T0136457

Against the Grain

L. Anders Sandberg
Peter Clancy

Against the Grain: Foresters and Politics in Nova Scotia

UBCPress / Vancouver

Printed in Canada on acid-free paper ∞

ISBN 0-7748-0765-2

Canadian Cataloguing in Publication Data

Sandberg, L. Anders, 1953-
Against the grain

Includes bibliographical references and index.
ISBN 0-7748-0765-2

1. Foresters – Nova Scotia. 2. Forests and forestry – Nova Scotia.
3. Forest policy – Nova Scotia. I. Clancy, Peter, 1949- II. Title.
SD146.N6S259 2000 333.75′092′2716 C00-910100-4

This book has been published with the help of a grant from the Humanities and Social Sciences Federation of Canada, using funds provided by the Social Sciences and Humanities Research Council of Canada.

UBC Press acknowledges the financial support of the Government of Canada through the Book Publishing Industry Development Program (BPIDP) for our publishing activities.
Canadä

We also gratefully acknowledge the support of the Canada Council for the Arts for our publishing program, as well as the support of the British Columbia Arts Council.

UBC Press
University of British Columbia
2029 West Mall
Vancouver, BC V6T 1Z2
(604) 822-5959
Fax: (604) 822-6083
E-mail: info@ubcpress.ubc.ca
www.ubcpress.ubc.ca

Contents

Appendices, Maps, Tables, Figures, and Photographs

Appendices

Maps

Tables

Figures

Photographs

Preface

In the pages below, we develop a series of arguments about the political character of forestry. Perhaps the most basic proposition is that the activities of professional forestry yield a diverse range of interests and perspectives about science, management, and public policy. This book illustrates these possibilities with studies of seven Nova Scotia foresters who enjoyed lengthy careers over the course of the twentieth century. Our sample draws from many historical and occupational phases, encompassing interwar, postwar, and contemporary careers, as well as time spent in corporate, government, and third-sector work. The experiences of these foresters illustrate the variety of political dimensions to their profession. We argue further that there is much to be learned from the traditions of dissent found in professional arenas such as forestry. Each, in its own way, runs against the grain of orthodox thinking, though established structures remain strong.

As this study moved toward publication, we experienced a fascinating confirmation of this truth. It is offered here as anecdotal evidence that 'politics,' as we define it, continue to infuse Nova Scotia forestry in deep and surprising ways.

While gathering copyright permissions for the appendices that follow the chapters below, we had reason to approach Nova Scotia's provincial deputy minister of agriculture, who is himself a forester. Since the document in question was more than sixty years old, we had little expectation of difficulty. However, the response to our query was as surprising as it was relevant to our overarching premise. First, we were asked to explain our interest in the 'musings' of a departmental official from 1939. Our answer was similar to that found in the introduction that follows. Second, we were asked about the perceived implication of the title: 'that foresters and politics are on opposing sides.' In subsequent conversations, we explained that it was the plurality of political dimensions in forestry, not their oppositional character, on which the book focused. Nevertheless, this official remained adamant. He suggested that he would need to review the entire manuscript if we hoped to secure permission. We replied that this sort of bureaucratic vetting was not appropriate for independent academic work, nor was it commensurate with a request for a limited right of reprint. Our correspondent then declared that he was 'uncomfortable' with the title. Unless this was changed, he was not prepared to sanction the use of John Bigelow's material in the book! Happily, this position was reversed later by decision of the minister.

Readers will judge for themselves whether this incident bears any relevance to the broad themes sketched in the following work. But to us it is a fascinating, if at times discouraging, reminder of the ways in which professional power can be deployed in forestry debates and decision making. To us, the reaction described above has less to do with fair legal dealing than it does with editorial control, less to do with public debate and understanding than with professional and bureaucratic insularity. It suggests that many of the institutional outlooks documented in this analysis remain alive and well in Nova Scotia today.

Acknowledgments

This volume began as a conversation, grew into a monograph, and ended up as a book. Along the way, it outpaced our initial investigation into the political economy of Nova Scotia forestry. We hope that the original study will follow soon. *Against the Grain* is a collaborative scholarly effort in the fullest sense. Reflecting this partnership, we regularly alter the order of author names in our joint publications.

During the work on this project on foresters and politics, we have received assistance and support from many people and organizations. It is a pleasure to recognize them now. First we must thank the central subjects of the volume, the foresters about whom we write. Over the past eight years we interviewed and re-interviewed John Bigelow, Lloyd Hawboldt, Donald Eldridge, David Dwyer, Richard Lord, and Mary Guptill. Through these encounters, they generously explained and interpreted their career experiences in the Nova Scotia woods. Regretfully, we must mention that John Bigelow, Lloyd Hawboldt, and Donald Eldridge passed away during these years. To all six of the key informants we owe an immense debt of gratitude.

We are happy to acknowledge the support of a talented and enthusiastic group of research assistants who have contributed over the years. Our team included Carolyn Chisholm, Michelle Johnston, David MacIsaac, Karla Tate, and Peter Twohig. At the Political Science Department at St. Francis Xavier, we would like to thank Marcy Baker. At the Faculty of Environmental Studies at York, we thank Carina Hernandez for preparing the final manuscript. Carol Randall produced all of the cartographic work, save for Map 6 which was drawn by David Cassels. Thanks go also to Holly Keller-Brohman, Ann Macklem, and Randy Schmidt at UBC Press, for their patience and support.

We are also grateful to the Social Sciences and Humanities Research Council of Canada, which provided generous financial support under

Research Grant No. 410-91-0460. Two small research grants were provided by the Faculty of Environmental Studies at York University.

Finally we wish to thank our families, who have graciously accepted the distractions, absences, and preoccupations which go along with this sort of project.

Against the Grain

1
Introduction

> **grain**: (def.): (5) the general direction or arrangement of fibres, layers or particles in wood, leather, stone, etc.; (15) natural disposition, inclination or character (esp. in 'go against the grain').
>
> (*Collins Shorter English Dictionary*, 486)

Foresters are in the business of growing and harvesting trees. So perhaps it is appropriate to begin a book about the politics of forestry with a metaphor derived from the structure of wood. To go against the grain is to pursue a course (whether in the sawmill or in life in general) that runs counter to the prevailing direction. Equipped with the proper tools, working against the natural grain of the wood is relatively easy and in some cases necessary. But care must be taken in choosing the time and the place. Under pressure, the strength of lumber runs along the grain, not against it. Sanding against the grain guarantees the destruction of the finish. In such cases, there is clearly a natural potential in the wood that, once recognized, can be preserved and enhanced through treatment.

Can there be a similar 'grain' in social or political affairs? In this book, we contend that there is and that it is evident in the practices of professions such as forestry. Here too the world is structured in certain fixed ways. There are conventions for managing forests just as there are for working with wood. It is possible, of course, to work against the grain of orthodox forestry thinking by challenging its central principles and practices, but not without paying a price. Established structures have their limits in accommodating change. Social and political interests can be counted on to defend the prevailing grain in the face of challenge. So once again there is a question of time and place. But unlike in the case of wood, it is possible to achieve creative outcomes while working persistently against the grain. The subjects in this book are prime cases in point.

There is a deeper significance to this situation. Too often the ideas and practices of professionals such as foresters have been viewed as monolithic. This comes in two basic shapes. The first is expressed in the ideas, reports, and memoirs of foresters themselves. Their stories are typically about the struggle for professional recognition, voice, and control. They are about the terms of employment, the state of the resource, and the problems posed by balance sheets, politicians, and the public. They celebrate the prodigious forces of nature and the colourful personalities who confront them. Such studies tend to be heavily descriptive and avoid conflict. The debates that they do record are largely intramural to the profession.[1]

A second school displays the ideas and practices of forestry as perceived by various 'outsiders.' These accounts too are often monolithic. In them foresters have been taken to task for much of what seems to be wrong in the Canadian forestry sector. Foresters have been portrayed as supporters of the growth of the Canadian staples export economy and the rapid exploitation and degradation of the forest.[2] They have been described as working in close concert with business and government to boost company profits and state revenues.[3] Labour historians have identified foresters as more aligned with their bosses than with forest workers and therefore instrumental in the deterioration of the conditions of forest work.[4] Others have pointed to foresters' and workers' different views on the forest environment.[5] And then, of course, foresters have come under fire from environmentalists for endorsing the use of destructive methods in exploiting the woods.[6] Some of these critiques have extended into Nova Scotia. Foresters have been described as prominent agents in the sellout of Crown lands to foreign pulp companies (by manipulating forest inventories); in the opposition to forest management legislation (because it threatened the profession's monopoly on knowledge); in the resistance to pulpwood marketing mechanisms for private woodlot owners (since private tenure failed to fit with the reigning industrial paradigms); and in the uncritical support of chemical use and clearcuts as core management tools.[7]

Here we argue that forestry is a more diverse and complex activity than the relevant literature has shown to date. By exploring in detail the careers of seven professionals active in Nova Scotia throughout the twentieth century, we point to the political quality of the profession. Difference lies at the root of politics, and Nova Scotia forestry has been punctuated by fundamental debates on matters of science, policy, and management. Although such dissent is seldom an 'all or nothing' exercise, but selective and episodic, all of our subjects run against the grain, raising challenging issues in the pursuit of better forestry. Many of these challenges failed because the established consensus proved resistant. Nonetheless, the plurality of views and experiences that they expressed is an apt reflection of the inherently political character of modern forestry and of the need to search beyond the surface to understand the foundations of both orthodoxy and dissent. The balance of this introduction outlines the framework of the study and the context in which Nova Scotia forestry professionals have worked.

This study of foresters and forestry in Nova Scotia reveals that a rich tradition of alternative and dissenting practices is intertwined with the professional and political orthodoxies of the day. The title of this book is intended to underscore the pluralism of thought and practice that is a part of modern forestry while demonstrating that all threads are not equally influential or binding. At the same time, the very concept of a 'profession' implies restricted control of specialized knowledge that is accumulated, transmitted, and applied over time.[8] This presupposes a shared paradigm, or framework of knowledge and technique, that is available for purchase or hire. Thus, there are always limits to the range of challenges that can be absorbed and sustained within the main corpus. In formal policy and in field activity, many of the alternative threads are rejected or ignored. Here again a 'political' situation arises between researchers, teachers, practitioners, and professional regulators when the conflicting currents must be reconciled. In this book, our goal is to assert the inevitability of politics within forestry and to illustrate its impact in the scientific and social relations of modern forestry. These explanatory factors have been seriously neglected to date, to the detriment of understanding both forest policy and the forest industry.

We have come to this conclusion through an exploration of the eastern Canadian province of Nova Scotia (Map 1). In a comparative ranking of primary wood product output by volume, Nova Scotia ranks sixth among

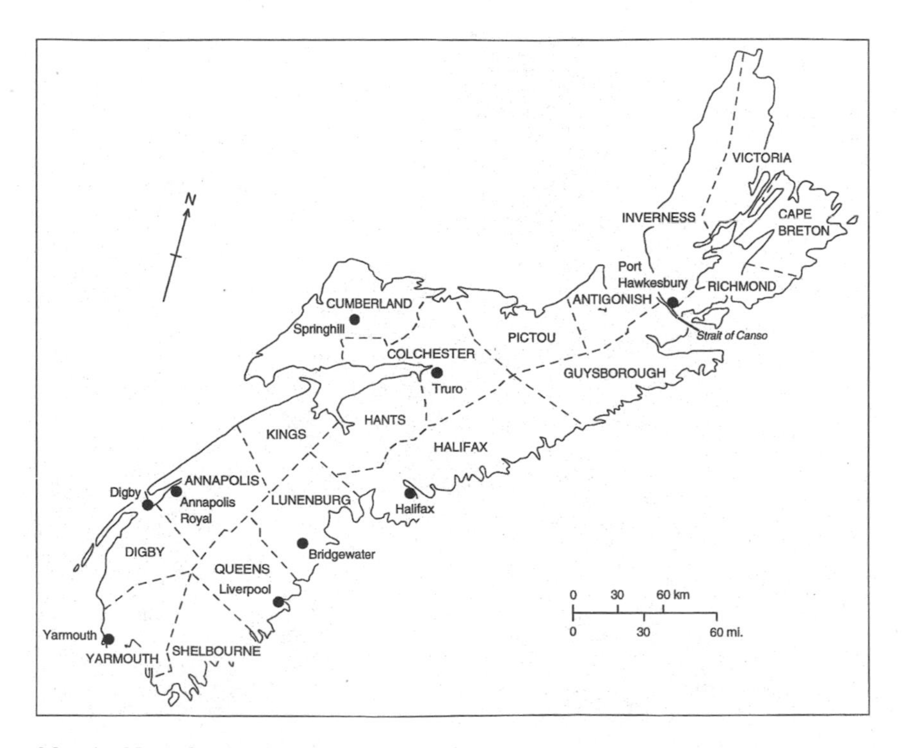

Map 1 Nova Scotia counties

twelve jurisdictions in the nation.[9] However, this relatively modest harvest level should not suggest that Nova Scotia forestry is marginal or unimportant to the wider scene. If anything, the reverse is true. The economic importance of the industry within the province is considerable, thus attracting sustained attention from both state and business. The forest resource itself is distinguished as part of the Acadian forest region (featuring mixed hardwood and softwood species), extending through Maritime Canada and parts of New England. As home to some of the earliest European settlements in the nation's history, Nova Scotia has witnessed more than two centuries of forest exploitation and a pattern of forest land tenure (unusual to Canada) falling three-quarters under private ownership, both large and small (Map 2).[10] There is much that is distinctive about the provincial scene. At the same time, its forestry has been cross-fertilized by national and international currents. In fact, it could be argued that as a smaller jurisdiction Nova Scotia has been shaped disproportionately by wider forces in the fields of forestry education, research, industrial production, and state policy. These will be explored extensively in the chapters below. Here we will present them as local applications and adaptations of more general currents rather than uniquely bred within the province.

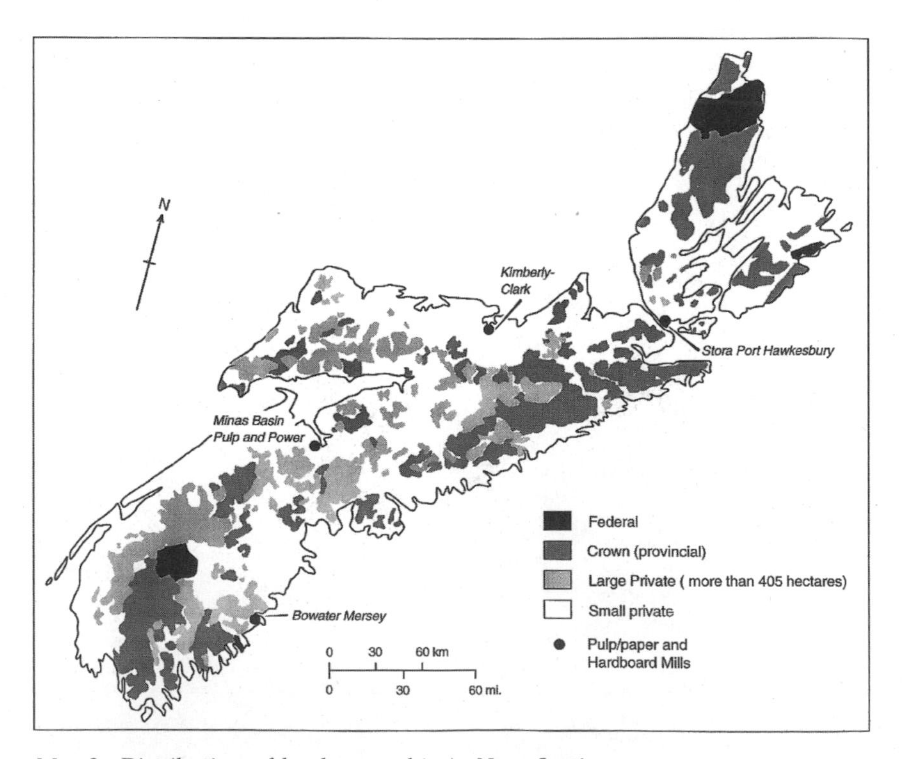

Map 2 Distribution of land ownership in Nova Scotia

Forest Ecology and Political Economy in Nova Scotia
In any setting, forestry practices are framed against the backdrop of two powerful and interacting structures. One involves the natural domain of forest ecology, while the other involves the social domain of political economy. Together they play a major role in determining the diverse interests, conflicts, and choice patterns that render forestry political. In this section, the defining features of these two structures are explored insofar as they shape and constrain the prospects for professional forestry.

The 85 percent of Nova Scotia covered by trees forms part of the Acadian forest region. Here the leading hardwood species are birch and maple, and the leading softwoods are spruce and balsam fir. Other trees that are strongly represented include beech and aspen or poplar (among the hardwoods) and hemlock, pine, and larch or tamarack (among the softwoods). Their exact distribution varies considerably across the province, according to site characteristics, which include, among others, soil, moisture, and temperature conditions. Ralph Johnson suggests that there are few uniform stands of more than fifty hectares anywhere in Nova Scotia. This variation is itself a distinguishing feature of the Acadian forest zone.

O.L. Loucks has identified six forest zones and twenty-four districts in Nova Scotia (they are illustrated in Map 3).[11] The most extensive is the spruce-hemlock-pine zone that covers most of the interior of mainland Nova Scotia and the shorelands of Northumberland Strait. The shallow rocky soils of the zone's extensive uplands, with low temperatures and high precipitation, harbour an exclusively coniferous forest. The lower elevations closer to the shore, with more moderate temperatures, lower precipitation levels, and better soils for tree growth, contain a more variable forest cover with sugar maple, beech, and yellow birch on higher slopes and red spruce, hemlock, white pine, and balsam fir on the lower slopes and valley bottoms.

A second zone, the sugar maple-hemlock-pine zone, extends from the valleys of the lowlands of the northeastern mainland to the central lowlands of Cape Breton Island, with an isolated extension in a southwestern section of the mainland. The pine, though cut heavily in the past, and dominant hardwoods (sugar maple and beech) occur most frequently throughout the zone, with hemlock, white and red spruce, and balsam fir common on lower valley slopes and valley bottoms. There is some variation within the zone. In the East River-Antigonish district, black spruce is common on poorly drained lands. In the Lahave district, red oak also occurs among the hardwoods and black cherry among the softwoods.

A third forest zone, the sugar maple-yellow birch-fir zone, is confined to four upland areas, the Cobequid Mountains, the Musquodoboit Hills, the Pictou Uplands, and the Cape Breton Hills. This zone stands out by the abundance of yellow birch, white spruce, and balsam fir in the dominant sugar maple stands. There is also a lack of hemlock on mixed wood slopes

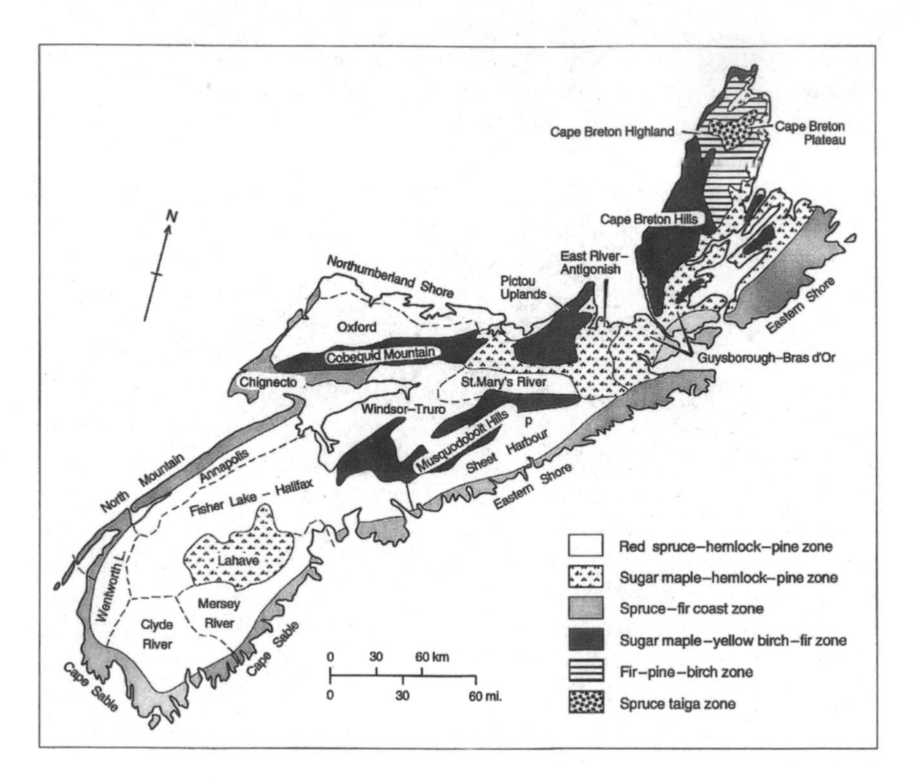

Map 3 Forest zones and districts of Nova Scotia

and a general restriction of white pine to sandy and gravelly soils in the valley bottoms.

The coastal fir-spruce zone is practically devoid of hemlock and white pine. It is divided into two distinct coniferous regions, one abundant with, and the other scarce of, red spruce. The Chignecto and North Mountain districts contain red spruce, though these stands are getting progressively scarce from north to south. The frequency of hardwoods exhibits the reverse pattern. In the Cape Sable and Eastern Shore districts, white and black spruce and balsam fir predominate. Bare bedrock is common, wind exposure is frequent, and the tree stands therefore tend to be open and the trees stunted. White spruce is most common in Cape Sable and black spruce in Eastern Shore.

The fir-pine-birch zone is confined to the Cape Breton Highland district. It consists of a balsam fir, white birch, and spruce association, where those trees exposed to the wind are seriously shortened. A belt of ridges and deeply incised valleys is marked by tolerant hardwoods. Finally, the spruce-taiga zone of the central portion of the Cape Breton highlands is composed of scattered stands of stunted black spruce, white spruce, balsam fir, and white birch. Various shrubs, lichens, and sphagnum dominate a barrens, where the climate is too severe for a closed forest.

Even without logging, forest ecosystems are dynamic entities, though change unfolds over the long term. Particular sequences can be discerned, beginning with pioneer species that adapt readily to a site. This is followed by a succession of associated species in the early, middle, and climax stages. Light and shade patterns, for example, can have profound effects on patterns of succession, as some (shade-tolerant) species can regenerate only under a canopy of older trees, whereas other (shade-intolerant) species require open sunlight to flourish. But the forest seldom changes according to a set pattern.[12] Major disturbances can accelerate or alter the paths of change, depending on their character. Fire, storm, insect infestation, and logging represent four prominent types of disturbance that can trigger abrupt change across wide areas.

The social relations governing forest exploitation and regulation are rooted in a combination of market and state. These have evolved dramatically over the course of provincial history, from Aboriginal society to military outpost to settler colony and finally to timber capitalism. Many of the pivotal events are presented in Table 1.

The Mi'kmaq reigned supreme in the forest until the mid-eighteenth century. Then the defeat of the French (with whom they were allied), the deportation of the Acadians (who supplied them with certain provisions), and waves of British and Loyalist settlements undermined their position.[13] Subsequently, the Mi'kmaq gradually lost access to fish and game, either through competition from market or sports operators or through legislative restrictions on their harvesting methods. They nevertheless achieved some measure of material prosperity as hunting and fishing guides and as skilful craftspeople with wood products.[14] However, the Mi'kmaq were severely affected by the economic decline of the Maritime economy in the 1920s and the Depression of the 1930s, leading to a period of unprecedented hardship and financial dependence on the state.[15] It is only recently that the Mi'kmaq have taken political, legal, and popular direct action to reclaim access to the forest and its various resources.[16]

Beginning with the colonial economy, the exchange value of timber products was defined partly by the natural supply of forest species and partly by the commercial demand in local and export markets. For example, early logging centred on masting timber (white pine logs) and ton timber (oak and pine logs) for export to Britain. In the late 1700s, the shipbuilding industry provided a domestic market for oak, birch, and larch. With the nineteenth-century sawmill expansion and the transition from water power to steam, pine and spruce lumber deals assumed greater importance in export to America and Europe. There was also considerable regional specialization of wood production in Nova Scotia after 1850, according to variations in forest

Table 1

Pivotal events in the history of forestry in Nova Scotia

Date	Key event in forest sector	Details
1728	British 'broad arrow' policy applied in NS	Reserved all white pine greater than 24" suitable for masts. Extended to private land in 1785.
1759	British land grants to settlers begin.	Crown lands granted and sold until Lease Act, 1899.
1899	Lease Act proclaimed.	Crown forest land conveyed by lease. Last land grants made in 1920s.
1926	Department of Lands and Forests created. Otto Schierbeck appointed Prov. Forester.	Merger of Dept. of Crown Lands and Dept. of Forests and Game. Attorney general serves as first minister.
1927	Tobeatic Park created.	First game sanctuary in NS. Liscomb, Waverly follow.
1928	Mersey Paper Co. signs Crown timber deal. Mill starts production in November 1929.	1,000,000 cords in Guysborough / Cape Breton over 30 years.
1930	Bill 151 (Embargo Bill) defeated.	Attempted to regulate export of pulpwood from NS.
1934	NS Forest Products Association formed.	

1946	Small Tree Conservation Act and Scalers Licensing Act proclaimed.	Diameter limits to protect young stock. First regulation of private forest land in NS.
1948	G.W.I. Creighton appointed deputy minister of DLF.	Staff included 12 foresters, 21 rangers, 7 game wardens, 4 land surveyors, and 2 scientists.
1953	Forest inventory begun by Belanger and Bourget (federal-provincial funding).	Four-year aerial survey and ground cruises described in Bulmer and Hawboldt, *Forest Resources of NS*, 1958.
1954	Hurricane Edna, 11 September.	Estimated 700 M FBM of lumber blown down. Years of salvage cutting follow.
1954	Nova Scotia Section of Canadian Institute of Forestry (NSS-CIF) formed.	Previously NS foresters were members of Maritime Section of CIF (established 1937). 40 members in 1954.
1959	NS Pulp Ltd. and GNS sign crown lease.	1.2 M acres on eastern mainland and Cape Breton.
1959	Organizing for private Woodlot Owners Association begins.	Supported by Department of Extension of St. Francis Xavier University.
1965	Forest Improvement Act passed.	Designed to replace Small Tree Act (rescinded).
1968	Bob Burgess appointed deputy minister of DLF.	

▶

◀ *Table 1*

Date	Key event in forest sector	Details
1972	Pulpwood Marketing Act passed.	Authorizes registration of private supplier groups and framework for negotiating contracts.
1976	GNS authorizes spruce budworm spray program.	Political controversy continues for two years.
1976	Canada and NS sign five-year Forest Resource Development Agreement (FRDA).	Two-year hiatus begins in 1982. Successor agreements signed in 1984 and 1989. Terminated in 1994.
1978	Donald Eldridge appointed deputy minister of DLF.	
1982	Royal Commission on Forestry established.	Report submitted in 1984.
1986	NS forest policy declared.	Formal response to royal commission.

composition and mill capacity. Finally, the emergence of the early groundwood pulp industry at the turn of the twentieth century boosted the importance of smaller-diameter spruce and fir stock. This also held out the promise of markets for inferior stock that might otherwise lack value. However, it took more than half a century before the pulp sector confirmed its political predominance over the sawmill segment in Nova Scotia forestry.

Even as market conditions drove investment and sales, the state also played a pivotal role in commercial growth. In part, this was due to its power to define the rules of property ownership and exchange. Despite the questionable legal basis of Aboriginal surrender by treaty, European Crown authorities asserted a strong presence from the outset. This began with policies such as the Broad Arrow (reserving tall pine for Crown military use) and land grants to settlers (transferring the majority of Nova Scotia forest lands into private rural tenures). Over the course of the nineteenth century, the prime sources of supply shifted from coastal forest regions into the interior. With the development of watersheds as log-collection networks in springtime, the state exercised powers over riparian rights of water use. Commercial tariffs and other trading rules were also instrumental in shaping the direction and depth of markets.

After Confederation, national and provincial authorities shared the relevant state powers, with Crown forest jurisdiction and land taxation at the provincial level, while trade, finance, and credit rested at the national level. In Nova Scotia, a decisive shift was marked by the discontinuance of the granting system in favour of Crown leases after 1899. This redefined the residual Crown forest from a dispensable resource to a permanent estate, whose management could be used as a development lever. This became increasingly strategic as pulp and paper operators came to rival sawmills as timber users in the twentieth century. However, Nova Scotia's preeminent policy problem was rooted in the privately owned forest sector, where the state lacked policy leverage to support the developmental plans of the lumber, pulp, or export business interests. The intensely competitive lumber industry could swing wildly in boom and bust directions. The highly speculative trade in timber lands impeded industry growth and proved impossible to contain. Attempts to prevent the export of raw logs were met with massive opposition from private timber owners and loggers. Furthermore, efforts by the province to buy back degraded forest lands raised the spectre of policy conspiracies against the private owner.

In many of these controversies, the Nova Scotia state found itself caught between irreconcilable sets of competing forest sector interests. The politically dominant big sawmill sector of the nineteenth century required little by way of state support in a business culture wedded to laissez-faire. Assembling large expanses of private forest and buying logs from the rest of the private sector, the lumber kings had little need for Crown lands. This preeminence was lost at the turn of the century when the big sawmills were increasingly challenged by rivals. From one direction came the pioneer groundwood pulp interests, competing for timber limits and carving out Crown leases. From another corner rose the small portable sawmills that high-graded small tracts into low-quality lumber before abandoning them to tax sales. Despite the efforts of the leading lumber families to organize their industry in the 1930s, through the Nova Scotia Forest Products Association, a structural transition from lumber to pulp was already under way.

Yet this transition was drawn out, and it remained incomplete for several generations. Many of the leading policy events of the mid-century can best be understood in this context of prolonged political tension and lack of hegemony. The Small Tree Conservation Act was promoted by progressive sawmillers during the Second World War as a defence against the degradation wrought by portable mills, yet the majority of small forest holdings was exempted from its terms. Nova Scotia's first systematic forest inventory in the 1950s was undertaken in the hope that pulp and paper investment could be lured by Crown lease to subregions lacking sawlog stock. However, the survey revealed that the most acute overcutting was in sawlog stock. Then the very success of the pulp promotional strategy in committing Crown forests for pulp at concessionary terms opened new crises. At a stroke, it deprived the sawmillers of direct access to Crown sawlog stock and triggered the collective organization of small private woodlot owners fearing plummeting markets for their own timber. In the 1960s, the provincial state found itself caught between the woodlot movement, the sawmill sector, and the pulp corporations. Halifax was unable to ignore the small owners' movement outright but unwilling to fully support it either. Halifax also sought to integrate the sawmillers into fibre exchange networks with pulp and paper, but at the cost of their subordination. Almost simultaneously, a popular citizen environmental campaign challenged the pulp industry's plan for aerial chemical spraying against the spruce budworm defoliation.

This prolonged political-economic transition posed continual challenges to the forestry officials within the provincial Department of Lands and Forests. For the first generation, the few professionally trained foresters struggled to legitimize their status within a bureaucracy of partly skilled patronage appointees. During this period, there was little call for management planning in the sense of balancing growth and harvesting rates. Rather, it was assumed that the fibre supply was inexhaustible. Then, after the Second World War, as the professional cadre began to reach critical administrative mass, it faced pressures and demands from diverse political directions. Small private woodlot owners called for commodity marketing and extension forestry support. Lumbermen called for conservation regulations. The pulp industry sought long-term Crown leases. And environmental advocates campaigned for alternatives to industrial forestry.

It is important to remember that the successive exploitation of various tree species for commercial forestry has been determined by ecological as well as political and economic factors. The forest itself has changed, and its own sometimes unpredictable dynamic has reinforced or altered specific forest uses. The selective cutting or high-grading of large-diameter pine and spruce in the nineteenth and twentieth centuries resulted in a precipitous decline of such trees. In the most distinctive forest of Nova Scotia, the red spruce-hemlock-pine zone, for example, vast tracts of red spruce and hemlock were depleted, leaving only remnants of the old-growth forest. The decline of large-diameter trees occasioned the growth of portable mills to gain access to such trees in the more remote locations. In some areas, such as the Northumberland Shore, Oxford, Windsor-Truro, St. Mary's, and Sheet Harbour districts of the red spruce-hemlock-pine zone, repeated cutting and burning have yielded witherod and rhodora shrubs that control sites so effectively that they exclude softwood regeneration. Once the best trees in the most remote locations were cut, the forest potential changed quickly to a pulpwood economy. The degraded forest, in short, played a definite role in shaping forest use.

The growth of agriculture and settlement in the nineteenth century also had an impact on forest use. This is because the forest reclamation of abandoned farmlands in the twentieth century has resulted in a different forest from the one cut by the early settlers. Abandoned farmlands have been invaded by dense stands of softwood pioneer species, predominantly white spruce and balsam fir. In the sugar maple-hemlock-pine zone of the Guysborough-Bras d'Or and East River/Antigonish districts, and in the Pictou Uplands district of the sugar maple-yellow birch-fir zone, for example, former cleared sheep pastures and abandoned farmlands have reverted to white spruce and/or balsam fir stands. The compaction of the soils in these districts has contributed to the presence of poor tree stands. White spruce and balsam fir also readily establish themselves on land clearcut for pulpwood. Nature has acted differently in the Lahave district of the sugar maple-hemlock-pine zone, where white pines form pure stands on abandoned fields.

The balsam fir and white spruce stands have reinforced the pulpwood economy because such species are suitable for little else. The dense and small-diameter nature of these tree stands has also encouraged the rapid mechanization of woods harvesting. While pulp cutting was done by woods crews with hand tools well into the 1960s, since then there has been a transition toward large-scale mechanical harvesting. Pulpwood contractors perform this task with skidders, forwarders, and tree harvesters, often working around the clock to make payments on their machines and to feed the pulp mills' insatiable hunger for fibre.

The growing presence of a balsam fir and white spruce forest has had other consequences for the forest industry. This forest is particularly vulnerable to spruce budworm infestations, and the province has seen an increase in the frequency and intensity of such infestations. In the mid-1970s, an unprecedented infestation destroyed a substantial part of the fibre supply in eastern Nova Scotia. The pulp companies argued strongly at the time for insecticide spraying, but the province resisted (for reasons explored later). The effect of the infestation nevertheless sped up the province's move toward industrial forestry, understood as the close control of the fibre supply by industrial techniques such as mechanical harvesting, broadcast pesticides, and planted monocultures.

There has also been a bias toward softwood utilization in both the sawmill and pulp and paper economies. This has resulted in the crude and selective use of the province's extensive hardwoods. The hardwoods have thus been degraded and transformed into a non-commercial composite of pioneer species. The large-diameter yellow birch and beech components were cut early for commercial purposes, but the regeneration has been uneven and has led to what many foresters and industry analysts describe as the 'hardwood problem.' Indeed, when birch and beech were exposed to disease in the 1940s and 1950s, the situation was cheered on by most sawmillers and pulp mill operators.

There are many other ecological processes that could be considered here, such as those resulting from fire suppression, a major task of forestry. Long a pillar of the forest service mandate, it aims to protect the wood supply. But it may also carry inadvertent negative effects, for the wood fibre supply as well as the forest ecosystem. Fuel loadings may build up and cause large destructive fires in the future. Fire suppression may also prevent ecosystem renewal, such as in the case of some pine species, which are dependent on fire for regeneration. Fire, then, may not be an external catastrophic event but an integral part of forest ecosystems. The same point may be made about periodic extreme storms (such as Hurricane Edna of 1955, which destroyed massive tracts of forest on mainland Nova Scotia) and spruce budworm infestations. These events may be integral parts of ecosystem processes that serve important roles and that need to be accounted for in planning forest harvesting.

This is not to suggest, of course, that all fires are natural. Human-set fires may play a destructive and/or ecosystem-altering role. Loucks observed, for example, that in the spruce-fir coast zone of the Cape Sable and Eastern Shore districts frequent burnings by the settlers may have encouraged the growth of even-aged dense stands of white and black spruce, balsam fir, and alder. Similarly, in the Cobequid Mountain district of the sugar maple-yellow birch-fir zone and the Clyde River district of the red spruce-hemlock-pine zone, frequent burnings to encourage blueberry growth have suppressed forest regeneration.

The Acadian forest is thus not a passive agent. It is an active and often unpredictable factor that has played its unique role in shaping the Nova Scotia forest economy. The specific ecological processes and patterns that have gone along with forest degradation and forest reclamation of abandoned farmlands have reinforced (even naturalized) an industrial form of forestry, based on clearcutting, pesticides (if not chemical insecticides, then biocides and herbicides), and monoculture plantations. A forest industry more in keeping with the forest ecology of the region, based on the maintenance and use of the diversity of the Acadian forest, has been distinctly absent.

The Foresters and Their Times

Another of our aims in this study is to capture the rich variety of Nova Scotia forestry practice. This raises issues of historical time span and workplace and professional specialization. An ideal sample would cover professional forestry from its birth to the present. It would include careers spent in industry, government, and the voluntary sector. It would also explore practices such as corporate woodland management, Crown land management, research, extension or private lands forestry, service to trade associations, and government policy making. This is a tall order, and it points to the need for careful choice. If the results are to do justice to the subjects – that is, to achieve the level of depth and detail desired, while at the same time permitting comparison and analysis – then the sample must be small but powerful.

In selecting subjects for the study, we were eager to highlight people whose career experiences were representative of the rich variation of Nova Scotia forestry. We also sought to include those whose exploits had not been well documented to date. Although the literature on Nova Scotia forestry is not vast, two prominent figures have contributed book-length histories. Ralph Johnson spent half a century as forester to the Mersey Paper Company (later Bowater Mersey) in Liverpool and played a leading role in the Maritime and Nova Scotia sections of the Canadian Institute of Forestry. His *Forests of Nova Scotia* was the first comprehensive study of the woods sector in the province.[17] Wilfrid Creighton is another senior professional. His thirty-five-year career was spent with the Department of Lands and Forests (DLF), first as Provincial Forester and later as deputy minister. In his history of the department, *Forestkeeping*, Creighton reveals much about his own career and philosophy.[18] The broad availability of these works permits their authors to speak for themselves, as it were, freeing us to search for equally revealing but less publicized subjects.

In the end, we settled on a group of seven subjects. Collectively, they cover Nova Scotia forestry from the First World War to the present day. Their careers reflect all of the significant phases and turning points in both industry and state policy. Furthermore, the overlapping experiences and even the direct interactions of these foresters result in multiple perspectives on key events or controversies. Our set begins with Nova Scotia's first professional forester, Otto Schierbeck, who arrived in 1926. Trained in Europe and experienced in Canadian pulp forestry, Schierbeck was charged with creating a forest service to undertake modern management. His experiences reveal the problems of building a balanced industrial structure, taking full advantage of the resource base, and reconciling professional methods with the cliental political framework of the day.

John Bigelow was a forestry graduate of the University of New Brunswick in the 1930s. Although his career was spent in Nova Scotia government circles, it was almost entirely outside the Department of Lands and Forests. Bigelow was an energetic promoter of forest improvement through increased and higher-value utilization. He was guided by a practical perspective on forest economics long before it became widely recognized as such. He was also an early and persistent advocate of forest sector organizations to represent the interests of sawmillers, woodlot owners, and others.

Following the Second World War, Lloyd Hawboldt joined the DLF, where he was to spend a distinguished forty-year career. A graduate entomologist, Hawboldt was a self-taught forester schooled through diversified service to his department. After developing an early research capability as head of the Forest Biology Division, he went on to build the extension program and eventually rose to the second highest position in the departmental service. Hawboldt played a crucial role in the provincial forest inventory of 1953-7, which opened the way for the modern management program. He was also a crucial participant in the policy deliberations over various treatment programs to combat the spruce budworm in the 1950s and 1970s.

Donald Eldridge was a member of the postwar generation of University of New Brunswick (UNB) foresters, and his career spans three quite distinct dimensions of Nova Scotia forestry. For more than fifteen years, he worked as forest lands manager for the Eddy Lumber Company. He was then appointed the first full-time executive director of the Nova Scotia Forest Products Association, an industry group representing sawmill, pulp and paper, and logging interests. A decade later, Eldridge moved to the apex of the provincial forest service with his appointment as deputy minister of the Department of Lands and Forests. More than any other figure in this collection, Eldridge speaks from the perspective of industrial forestry.

Another postwar UNB graduate was David Dwyer, who spent his entire career in the DLF. Dwyer declared a preference for the social side of forestry. He preferred working with forest people: woodlot owners, rangers, sawmillers, logging contractors, and other government officials. For more than thirty years, he promoted small, private land woodlot forestry, which made up more than half of the forested land base of the province. This commitment placed him outside the mainstream, where interest centred on the exploitation of vast Crown timber limits by corporate leaseholders.

Rick Lord was drawn to many of the same woodlot issues as Dwyer, though he pursued them from a non-governmental and non-corporate base. Lord came to Nova Scotia to build organizational capacity for the small private landowners, who faced the lowest returns for pulpwood in the country. First as an organizer and later as the manager of the Nova Scotia Woodlot Owners Association (NSWOA), Lord was embroiled in prolonged political battles in pursuit of organized fibre supply and woodlot forest management. His organizational acumen carried on to the provincial and national Christmas tree growers' associations once the woodlot movement reached its denouement in the 1980s.

In several respects, the 1970s was a transitional decade for forestry. Mary Guptill's experiences are an apt reflection of this change. When Guptill entered forestry school in 1973, many of the basic tenets of industrial silviculture were only beginning to be applied in Canada. By the time she joined the workforce five years later, the face of forestry had been transformed, in Nova Scotia as elsewhere, by the new generation of federal-provincial forest development agreements. She went to work as a field forester for La Forêt Acadienne, one of the newly conceived 'venture group' enterprises in Nova Scotia. Her experience offers a fascinating glimpse into the challenges of delivering management plans on the microproperties of the province's French Shore.

Altogether, then, our set of foresters contains figures from the interwar, postwar, and contemporary generations. The majority of our subjects divided their careers between several sectors. Schierbeck, Bigelow, Eldridge, and Lord spent time in both business and government service, while Bigelow, Eldridge, and Lord also worked with trade and industry associations. Two subjects, Hawboldt and Dwyer, spent careers exclusively in the public service, while Guptill has spent most of her career to date as a field forester in the not-for-profit sector. Hawboldt is not strictly a forester (i.e., not a formally accredited forestry graduate) but an entomologist. However, his rightful place in this study will be evident from the sweep of his career. Bigelow, Eldridge, Dwyer, Lord, and Guptill were trained at the University of New Brunswick in Fredericton, while Schierbeck studied in Denmark and Hawboldt at McGill University in Montreal. As will be evident in the detailed profiles below, this only begins to capture the significant variations within the sample. However, for the sake of comparison, some of the summary features have been captured in Figure 1.

Finally, it is appropriate to indicate our methodological approach to this work. Despite its deep narrative base, this is not a work of oral history in the

Figure 1

Time line for foresters

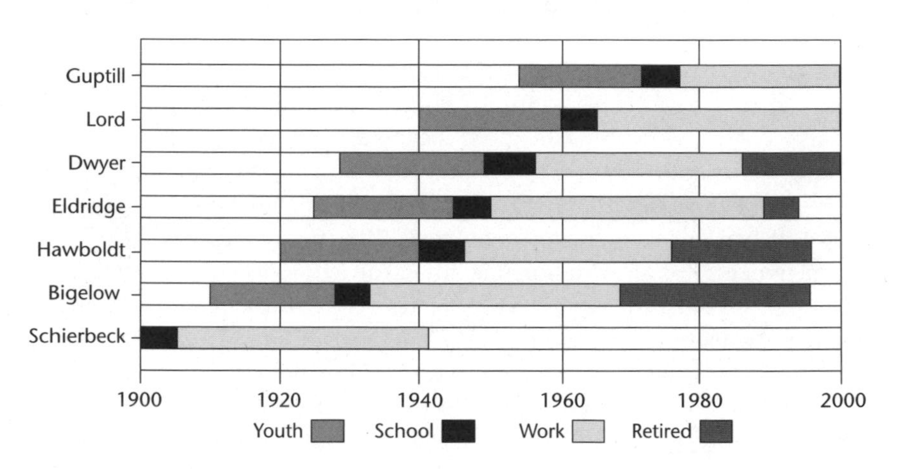

classical sense. Such studies are not unknown in Maritime Canada. One example of this technique is Mike Parker's *Woodchips and Beans*, a collection of 'personal experience narratives' on early Nova Scotia logging.[19] Another volume, *In the Mersey Woods,* explores the world of company bush camps.[20] While there is a unique authenticity in first-person accounts, oral history faces a potential challenge in ordering and interpreting the disparate accounts on which it rests. In postmodern times, this may be regarded as a strength rather than a weakness, since a plurality of partial narratives offers the only legitimate analytical prospect. We too have relied extensively on interviews with our subjects to compile the accounts below. Here, however, we seek to weigh, compare, and interpret the experiences of the foresters outside their own views. For this, neither oral history nor postmodern premises will suffice. At the same time, we share the desire to let the foresters speak for themselves. Consequently, each chapter concludes with a primary document reflecting signature themes from that person's career.

An intermediate approach combines commentary and interview transcripts. In forestry, this is represented by Ray Raphael's two volumes on the forests of the American northwest, *Tree Talk* and *More Tree Talk.*[21] We sympathize with Raphael's aspiration for 'a running narrative ... punctuated by individual portraits intended to personalize the issues, to translate both the political and academic aspects of forestry into human terms.'[22] However, our analytical approach goes beyond the search for the human face behind the abstract forces of social life. First, through the experiences and impacts of our subjects, forestry stands revealed as a complex professional pursuit. In the chapters below, we will have much to say about the succession of ideas, principles, and practices that have guided professional foresters. Second, the careers of our subjects open a window onto the history of forest activity in twentieth-century Nova Scotia. Since these foresters are affected by virtually all of the modern milestones of forest development – physical, commercial, and political – their story is its story as well. Finally, this study demonstrates that the foresters of Nova Scotia are a diverse and original lot who have been vital protagonists in the politics of forest practice. We are concerned to establish the character and significance of this politics as a positive rather than a negative attribute. This begins immediately below, with a general reflection on the relationship of forestry to politics, outlining some concepts and frames of reference that will figure in subsequent chapters.

Forestry and Politics

As the twentieth century draws to a close, the forestry field is being swept into expanding circles of controversy. For this there are many reasons: dramatic evidence of deforestation both around the globe and in our own backyards; suspicion that corporate exploitation will not rest until the last tree has passed through the blades of the last saw; and anxiety that public authorities have failed in their regulatory and stewardship responsibilities, to name only a few. Collectively, they suggest that a key renewable resource has been grievously mismanaged. Professional forestry is not without a reply to this charge. Its roots run deep in the turn-of-century North American conservation movement. Among its father figures are Gifford Pinchot, who founded the Biltmore Forest and the United States Forest Service, and Bernhard Fernow, who set up Canada's first forestry school. Foresters bow to no rival in their commitment to rational management. From the early 1900s onward, the primary goal was to harness scientific knowledge to wise resource use. For more than half a century, this goal lay at the heart of the modern forestry management paradigm, linking inventory, growth, and harvest in a regulated yield equation. This is the framework of sustained (and maximum sustained) timber yield, which aims to ensure a permanently available forest resource by closely linking annual cut to annual growth.

While its logic is unassailable, sustained yield forestry has suffered from a host of practical problems. Some arise from the politics of modelling, in which the parameters for estimating inventory and growth are manipulated in unwarranted ways and thereby distort the authorized harvests. Others arise from the overcommitment of the resource to manufacturers, as part of the search for rural investment and employment. Still others stem from the failure of regeneration programs to meet their targets under conditions of intensive silviculture. Whether these failings are inherent or incidental to the management paradigm is arguable. Either way, however, the result leaves professional paradigms open to harsh attacks from critics.[23]

More recently, in response to ecological concerns, North American foresters and forest agencies have altered their orientation significantly. Most now embrace a 'new forestry' defined in terms of ecology and sustainability, where the emphasis is on managing forest ecosystems rather than forest fibre.[24] But in spite of this shift from industrial to environmental goals in recent years, the old forestry paradigm still remains very much alive. Much of scientific forestry is highly instrumental and manipulative, where experts or technocrats (now equipped with computers) map, measure, and model the minute details of the forest landscape and then take active steps to accommodate human demands for timber, wildlife, species diversity, recreation, and old growth.

Expert forestry opinion does not always respond effectively when confronted by public criticism of its technocratic bent. One answer sees the public as misinformed and prone to exaggerate problems. The antidote is seen in more effective 'education,' and environmental roundtables and stakeholder forums are advanced toward this end. Often this amounts to little more than an urge to have public opinion brought more closely into line with expert opinion. On the other hand, it is argued that expert knowledge is blocked or even deformed by the considerations of 'political' expediency or necessity. Here elected officials are seen to tie the hands of the technocrats by inadequate resourcing (preventing full implementation) or failure of nerve (conceding to popular prejudice) in implementing forest policy.[25]

For eighty years a strength, this paradigm is increasingly subject to challenge. In an emerging age of ecological consciousness, environmental protest, and interdependent networks, a management regime built on the expert manipulation of timber values is increasingly viewed as untenable. In effect, the entire knowledge base of forestry has come into question, for, as Maser observes, 'Ecological understanding is a nonexact, nonstatistical subject. Cumulative effects cannot therefore be rendered statistical, because ecological relationships are far more complex and far less predictable than our statistical models lead us to believe. We cannot foresee the moment when cumulative effects become irreversible.'[26]

It is difficult to dispose of the criticisms of modern forestry. However, our sense of forestry 'politics' extends far wider and deeper than that described above. Given the complicated and contested nature of its subject matter, modern forestry can be little else but political. This should be acknowledged and even welcomed as a progressive development. Differences of interest, of perspective, and of program are inevitable, and little is served by ignoring or denigrating them in an effort to build or defend a single unified knowledge base.

Like many professions, foresters have had difficulty in acknowledging the political aspects of their position in modern life. They are apparently more comfortable with an engineering or technocratic sense than with a social or political sense of their vocation. This is reflected in much of the professional literature. There is no lack of studies on forest management practices as applied to different physical settings, political jurisdictions, and periods of time. But it is common to find the forester portrayed as an intermediate actor, the agent of broader and higher forces such as the corporation or the state or the public interest. There is a tendency here to see foresters as functional operatives, vital and necessary to be sure, but acting in tightly circumscribed arenas where the focus is applied science.

Perhaps unwittingly, the profession has added to this impression. This derives in part from the specialized language and somewhat esoteric concerns that dominate most professional deliberations. The terms invoked are typically about manipulation and control of a promiscuous Mother Nature. The forester's role is to 'protect' trees from various 'enemies,' such as insects, fires, and storms. In all likelihood, the professional outlook is also a product of the diffidence shared by most foresters, reluctant to step squarely into social or political controversy. In fact, it might be suggested that the profession has displayed an outright aversion to, and denial of, political realities. How often is it proclaimed in the professional journals that forestry and politics do not mix, that the rotation ages for trees and the electoral cycles for politicians will never coincide, that the formulation of policy (presumably at the 'political' level) and the practice of management (presumably at the professional level) are appropriately kept separate? These are contentions that we find both highly questionable and highly revealing, and we will examine them extensively in the chapters below.

The profession's long-standing awkwardness with the political process has lately turned to one of crisis. For more than a decade now, forestry has found itself on the defensive. The expansive confidence of the postwar generation seems to have evaporated, in part through the ravages of recent recessions on the job market, but also because of public challenges to professional practices and integrity.[27] To many of its critics, the forestry profession is seen to be hopelessly co-opted, or even corrupted, by the recruitment of so many foresters to employment in large resource-processing corporations. Neither have government foresters escaped this charge, because they are often viewed as the handmaidens of corporate power. The most fundamental questions are being raised about the objectivity of the forestry outlook in a commercial age and its ability to deal fairly with the needs of the resource when they conflict with the need for the cheapest possible wood supply. There is, not surprisingly, a spirited response from the professional mainstream. It contends that contemporary forestry practice is founded on a century of scientific advance and field application, geared to North American ecologies and business realities. This has brought unprecedented knowledge of forest biology, wood inventory, silvicultural strategies, and logging techniques.[28] Foresters insist that this knowledge is highly relevant in the age of environmentalism and forest ecosystem integrity. Many go further, constructing a new vocabulary and practice of 'sustainable forest management' to suit the 1990s.[29]

In spite of such claims, orthodox professional forestry is still subject to a withering critique and dismissal by many environmental activists. They view its central corpus as unacceptably limiting, tied to a commodity-based approach to the natural world when an ecological approach is required. From this perspective, it can be little more than a technocratic assault on nature. Even debates within the forestry profession reveal a profound uneasiness about the new public and political perspectives on organized forestry and a striking ambivalence about the appropriate response. There was a time when the views of 'non-foresters' could be relegated to the margin if not dismissed outright. But with public perceptions now solidifying, the organized face of the profession is committed to respond.[30]

In our view, these are fascinating debates, being conducted at a critical time for forestry. They are timely, appropriate, and pressing. The critique is understandable, though it is accurate only in part. The predicament for forestry and foresters is that its professional practices have become a lightning-rod for public discontents about a far wider set of forces and problems. Ironically, however, the profession's misperception of forestry's political character has compounded its difficulties in responding.

This study does not seek to advocate or confirm any particular perspective. Rather, we wish to explore the politics of forestry, in several different senses, and to establish that the political character of forestry is not new. While the unprecedented level of public interest and debate is relatively recent, forms of intramural and interagency politics have pervaded twentieth-century forestry, in Nova Scotia as much as elsewhere. The profession has seldom in the past achieved a full consensus on many matters. It is more typically characterized by competing clusters of majority and minority views, regional geographic variations, and generational differences. It is this range of interests, agendas, and encounters that we wish to address.

In the pages below, we will stress the complexity of forestry as a knowledge system. Indeed, much (though certainly not all) of the confusion and acrimony that define today's situation has resulted from overly simple approaches that confuse separate phenomena by lumping them together. As a result, there are moments when 'forestry' is blamed for the excesses of national and international corporations pursuing maximum profits. At other times, 'forestry' is blamed for the timid and even apologetic policies of state agencies charged with managing public resources and promoting forest enhancement. In still other cases, 'forestry' is held responsible for the arrogant invocation of science and expertise to legitimate or excuse the 'necessity' of what are often highly controversial activities. There is no question that modern forestry, in theory and in practice, has much to answer for, but we are convinced that the most sensible and rewarding path to its understanding is an analytical one that starts with concrete experiences and practices on the ground. However, before moving to the primary subjects, we will consider certain political and social dimensions of forestry at large.

In the forestry domain, politics are likely to surface at many locations and to take several forms. Politics are present in any walk of life in which a diversity of interests calls for decisions to be taken collectively. In each case, competing interests will bid for priority, and leaders (often with no party political involvement) face the task of forging responses that are broadly acceptable to participants. Consider the political premises of the following three comments.

> By emphasizing one function over others, by aggressiveness or passivity, by inventiveness or adherence to the *status quo*, by risking the displeasure of superiors or colleagues or neighbours or by following the path of least resistance, by enthusiastic or indifferent or reluctant performance, the Rangers in effect modify and even make policy – sometimes without knowing it. (Herbert Kaufman[31])

> The Forestry profession persistently restricted its orientation to the needs of the tree while avoiding consideration of the economic and social needs of the owners of these trees. Yet it is on the Profession that our governments rely for the expertise in planning development policy for our forest industry. Ultimately, of course, it is our Universities which presume to offer degrees in forestry which are the real culprits in this sorry story. (Alexander A. MacDonald[32])

> In a public opinion poll commissioned not long ago by the Federal Government, 97 percent of the lay-public respondents agreed with the statement that 'clearcutting is a poor forest management practice.' Seventy-nine percent of foresters questioned disagreed with the same statement, and only 7 percent of them agreed 'strongly' with it. Clearly on this issue there are two solitudes – the popular and the professional – which foretells of serious problems ahead. (Edward S. Fellows[33])

Kaufman points out the crucial role of field staff in delivering programs to the public, MacDonald underlines the inevitable links between forestry and society that are transmitted through professional training, and Fellows notes the growing discrepancy between technical and popular outlooks on forest matters. These are all primary domains for the politics of forestry. Consequently, politics is as likely to figure in the affairs of the International

Woodworkers of America union, the Bowater Mersey corporation, or the Canadian Institute of Forestry as it is in the Nova Scotia electoral process or on the Legislative Order Paper.

In a similar sense, much of the forester's professional practice is permeated by political relations. It is true that to many people, foresters included, evidence of 'political' considerations is a cause for regret. By this view, the political represents an unwanted intrusion by arbitrary or unknown forces. It is manifest where partisan loyalties and animosities offer the basis for conferring or withholding benefits, where croneyism and clientele networks serve as means of controlling access to resources and power. Prior to the Second World War, for example, it was routine for the entire staff of Nova Scotia forest rangers to be dismissed after an electoral change in governing party, to be replaced by friends of the incoming group. Chief Forester Otto Schierbeck experienced the harsh consequences of rural resentment channelled through the Conservative government, which led to his firing in 1933. Not long after, Nova Scotia witnessed the 'Woodpecker Election' of 1937, featuring fierce recriminations over the dispensing of Crown timber access by a Cabinet minister.[34] Viewed in this way, the politician is an external agent who imposes a powerful but unpredictable stamp on forestry policies and practices. Consequently, the rueful rationalization 'it's a political matter' conveys an awareness that an external authority has taken control of matters out of professional hands, with the prospect of arbitrary and even irrational rulings ahead. This is a continuing reality in many provinces, including Nova Scotia.

But might professional forestry also be inherently 'political' in a more positive and reasonable way, according to our alternative definition? To the extent that foresters, individually or collectively, articulate and pursue interests that are at odds with other social groups, they are engaged in a legitimate process that is inherently political. This may involve efforts to define and implement a shared interest of foresters as professionals, or it may involve foresters lending their expertise to the initiatives of others. Three aspects of this alternative politics of forestry are outlined below.

Professional and Associational Politics

One type of politics, central to the subjects of this book, concerns professional training, governance, and associational representation in public affairs. The formal course of training that makes up the bachelor of science in forestry degree (BScF) marks an entry point into a carefully controlled occupational specialty. But like all accredited guilds, foresters share a bundle of professional interests. Which consequences flow from the successful creation and continued defence of the forestry profession? It is important to appreciate that there was nothing natural or inevitable about this development. In fact, it was the product of prodigious efforts by a vigorous cadre of leadership figures who founded and maintained the forestry faculties and the professional associations that seem so familiar today.

Canada's first forestry school was founded at the University of Toronto in 1907. New Brunswick followed one year later, with Laval (1911) and the University of British Columbia (1917) rounding out the original four.[35] In the early decades, the faculty complements remained small, as did the student enrolments. At the University of New Brunswick, the graduating class averaged six students in the years 1910-30. Initially, the federal forest service was the main employer of graduates. Indeed, for the first several decades, both industry and provincial governments remained sharply sceptical about the necessity of maintaining permanent foresters on staff. In Nova Scotia, there were likely fewer than a dozen foresters employed in any professional capacity as late as 1945.

The first professional association, the Canadian Society of Forest Engineers (CSFE), appeared in 1908. Not surprisingly, it was University of Toronto's dean of forestry, Bernhard E. Fernow, who brought together the twelve founding members.[36] Modelled after its American predecessor, the 'forest engineering' label underlined a professional aspiration geared to organizing complex systems. In this case, forest engineering involved getting the wood out of the forest and into the mill by cruising and mapping timber stands, building roads and railways, damming rivers, and mechanizing extraction of wood where possible. The purpose of the organization was to bring together foresters of various backgrounds to advance scientific knowledge while promoting improved forest practices.[37] Annual meetings shifted to a more technical basis after 1921. In 1925 the society launched its own journal, the *Forestry Chronicle,* as a forum for professional discussion.

Significantly, the CSFE was open to persons other than graduate foresters and continued to function as a social club as much as a professional body.[38] However, the situation changed following the First World War, as membership criteria were tightened. A professional Code of Ethics was developed, one of its articles containing a provision that 'He [the forester] will not subjugate his professional principles or judgement to the demands of employment.'[39] As we will see later, honouring this commitment could be anything but simple.

After the Second World War, the context for forest industry expansion was more suited to professional growth. In the rush to secure forest concessions for long-term supply, both government and industry required expert staff, and returning veterans swelled the student ranks of the forestry schools.

Already the CSFE recognized some of the problems posed by a dispersed national membership within a highly regionalized industry. In 1939 it introduced a new organizational layer of regional and provincial 'sections' to convene meetings between annual conferences. The Maritime Section

The graduating class in forestry at the University of New Brunswick, 1948. Mona Roy, pictured in the foreground, was the first female forestry graduate in Canada.

drew together members from the three provinces (and after 1949 from Newfoundland). Since most regional members were graduates of the Faculty of Forestry at the University of New Brunswick, the Maritime Section must have functioned as a veritable UNB alumni club. At the regional meetings, the associative relationship with industry was reinforced in at least one respect, because pulp and paper companies regularly made donations in support of the gatherings. In 1954 a separate Nova Scotia Section was formed, with fifty members. (By 1980 it had almost doubled to ninety-two.[40]) In 1950 the association adopted a new name, the Canadian Institute of Forestry (CIF), for its 1,127 members.

Despite the contributions of the CSFE and the CIF, there were some respects in which the professionalization of Canadian foresters remained incomplete. Even with the CIF membership being confined to graduate foresters holding the BScF degree, still missing was the special certification of expertise, not by the universities alone but by a form of professional 'registration.'[41] In this Quebec took an early lead. Well prior to the 'Quiet Revolution' of the 1960s, there existed a program of study and examination (associated with Laval University) leading to certification as an *ingénieur forestier* (*ing.f.*). Other leading forestry provinces followed by instituting the credential of registered professional forester (RPF).

Yet no parallel initiative occurred in Nova Scotia. While the issue arose periodically in the councils of the CIF(NSS), it was not until the advent of the Canada-Nova Scotia Forestry Sub-Agreement in the mid-1970s that serious concern arose with professional certification. This was on the eve of another major hiring explosion and turned on the question of who would be authorized to approve the massive new wave of publicly funded silviculture works. Ambivalence about the effect of dividing loyalties between two professional groups appears to have scuttled the proposal.[42] Instead, the CIF(NSS) was content to urge its executive council to become an active force in public forestry issues. More recently, it is widely perceived that the growth in the ranks of forest technicians, also organized into their own association, has complicated the task of securing provincial state consent for a registered foresters' association.

For generations, the battles between organized groups in the forest sector have shaped the growth of the profession. As professionals, foresters have been active in other organizations to advance their cause. As a broad public alliance of resource owners and users, the Canadian Forestry Association followed its American counterpart in sponsoring the conservation movement after 1900.[43] Although important at the time, the CFA (and its provincial affiliates) have declined dramatically in influence during the twentieth century. The Commission of Conservation, formed in 1907 to take stock of the management of Canada's natural resources, also constituted a forum to present the skills of the forestry profession, but it folded amid squabbles in 1921.[44]

As an expert group, foresters have sought to influence state and corporate policy in particular directions, and inevitably they have confronted advocates for other interests. In 1941 the CSFE moved into national forest policy advocacy with its 'Statement of Forest Principles.' Beginning in 1944, the Nova Scotia foresters did likewise. Their brief, 'Forestry, Economy, and Post-War Reconstruction,' outlined a comprehensive program of action 'towards better use and protection of one of Nova Scotia's greatest assets.'[45] This launched a tradition of periodic professional comment on provincial forest policy issues. The CIF(NSS) appeared in 1954, evolving into an effective (social and policy) vehicle for bringing together corporate, government, and consultant foresters, contributing solidarity to a group whose diversity of employment threatened a fragmentation of interest.[46] Constituted around working committees and an annual social/technical meeting, the CIF(NSS) became an important point of reference. As a non-business and non-government body, it enabled foresters to articulate concerns *as professionals* that it might not be possible to raise *as employees*. Thus, the Nova Scotia Section petitioned the government in 1959 to replace the Small Tree Act, which it criticized on silvicultural grounds. The subsequent Forest Improvement Act, modelled on Swedish legislation, was strongly endorsed. The section lobbied for almost twenty years for the reform of forest taxation, which was finally acted on in 1977.[47] However, its

Participants at the annual meeting of the Nova Scotia Section of the Canadian Institute of Forestry, 1959.

most ambitious policy intervention may have been the 1971 proposal *A Forest Policy for Nova Scotia*, a comprehensive eighty-page report.[48] At the same time, its organizational 'plasticity' is evident in at least two respects. First, with section duties having to be fit into heavy career commitments, it was often necessary for NSS executive officers to delegate responsibilities to willing and available members in a somewhat haphazard manner. Second, as a recognized stakeholder in the provincial 'forest sector,' the section executive enjoyed the prestige of a professional and expert body. Yet given the frequent overlap of members in representing employer, trade association, and professional interests, there were times when these hats appeared to be juggled entirely too casually.[49]

Thus, professional foresters were and are involved extensively, through their associations, in the ebb and flow of interest group politics. The chances of maintaining a consensus proved greatest when the CIF confined its attention to matters of technical forestry, and the prospects of internal schism grew as advocacy transgressed on occupational or ethical commitments. One of the foresters in our collection, David Dwyer, withdrew in protest from CIF(NSS) activities in the 1980s. Despite several decades of service to the section, he could not accept its repudiation (at the urging of industrial foresters) of the forest Group Venture program that it was his job to promote. Normally, however, the CIF(NSS) was sensitive to the core commitments of its members' paymasters in business and government when it chose the grounds for political engagement.

The Politics of Institutional Affiliation

Foresters' professional outlooks are heavily influenced by their immediate institutional affiliations with the work world. Like all expert groups, they face the challenge of adapting their skills and practices to the interests of particular employers. This plunges foresters into authority structures that follow wider commercial or administrative imperatives. Such practical accommodations may move them a long way from the scientific optima and best practice formulas taught in faculties of forestry and prescribed in research journals and professional codes of conduct. A few examples may suggest the considerable variety of potential accommodations.

One of the most familiar distinctions is that between 'industry' and 'government' foresters, and there is much to confirm this distinction. Forest products corporations look to foresters for expertise in measuring and classifying the available wood volumes, as well as planning the extraction and renewal of the forest base. State forest agencies employ foresters to administer and manage public (i.e., state-owned) forest lands in order that superior stands are available to meet public policy needs. Until relatively recently, this has meant making them available for use by private business, according to lease and in return for a royalty payment known as stumpage.

This relationship between forester and employer is complex, and it involves many variations. For example, within the corporate category, there are differences between employment by lumber firms, pulp and paper firms, and forest land companies. It has been suggested that the explosive expansion of the Canadian pulp and paper industry in the 1920s was critical to solidifying the place of foresters in business. All but the largest lumbermen remained part of the competitive economy, in which capital needs were relatively modest, entry into and exit from the industry was easy, and levels of activity could be adjusted according to flexible sources of timber supply. By contrast, pulp and paper producers faced far higher capital costs for their elaborate production facilities, were financed over a longer term at fixed sites, and used wood more intensively. All of these factors accentuated the significance of long-term guaranteed wood supply in corporate planning, finance, and operation. To the pulp operator, foresters offered unique and indispensable talents. Over time, the term 'pulp forester' took on distinct connotations of even-aged softwood plantations, clearcut harvesting, and relatively short (forty-year to sixty-year) rotation cycles. By contrast, many 'sawlog foresters' worked with a wider species mix and uneven-aged selection harvesting. In British Columbia, the H.R. MacMillan Company argued against the postwar tree farm management regime on just such grounds.[50]

Pulp forestry developed its own professional institution as early as 1918, when a Woodlands Section was established within the Canadian Pulp and Paper Association (CPPA). At the founding meeting, it was agreed that 'the imperative problem is to provide a permanent wood supply at minimum cost.'[51] For the next thirty years, the CSFE annual meeting was scheduled in conjunction with that of the CPPA Woodlands Section, thereby bringing foresters of all backgrounds, industry and government alike, together with company woodland managers. The foresters were not unmindful of the conflicting pulls of the workplace.

In pulp and paper firms, most foresters were attached to a woodlands division, in which they worked under the authority of the woodlands manager. Here there was definite room for tension between the forest development mandate of the professionals and the fibre production imperative of the managers. This could crystalize in any number of issues: the range of forester duties that went beyond inventory cruising; the choice of harvesting by clearcut or selection cut; the extent to which the ease of cutting crews would be qualified by the needs of site regeneration; and the disposal of slash and other wastes left at the cutting site. In the era of vast, cheap, virgin forests, it was virtually impossible to build support for intensive silviculture practices. John Bigelow looked to dynamic commodity markets to provide the price and equity incentives for improved forest management, though his was a voice before its time.

We should stress that lumbermen were not at all blind to the advantages of technical forestry. Indeed, one of our subjects, Donald Eldridge, began his career with the Eddy Lumber Company as a surveyor and forest land buyer. But here there was more room for variation. Where pulp companies were obliged to secure extensive forest acreage prior to financing and start-up, most lumber companies continued to acquire quality sawlog properties over their operating lifetimes, constantly engaged in the buying, selling, and leasing of timber stands. Since the larger firms had wood volume needs closer to that of the pulp sector, and had to plan accordingly, it was here that foresters were most likely to find employment. But for many other mills, the job of finding and cruising timber went to practical woodsmen whose knowledge was experiential rather than technical.[52] Once again the practice of the craft was filtered through a commercial screen, though one of different shades. Sawlog forestry involved a wider range of species (including hardwoods) and longer rotation ages (80-120 years). The variation in cutting practices was extreme, reflecting great differences in stand characteristics, enterprise philosophy, and state regulation. They extended from high-grade logging to clearcutting to selection management for long-run sustained yields.

On the other side, 'government' forestry was less explicitly commercial in orientation. In principle, its goal, long-run management of public forests for maximum growth and quality, coincided closely with the classic imperatives (and codes of conduct) of the profession. No doubt many enthusiastic graduates saw government forestry as an alternative calling, free from the relentless commercial dictates of business. David Dwyer spent his career in the Nova Scotia Department of Lands and Forests, first as a District Forester, then as an Extension Forester, and finally as a coordinator of private forest management ventures. This institutional setting followed its own set of norms, and quite a separate policy framework, from that of industry. However, it also imposed its own limitations, as newly hired foresters soon discovered. The scale of resources available in the public administration often paled beside its private sector counterpart. This could be reflected in salary scales, equipment, support personnel, and operating budgets. In many parts of Canada, government forestry was also burdened with a long-standing obligation for fire protection services. This obligation carried over from the pre-professional era, and, while it was always regarded as integral to forestry work, it could preempt large blocks of funds at the expense of silvicultural and management work. Finally, there was the ubiquitous 'political' factor, which generally meant interventions by elected politicians and Cabinet ministers in search of jobs, permits, and leases and general-purpose preferences for local constituents. In Nova Scotia, the forest ranger system began in 1904, with the appointment of a Chief Ranger in each rural municipality. While the numbers rose and fell over the years, these jobs exemplified old-style patronage.

When foresters first entered government organizations, they encountered a field staff of woods-wise rangers with little formal training but considerable practical experience as well as partisan connections to the government of the day. It was also a time when ministers of the Crown could take a firm grip on the minutiae of the forestry department. Consequently, state foresters had to operate in a competitive environment. They encountered a system already deep in hierarchies and fixed outlooks, and their expertise was potentially destabilizing. In turn-of-the-century Ontario, for example, the politicians 'had been convinced that foresters wanted to go too far too fast.'[53] In New Brunswick, the Forest Service was established in 1918, but electoral politics undermined its early momentum, and progress was not restored until after the Second World War.[54] The presence of a senior provincial forester proved to be one of the key variables affecting the timing and extent of professional advancement. Otto Schierbeck, the first Chief Forester appointed in Nova Scotia in 1926, is profiled below. He faced constant challenges in his efforts to adapt European-inspired practices to the ranger service that he led. After his dismissal, his successor, Wilfrid Creighton, proceeded far more cautiously, recalling that 'over the next few years I learned by trial and error what my position in the Department was supposed to be.'[55] Even though the MacDonald government established a Civil Service Commission in 1935, more than twenty years elapsed before partisan hiring was seriously curtailed under Stanfield, and even then part-time employment in highway maintenance and liquor sales was exempted.

Another institutional variable that closely shaped the circumstances of government forestry was bureaucratic location. This refers most generally to the place of an agency, such as a forest service, within the wider complex of administrative departments and agencies. Much of the intellectual climate and operational mandate of a service springs from location. This theory of 'bureaucratic politics' as a determinant of policy behaviour is summed up in the epigram 'Where you sit [i.e., where your desk or office is located within the administrative state] is where you stand [i.e., the perspective held on the issue under consideration].'[56] It will also determine the place of the branch, agency, or department in the hierarchy of state institutions, the patterns of bureaucratic alliance and rivalry within the state overall, and degrees of administrative autonomy open to the forest service.

One well-documented instance involves the US Forest Service. It was established in 1905, on a tide of conservation thinking supported by President Theodore Roosevelt.[57] Significantly, the Forest Service was attached to the Department of Agriculture, given the evident analogies between farm and forest crops as renewable resources, along with the fact that both land bases were held largely (at the turn of the century) in private hands. This choice of site was to prove critical for the Forest Service. It was influenced subsequently by the wider policy philosophies of the farm bureaucracy, including the techniques of working with private landowners, and the integral relationship of resource conservation and use. Over a period of time, these were blended with the corpus of technical forestry, including its concern with the management of public lands. Then, in the early twentieth century, a bureaucratic rivalry began to emerge that carried policy and professional consequences of the first order. Virtually from the moment of its inception in 1905, as the administrator of vast federal lands in the western United States, the Department of Interior sought to bring the Forest Service under its umbrella. In the far more commercially driven environment of Interior, which elevated the priority of exploitation above that of conservation, forest management would have been conditioned by a far different set of norms. Significantly, the Forest Service mounted extended campaigns to resist incorporation by the department, carrying these on occasion to the highest political levels.

The pattern differed considerably in Canada. Here the federal state surrendered its imperial grip on prairie lands in 1929, with important consequences for the federal forest service. Having lost a proprietorial base, its mandate was redefined in terms of forest research and commercial development.[58] Since provincial authorities hold prime jurisdiction over forest management, a different pattern developed. Here the design of the public administration tended to separate the mandate for arable farming (a private tenure resource) from that of forests and mineral resources (predominantly

Crown-owned resources). The Canadian provincial tradition has been to locate forest services either in distinct departments of forests or in subdivisions of consolidated departments of lands and forests (including wildlife and parks) or departments of natural resources (sometimes including minerals).[59] Since 1926, Nova Scotia has followed the middle pattern closely under a combined Department of Lands and Forests.[60]

Yet Nova Scotia broke with the provincial norm in one critical respect, since Crown forest holdings constituted less than one-quarter of the provincial forest area, while the preponderance was privately owned. Nevertheless, the frameworks and philosophies of government forest administration predominated from the outset. Private forest management, whether for farm or non-farm woodlots, large tracts or small, has been a distinctly secondary policy concern for most of the twentieth century. The forest service developed its operational framework in virtual isolation from primary forest product marketing (advanced, significantly, by the Department of Agriculture) and forest manufacturing (handled by the Department of Trade and Industry). As illustrated in several chapters below, this arrangement congealed over time into an informal division of labour (and rivalry) between the Department of Lands and Forests and the Department of Agriculture, with the latter demonstrating a far greater commitment to small private forest owners as a productive and management segment. It testifies to the potential impact of bureaucratic politics.

The Politics of Internal Hierarchies

We need to note one final political dimension. It concerns the experience of individual foresters according to their locations within formal organizations. Any complex agency, whether corporate or public administration, is founded on the principle of hierarchical authority. In this way, policies are formulated and executed, and specialized talents are organized and applied. The economist John Kenneth Galbraith has captured part of this phenomenon in his concept of the 'technostructure,' a configuration of information specialists who provide an indispensable intermediate layer of expertise for planning in large organizations facing complex problems.[61] This technostructure is an essential element of both public and private bureaucracies. Furthermore, it can be argued that these hierarchies are political in our stipulated sense of the term, since they succeed in generating solutions in situations of difference. Most significantly, foresters are caught within such webs of hierarchical authority, which play a crucial role in determining their actions in the short run.

Consider the pulp and paper firm as an authority structure. A diversified organization, it combines a series of complex operations (wood production, transportation, single-stage or multi-stage processing, marketing, and sales) in order to realize commercial income. While the woodlands operations, where foresters operate, are crucial in the planning stages of the venture, it is equally clear that once production begins the mill dictates the crucial parameters. The rated capacity of the pulp and paper machines sets the general target for timber production, while actual levels of production will be adjusted to market conditions. Woodland managers are furnished with fibre volumes that must be met come what may. Since foresters tend to operate under woodland managers, their forest development efforts are doubly subordinated – by company-wide constraints and the immediate economics of wood supply. This is illustrated well by Ralph Johnson, the longtime forester for Mersey Paper Company in Nova Scotia:

> Unfortunately, at Mersey the woodlands manager was in charge of both the woodlands department and the forestry department until 1958. The logging superintendent wanted only clearcutting and did all he could to oust any silviculture from Mersey's forest operations. Roadways were deliberately cut three times wider where they passed through coniferous forests than elsewhere, and other openings – such as log brow sites – were made unreasonably large. The woodlands manager did nothing to stop this, and as a result of the large openings in the forests there was more storm damage than there should have been. Logging contracts and stumpage sales were under the supervision of the forestry department and in these, selection cutting was quite successful because the openings were kept small.[62]

Similarly, in government forest services, there are varying levels of authority descending from the deputy minister through the provincial forester to regional and district forestry offices. Staff outside headquarters seldom have a direct impact on policy development (as distinct from policy implementation), which is normally confined to a small group of senior officers. This helps to explain why 'political' intervention by industry interests will focus at these very levels. At the same time, line foresters may face severe sanctions for ignoring or defying the command and control mechanisms of the organization.[63] The most blatant forms of dissent may be met by reprimand or outright dismissal, as in the case of Donald MacAlpine. A unit forester in Nipigon, Ontario, MacAlpine was fired in 1982 after refusing to issue timber-cutting licences on tracts where official inventories misrepresented the available stock.[64] In situations lacking a major infraction, more subtle forms of discipline can be imposed. Field staff can be transferred, or threatened with transfer, to distant localities or to 'punishment' positions. Algonquin Unit Forester Don George was a tenacious opponent of clearcutting in the Pembroke region of Ontario in the 1970s and 1980s. He was personally threatened with a transfer to the outer reaches of northwest Ontario by his regional office, and he observed a colleague who was uprooted to Sudbury in similar circumstances.[65]

Forestry as Ideology

The politics of forestry as expressed at the various levels identified in the previous section suggest that forestry is thoroughly ideological. The term 'ideology' carries uncomfortable overtones for many people. It may suggest narrow and dogmatic thinking, an unwillingness to accept awkward realities, and a penchant for rigid planning. It may also conjure up the battle of the 'isms': conservatism, liberalism, socialism, and so on. While each may have a limited basis in reality, they are unnecessarily limiting and should not be allowed to stand in the way of a potentially useful concept. Applied in a different sense, ideology offers an extremely useful tool in analyzing the conflicting outlooks found in many fields, including forestry.

We approach ideologies as systems of ideas that are formulated to help make sense of a complex situation. They are forms of intellectual shorthand that allow people to discuss and organize action in modern society. These ideas are not associated at random but are related according to the particular problems at hand. They need not be rigid and unyielding; in fact, most ideological outlooks are quite supple and capable of evolution over time. Not only do ideological perspectives help to explain situations, but they also offer prescriptions for action, and as a result they are often front and centre in political controversies. Ideologies are not confined to experts either. They can be fashioned from a variety of raw materials, such as folk values, traditions, scientific ideas, and philosophical ideas. Each instance of ideology is also of interest since it tends to spring from particular social and economic (or class) concerns.

Seen this way, ideology enters the forestry field in a particular fashion. It is distinct from the professionalized pursuit of forestry in the sense that it is accessible to a far wider range of social and public interests. It is inherently political, but its core constituency is far more broadly drawn. It offers a means for such diverse groups as forest landowners, woodsworkers, rural residents, wood-harvesting and -processing enterprises, outdoorspeople, environmental advocates, and any number of additional public interests to communicate and to act. It should be evident that we see ideology in a positive light, as a necessary and inevitable part of the forestry scene.

Another fascinating property of forest ideologies is that they may be expressed both in very simple and in very complex terms. In the midst of a confrontation between loggers and old-growth forest preservationists, the dialogue (or, as it is sometimes labelled, the discourse) can be sharp and blunt. In some cases, it may be seen as a relatively simple choice of 'jobs or nature.' Both sides may be willing to accept this formulation, while differing on how to resolve it. In such a case, it helps to make sense of an intractable problem, with each side invoking worthy preferred outcomes. In ideological shorthand, the confrontation offers powerful symbolic fuel for social solidarities. Under such banners, a woodsworking coalition (which may be supported by trade unions, employers, local business chambers, or others) advances one action plan, while a preservationist coalition (possibly aligned with wildlife advocates, nature groups, recreational users of the forest, tourist operators, or others) advances another. Despite the evident simplicity, this sort of ideological statement taps into social beliefs and concerns at many levels.

In this book, the concept of ideology is especially useful when examining two concepts of professional forestry: its scientific and its social dimensions. Ideological notions abound in the specialized vocabularies of the science of forest management. Consider, for example, the prospective treatment of forest insects destructive of forest biomass. Biological and botanical science has always been close to modern forestry, both in theory and in practice. Not only is it a foundation of the university curriculum, but it also animates much of the field research for modern silviculture. This is both a strength and a weakness, since science is both selective and dynamic. Important questions need to be asked about the state of botanical thinking at the time when modern forestry congealed as a subject of study as well as about the capacity of forestry theory to take account of subsequent advances.

For example, how closely were related disciplines such as entomology (the study of insects) tied to forest biology? Moreover, what was the state of entomological understanding? After the Second World War, the chemical industry promoted major advances in insecticides and herbicides, and both found their way rapidly into farm and forest applications. In this process, entomologists' experimentation with and application of 'biological' antidotes to forest pests by identifying and releasing their parasites into infested areas were rapidly marginalized. The same was true for the widely held beliefs that birds and silvicultural methods could serve to mitigate insect damage.[66] Thus, several approaches were available in principle to control tree damage from insects. How was the choice resolved? Evidence from the philosophy of science suggests that organizational factors may play a major role in deciding such questions. Social and business ties may figure prominently here, in the field sometimes labelled the 'chemical-industrial complex.'[67] As disciplinary affiliations and bureaucratic rivalries reinforce the preference for one research paradigm over another, a commitment grows over time as research funds are invested and an agency's 'stake' in a particular strategy intensifies.[68] As we will see below, these factors were prominent in a number of Nova Scotia settings, particularly on questions arising from spruce budworm damage. Hawboldt's text below sheds light on the province's advocacy of silvicultural solutions and its resistance to the spray option in addressing the spruce budworm invasion of the 1970s. This is in contrast to the chemical spray option so thoroughly embraced in neighbouring New Brunswick.

Consider also the phenomenon of clearcut logging. In general, it is the practice of cutting all the trees in a forest stand, with the expectation that the site will be regenerated by natural seeding or by the planting of seedlings. Adjusted to site conditions, it has been one of many harvesting and silvicultural techniques practised for more than a century of Western forestry. Yet in modern public discourse, clearcutting has acquired a far more specific connotation as the standard industrial forest harvest in coastal rainforest and boreal forest stands. It is graphically illustrated by the ground and aerial photographs of vast denuded landscapes taken soon after the harvests.[69] Apart from the now-famous West Coast sites of the Carmanah Valley and Clayoquot Sound, the public consciousness is now sensitive to vast clearcut images, from the Quesnel Valley of the BC interior to the Keppoch Plateau of Cape Breton Island, Nova Scotia. Now generally regarded as an industry standard, these massive clearcuts have become emblematic of rapacious corporate forestry.

On the immediate level of pictorial image, there has been a debate on the question of authenticity. Both industry and government foresters contend that such pictures are extremely misleading since they unfairly freeze a single image in the public mind. Typically, the newly planted seedling forest is visibly insignificant against the stark barrenness of the clearcut landscape, and subsequent pictures of the same forest after twenty or forty years of growth are seldom available to balance the context. On this particular point, there can be little question that opponents of clearcutting have scored a massive tactical advantage. It would seem that not only is a picture worth a thousand words but also that the first picture is worth more than any subsequent pictures to the contrary.

Quite apart from the photograph itself, it is notable how the image opens the way for a wider debate that is itself redolent with ideological significance. How large does a patch cut (which will reseed naturally) have to be to be labelled a clearcut?[70] There is also the question of regenerative efficacy, or whether seedling forests achieve adequate coverage to start the replacement forest. In the public mind, a positive connotation attaches to tree-planting efforts, leading both government and private agents to launch triumphal celebrations of 'million tree' or 'hundred million tree' thresholds. Also relevant are the consequences of replacing a diverse, uneven-aged forest with an even-aged monoculture. More specifically, this practice raises questions about the impact of narrowing the forest gene pool and whether it leaves any future forest more vulnerable to insects and disease. Finally, there is the question of a net depletion of forest volumes in the event of large-scale failure to regenerate. We have dwelled in great detail on the ideological status of the clearcut. Yet similar explorations are possible on any number of forest management concepts and practices, including the 'tree farm,' the 'annual allowable cut,' and the 'sustained yield.'

The second ideological aspect on which the subjects of this book shed some insight is social. Consider, for example, the central position of forest tenure. By far the greatest proportion of Canadian forests lies in state (Crown) hands. This has shaped approaches to forest management in innumerable ways, since Crown ownership conveys powers to government foresters well beyond those that they could mobilize to influence forest management by private landowners. Only in the eastern provinces of Quebec, New Brunswick, and especially Nova Scotia do private forest lands account for a major proportion of the whole. This is closer in character to the situation in parts of central Europe and Scandinavia.

Consequently, it is relevant to inquire about the sort of adjustments made by government foresters in a significant 'private land' jurisdiction. Nova Scotia's unusual tenure pattern seems to have been considered an anomaly best ignored. Government foresters have viewed their restricted Crown share as a crushing handicap. To paraphrase Sigmund Freud, it might be said that the Department of Lands and Forests has suffered from the syndrome of 'Crown land envy' when it has met its counterparts from other parts of Canada. One memorandum aptly captured this spirit in referring to 'the complexities, if not vexations, that it [privately owned forest land] poses for any agency of Government responsible for the management of the resource in the public interest.'[71] As a result, programs of private land purchase have been pursued, with different levels of vigour, since the 1930s.

This special feature seems to have been little recognized in forestry education, which remains focused on managing Crown holdings. To the extent that private woodlands were part of farm woodlots, they could be delegated to agricultural college programs. Alternatively, the private owners became the responsibility of Extension Foresters.

Although the extent of private woods holdings in Nova Scotia became clear as a result of the 1953-7 forest inventory, it was not until 1971 that the Department of Lands and Forests made its first empirical investigation of small private landowners as a group. This underlines the low priority attached to this vast group of small owners in the postwar period.

In Nova Scotia, the predominantly private ownership of the forest estate led to an ongoing concern with the security of forest property rights from state encroachment. This figured in debates about forest land taxation, poorly surveyed property boundaries, regulation of the export of raw wood, prescribed forest management practices on private lands, and the Crown purchase of private forest lands, engendering deep suspicions by rural woodlot owners about government designs on their lands.[72] In some cases, this was reinforced by a perceived bias of government policy in favour of large corporate interests. By the 1960s, the cumulative impact of these debates led many small rural property owners to fear all government interventions as inherently threatening. Thus, during the campaigns for state-supported silvicultural and pulpwood marketing legislations, opponents of the bills argued (effectively in some instances) that silvicultural programs and commodity marketing were ploys by the provincial government to wrest control of private woodlot products.[73] This theme of the 'tyranny of state regulation' forms an enduring part of forest policy discourse in the province.

One final instance of the ideological dimensions of forest management concerns educational programs directed to the lay public. This was closely interwoven with the twentieth-century movement for resource conservation that spilled into Canada from the United States during the Theodore Roosevelt years. While most of the disciplined regulatory interventions fell to state authorities, it was recognized that a public educated and committed to the conservation ethics of forest renewal and optimal utilization was an essential adjunct.

Consequently, the Canadian Forestry Association was established in 1900, with its respective provincial branches, to promote woods safety (particularly to the danger of fires) as well as the worth of tree planting as a forest renewal measure. It was also hoped that these associations would act as effective pressure groups in favour of conservation policies. (This is aptly captured in a comment attributed to Franklin Roosevelt after meeting a conservation lobby group: 'Okay, you've convinced me – now go out and bring pressure on me.'[74]) Through these efforts, the public was conditioned to the symbolic worth of tree planting, even one plant at a time. This continues to be a standard of efficacy, where the corporate slogan of 'one tree planted for every tree felled' attracts a certain credibility, without any concern for species, age, and overall forest structure.

Ideology, then, strikes at the heart of a fuller understanding of why forestry has come to be what it is. It also helps us to see alternatives to the status quo and to explore dissent and forest views that go against the grain. A diversity of opinions and free expression of these opinions are necessary in promoting change. The late forester Jack Westoby had it right when he wrote that 'Living controversy, with full freedom of discussion, is the only way in which science can advance; and ... it is also the precondition of forestry policies which will fully serve the people.'[75]

2
Otto Schierbeck: Nova Scotia's First Chief Forester[1]

The transition from a lumber to a pulp economy introduced professional forestry to eastern Canada. As elsewhere, forestry assumed an instrumentalist view toward the use of forest resources and an elitist orientation that routinely overlooked the forest practices and needs of non-industrial groups: small woodlot owners, poor settlers, and Aboriginal populations. As Vivian Nelles put it, 'the business and professional middle classes, the opinion-making groups in distant urban centers, were the ones who initiated the demand for the "business-like management"' of Canada's forests.[2] In a decentralized national economy based on staples exports, where provincial ownership of forest lands was dominant, forestry was further influenced, even compromised, by the provincial government's clamour for pulp and paper industry investment. Provincial governments here faced double-barrelled demands from the pulp and paper industry, which exploited the old partisan political system for favours while at the same time pressuring governments to form professional bureaucracies to accommodate and service their long-term needs for cheap fibre and hydroelectricity. This occurred in contrast to the short-term demands of the lumber industry for sawlogs, demands that were more frequently dependent on partisan connections. The rise of forestry and the forest profession in Canada was closely tied to the growth of the pulp and paper industry; for example, forestry schools were sponsored by the industry, and the annual meetings of the Canadian Society of Forest Engineers took place in conjunction with the Woodlands Section of the Canadian Pulp and Paper Association.

Here we explore how one individual, Otto Schierbeck, the first Chief Forester of the province of Nova Scotia, fared in and coped with this situation. Schierbeck was a Danish forester committed to forest conservation and the defence of professional practices against political interference and public apathy and resistance. As such, he exhibited the typical strengths and weaknesses of his profession at the time. But he was also unique in the sense that he challenged the conditions under which Canadian lumber and pulp and paper businesses conducted their mill and forest operations. Schierbeck questioned the staples economy, the political institutions that had grown up to support it, and the successive exploitation of staples for export. He was thus critical of the practices of both the lumber and the pulp and paper industries and the various private and Crown land and forest management policies associated with them. He envisioned a forest economy in which forestry and professional foresters were at the cutting edge of planning and manipulating the forest for the greatest human use. This involved a more active role of foresters in forest management and in the development of a more diversified forest economy.[3]

Schierbeck's career provides an excellent medium for exploring the Canadian staples economy by bringing into focus the conflicts and the contradictions between Canadian forestry and foresters and industry, government, and society. Schierbeck not only held a position of the highest authority in the Nova Scotia Department of Lands and Forests for seven years, but he was also uniquely outspoken about Canadian forest practices, something that may have stemmed from his Danish background, for in Denmark foresters held a higher social and independent status than in Canada.[4]

A Forestry Professional in the Political Wilderness

The progressive reform movement in the United States spilled over into Canada in the early twentieth century. Following the tenets of American forestry as propagated by Gifford Pinchot, the Canadian Forestry Association and the Commission of Conservation, both national associations, lobbied for the recognition of forestry as a science and profession free from political interference. The impact of these organizations typically remained weak, and, in the face of jurisdictional conflicts and jealousies between federal and provincial governments, the Commission of Conservation folded in 1921, while the Canadian Forestry Association maintained a purely educational role.[5]

Partisan politics and patronage remained prominent in Canada in the 1920s, and they remained important in the sale and lease of Crown lands. They also proved instrumental in Schierbeck's rise and fall as Chief Forester. Legendary forest conservationist, industrialist, and politician Frank John Dixie Barnjum played a key role in Schierbeck's appointment. Barnjum was Canada's most well-known forest conservationist and an unabashed gilded-age entrepreneur who used politics and forest conservation to advance his own business interests. By the 1920s, he had become the most prominent advocate of an export embargo on Canadian pulpwood from privately owned lands on grounds of conservation, contending that the increasing exports of pulpwood to the United States were devastating and depleting Canada's forest resources. Barnjum's lobbying efforts occasioned the Liberal government to appoint a royal commission to investigate the pulpwood situation in Canada. During its hearings, Barnjum's conservationist message was questioned, his critics arguing that Barnjum stood to profit handsomely from the embargo through the sale of his extensive pulpwood lands in the United States and the purchase of pulpwood for his pulp mill in Nova Scotia. His critics faced equal suspicion about their concern for conservation. The most active opponents were American pulp and paper companies and Canadian pulpwood exporters interested in the free flow of pulpwood between the two countries.[6]

When Barnjum failed in his bid to have a national export embargo passed, he switched political affiliation from the federal Liberals to the Nova Scotia provincial Conservatives. In the 1925 provincial election, he gained a seat in the legislature. The following year he was instrumental in the appointment of Schierbeck as Chief Forester of the newly formed Nova Scotia Department of Lands and Forests. Schierbeck had previously received a $5,000 prize in an essay competition arranged by Barnjum on how best to combat the spruce budworm and spruce bark beetle. Barnjum then recruited Schierbeck in 1923 to become forester for his Nova Scotia holdings. Barnjum's unique position allowed him to delegate extensive powers to Schierbeck, whose duties involved far more than managing forest resources; the Chief Forester was also the Chief Game Warden, the Chief Provincial Land Surveyor, the chairman of the Board of Examiners for Provincial Land Surveyors, and the chairman of the Board of Land Surveyors.[7]

Barnjum possessed clear motives for promoting Schierbeck. As in Barnjum's other dealings, a mixture of private gain and public service figured in the calculations. Subsequent to the appointment, Schierbeck and Barnjum maintained reciprocal personal and business ties. Barnjum used the names of Schierbeck and William L. Hall, the attorney general and Lands and Forests minister, in his prospectus for a pulp mill development in the province. Schierbeck informed Barnjum on forest inventories and the attempts by rival interests to establish a pulp and paper mill in the province. Throughout his appointment, Schierbeck remained critically aware that 'the possibility of a change in government ... might lead to my dismissal in favour of a man from the Liberal party.'[8] Schierbeck's allegiance to Barnjum thus remained strong, even after Barnjum's death in 1933. By that time, Schierbeck had begun publishing a magazine, the *Forest Crusader*, which included a Frank J.D. Barnjum page and a dedication to Barnjum's memory and 'to the advancement of the cause he sponsored.'[9] Despite his ties to Barnjum, Schierbeck on several occasions opted for professional responsibility over personal allegiance, proving that he was not simply a 'yes man.' He pushed Barnjum to pay overdue taxes to the province, he charged him for services rendered, and he urged him to donate money to the Boy Scouts and to buy fire pumps for his department, requests that Barnjum declined.[10]

In keeping with the times, Schierbeck held an instrumentalist view of the forest environment. He had little notion of or respect for nature's own dynamic and other systems of knowledge rooted in local populations and communities. Nature existed to be controlled, improved upon, and managed by professionals 'for the greatest good of the greatest number for the longest time.'[11] Reforestation by planting provided a way of 'getting quicker growing and better formed trees ... [Natural reproduction] is slow, and unfortunately old mother nature is rather disorderly in her methods of regeneration.'[12] Referring to European forestry, he asserted that, 'when planting, the forester or landowner has complete control of the new stand. He can be assured that it becomes uniform and regularly spaced without openings or weed trees.' He similarly stressed the urgency of forest fire suppression and emphasized the importance of a continuous forest cover for industry and for protection of water reservoirs. He also pioneered efforts to halt the spread of the spruce budworm in Nova Scotia by cutting buffers and dusting the forest and predicting a 'sweeping victory ... over the insect pests.' Schierbeck, along with most of his contemporaries, possessed clear and supreme confidence in his ability to manipulate and control nature.[13]

One aspect of Schierbeck's mission to control nature and to promote better forestry practices involved the development of a knowledgeable and

Otto Schierbeck with a favourite hound. The dog once saved his life after he fell through ice on a river.

professional Forest Ranger Service, whose major job was to prevent and suppress fires. This was an exceedingly difficult task in a province that included no practising professional foresters. Prior to Schierbeck's appointment, forest rangers served only on a voluntary and patronage basis. On Schierbeck's initiative, the province engaged six Chief Rangers on an annual basis and six to eight subforest rangers serving full time in each of six districts during the fire season. At that time, more than 560 volunteer forest rangers complemented the fire-fighting and -prevention force.[14]

Just as Gifford Pinchot had done for the US Forest Service some twenty years earlier, Schierbeck tried to build up a knowledgeable and quasi-professional field staff that understood various forest sector matters. He described the work of the forest ranger or game warden as highly specialized: 'First of all, the official must have a thorough knowledge of the topography of the country he is going to protect. He must know all the roads, trails, lakes and streams in the region. He must know where all the sporting camps, lumber and mining camps are located ... He must be able to write a report and must have a working knowledge of surveying, cruising and scaling.'[15]

Two factors seriously compromised the prospects for such a field service: the Canadian partisan political system of patronage appointments to the Civil Service and the fact that local rural forest practices often did not correspond to the dictums of scientific forestry. Accustomed to such political appointments, with associated favoured status and the ability to dispense and profit personally from patronage, his rangers did not always follow orders. One anecdote speaks of a ranger telling Schierbeck off, proclaiming 'I take my hat off for a lady and I kneel for my God, but I don't take my hat off for a goddam man.' Schierbeck, on his part, could respond in equally harsh terms. When four of his rangers neglected to send him reports on the status of fire-fighting machinery in their districts, he chastised them for laziness in their conduct of official business and for undermining the relationship between field operatives and the central office. Schierbeck's zeal must at times have burdened his rangers, such as when Schierbeck asked them to keep account of and report on the distances that they travelled or when he implored them to always 'be on the go' in their search for forest fires.[16]

The success of Schierbeck's field staff was compromised not only by partisan politics but also by the general hostility of the conservation movement toward certain groups in society. Settlers and farmers were often subjected to criticisms. In New Brunswick, Quebec, and Ontario, settlers were blamed for settling on lands unsuitable for agricultural purposes. They were accused of wanton destruction of the forest through cutting trees and setting fires. Lumber and pulp men thus called for the establishment of forest reserves, exclusive areas for forest extraction where no settlements were allowed. In Nova Scotia, where three-quarters of the Crown lands had been granted to settlers and lumber companies, and where human settlements were relatively dense, the establishment of forest reserves was less relevant than coping with the use of fire by settlers. The long-established practices of clearing lands, burning barrens to promote blueberry growth, and burning brush, garden, and house trash, from which fires allegedly spread to destroy forests, were cases in point. Schierbeck attempted to establish blueberry associations to manage and prescribe burns to particular tracts of lands under the supervision of his rangers. He set up demonstration lots in an attempt to show that mowing or grazing by sheep served better than burning in promoting blueberry growth. He tried to identify which lands were most suitable for the blueberry industry, leaving most other lands for forestry. He also introduced the practice of issuing permits for burning.[17]

Sawmill operators also needed permits and were regularly inspected to minimize the fire threat that they posed to the forest. Indeed, most of the young department's efforts and expenditures went to fire control. The rangers took part in building fire towers to spot and locate fires. The larger forest products companies supported and assisted the department in this respect by providing building material or assisting in the installation of phone lines to the towers. The companies also lent their equipment to the rangers when fighting local fires. The rangers even visited the Boy Scout Associations to promote forest education and to instruct them in the forest regeneration of the province's extensive fire barrens.

Efforts to suppress forest fires were often not appreciated by rural landowners, and local residents commonly viewed prevention measures with suspicion. The blueberry associations never met with success, and the burning stipulations were routinely ignored. In cases of infraction, Schierbeck instructed his rangers that all 'must, without exception, be prosecuted.' In the 1927 season, he mounted an investigation into the burning of the blueberry barrens without permits that resulted in sixteen convictions. Local residents also thought that fire protection and fire fighting benefited the large forest companies more than them. Fire protection, in short, took place at government expense for the largest landowners. Some also resented the power given to the rangers to commandeer local residents to fight fires. Others set fires deliberately to 'create' local employment. So common was this practice that Schierbeck's successor, with far less authority and far fewer resources to patrol the forest, kept the pay for casual firefighters low because 'higher wages could have resulted in more fires.'[18]

Another aspect that set the conservation movement apart from many of the less wealthy elements of the rural constituency was the lack of support for small woodlot owners in the pulpwood market. In the Canadian provinces, the concern for forest regeneration and farm woodlot management stood apart from Crown lands administration and the Crown land leases. In Nova Scotia, where small woodlot owners held more than 50 percent of the forest lands, this issue had particular gravity. Schierbeck understood the urgency of introducing forest management on small woodlots but somehow failed to connect it to the broader questions of political and economic power. In 1928 he embarked on a collaborative program with the Dominion Forest Service for the thinning of farm woodlots. Ten farm woodlots of immature white pine, red pine, balsam-spruce, and poplar were chosen in Kings and Annapolis Counties. The program sought to demonstrate the financial advantage of the thinning process by providing a marketable crop and then showing 'the subsequent increase in quality and merchantability of the remaining stands.' Control stands were located beside the thinned plots, and both were selected at sites adjacent to well-travelled highways for demonstration purposes. During a mere four-year period, the difference in the growth of the standing trees between the thinned and the unthinned plots was insignificant, but the total volume (including the original thinnings) produced in the thinned stand was significantly higher. Mortality was also lower in the thinned stand.[19]

But Schierbeck could not complement the forest-thinning program with an improvement in the prices and markets for wood products. To his credit, he understood the critical importance of prices and markets for the success of forest management, arguing that 'proper forestry management will never be started in our forests as long as it is advocated on the basis of good citizenship, national economics, and philanthropy. Forestry will only be practised when the interested parties are convinced that it is a paying proposition.' But he was naive in promoting such a cause. In favouring the pulpwood export embargo, Schierbeck believed that he could help farmers to market their wood at better prices locally. During the embargo debate, he lobbied aggressively against absentee ownership of forest lands and the export of pulpwood, noting that Nova Scotia was 'sorely beset by a bad stepmother, the pulpwood exporter,' and awaited 'her fairy prince, the Canadian or English pulp and paper manufacturer, so that she can take the place she deserves in the royal castle of the Canadian paper industry.' Once the prince arrived, in the shape of the Mersey Paper Company's newsprint mill in Liverpool in 1928, Schierbeck thought that it would provide farmers with an important market for pulpwood and that it would be 'the wedge that will open the way for other similar industries which will manufacture Nova Scotia wood in Nova Scotia mills, and by Nova Scotia workmen.'[20]

But such conditions never materialized. Pulpwood exporters, in fact, contributed to making a less than perfect market somewhat more competitive. The Mersey Paper Company maintained a firm monopoly in the pulpwood market, and the exclusion of wood products from the province's primary products marketing legislation made organizational efforts by small woodlot owners difficult. These were seriously limiting conditions. At best, Schierbeck's thinning program was seen as a government handout. At worst, it was seen as irrelevant to the needs of woodlot owners and a mere subsidy to the Mersey Paper Company and the smaller groundwood pulp mills in the province.[21]

Partisan politics and patronage issues provided the immediate cause for Schierbeck's dismissal in 1933. Some of his rangers appeared to be 'captured' by local populations. They either shared the resentment of local residents toward any dictates from professional men in the emerging central bureaucracies of the provincial capital or owed allegiance to the political forces (sawmillers and pulpmen in particular) that opposed Schierbeck's policies, seeing them as a threat to their continued practice of dispensing patronage. In 1933 pressures were brewing among the discontented faction of Schierbeck's field force. While Schierbeck was on an official visit to promote Nova Scotia lumber products in the United Kingdom, a group of rangers met in Halifax and staged a 'palace revolt' in the Department of Lands and Forests, demanding his dismissal. A publicity campaign against Russian pulpwood imports to Canada orchestrated by G. Howard Ferguson, the Canadian high commissioner to Britain, did not help Schierbeck's situation. While in the United Kingdom, Schierbeck agreed to add his name to an article describing starving people and deserted villages in Canada that attributed their plight to the Russian pulpwood imports that undersold local products in various St. Lawrence ports. The article did little for Schierbeck's popularity in Nova Scotia, where the impact of the Russian wood was negligible.[22] Embarrassing to the government, the incident was used to justify his dismissal on his return.

Partisan concerns, popular resistance to the implementation of scientific forestry that seemingly favoured some at the expense of others, and the waning popularity of Schierbeck in the face of a pending election prompted the Conservative government to dismiss him in 1933. The official announcement claimed that the position of Chief Forester had been abolished 'owing to the necessity of effecting economies in the public service.' Later in the same year, following the provincial election and the return of the Liberal Party to power, the staff of the Department of Lands and Forests, whom Schierbeck had hired and trained, were dismissed, and the department was reorganized.[23]

Schierbeck's dismissal reflected in part the tensions between the new forestry science and professionalism and the old political partisan system and the different ways of seeing and using the forest in the old rural economy. This was a common pattern wherever scientific forestry confronted the old rural society. But these conflicts were not sufficient to occasion the dismissal of Schierbeck. He remained unique in that he did not bow to the 'business' in 'business-like management.' He continuously challenged and questioned the methods and terms under which the staples economy operated in Nova Scotia and Canada.

Challenging Forestry in a Staples Economy

The lumber trade in Nova Scotia had a long colonial history at the time of Schierbeck's appointment as Chief Forester. The mixed coniferous-deciduous Acadian Forest Region of Nova Scotia had been high-graded for the best pine and spruce trees, which were sawed crudely and exported to imperial markets. By the 1920s, as the international lumber market collapsed, the Nova Scotia forest was seriously degraded. From the perspective of lumbermen, pulpmen, and speculators, this did not cause too much alarm because the reconstituted forest provided a ready source of pulpwood for the emerging pulp and paper industry. The degraded forest was simply another staple for which demand rose. To Schierbeck, however, the successive exploitation of different species of smaller-diameter stock and their use and export as crude staples constituted a crime. He saw, no doubt through Danish eyes, possibilities for seeking out new markets, improving production facilities, creating value-added industries, and renewing the forest. In this context, he took on the province's lumbermen and pulpmen and Canada's compliant forester community, challenging many of their standard operating practices.

Schierbeck argued that the Nova Scotia lumbermen's methods of sawing were crude and that no one knew much about the nature of foreign markets. This condition especially defined the British Empire market, where, he believed, considerable room for expansion existed. He advocated better sawing and grading of lumber and the coming together of the industry to promote production and marketing.[24] Schierbeck also argued that the quality hardwood stands of the province were underutilized. In strong language, he claimed that 'the amount of wood manufactured into woodenware, flooring, etc., is ridiculously small and unfortunately, a large amount of splendid hardwood lumber is every year cut up into fuel wood.' In 1930, looking at the sawmilling industry generally, he insisted that 'the hardwood forests of the province of Nova Scotia have been hardly touched.'[25]

In the spring of 1933, Schierbeck travelled to England as a lumber envoy to explore the market for Nova Scotia lumber. There he found that eastern Canada was the least aggressive of all the supplier regions competing for the English market. He also alluded to some of the idiosyncrasies of the British market that needed to be considered by the industry. Canadian white wood (spruce) had a very bad name in the market, a fact that Schierbeck attributed to the mixing of spruce and fir in shipments, fir being less durable than spruce. He also found a great antipathy toward hemlock in spite of its large sizes and resistance to rot. Schierbeck speculated that this might be related to the fact that the term 'hemlock' referred to a poisonous weed in England, and he recommended that another term, such as *'grey fir,'* be resurrected from the past. He also reiterated his belief that, with 'the present method of manufacturing lumber in Eastern Canada, with our short length, poor sawing, and poor sorting, we cannot expect materially to increase our export of softwood sawlogs to England.'[26]

Yet Schierbeck stressed that the preferences of the British Empire opened up opportunities for the Nova Scotia lumber trade. For example, one of the largest consumers in the United Kingdom, the London County Council, had passed a regulation directing the specification of 'Empire timber wherever it is available in all their construction work.' The key here was the expression 'available,' and Schierbeck urged Nova Scotia lumbermen to undertake a concerted effort to improve production and marketing to compete in this market. His reasoning reflected his recent observations: 'When all these questions are taken into consideration it must be understood that the British lumber consumer is not at all keen on entering into a comparatively unknown market; to cast his old customers overboard, even if he is strongly urged to do so, for reasons of Imperial trade preference; and certainly he will only do so if it is strongly urged upon him.'[27]

Nova Scotia lumbermen did not take kindly to Schierbeck's criticisms. One observer complained that Schierbeck was now 'over the radio telling the Hardwood lumbermen that they do not know their business ... Barnjum bequeathed a heavy load to the Conservative party when he forced him on us.'[28] Another complained that, while Nova Scotia still had large quantities of hardwood, most of the accessible stands had been cut.[29] The critics, however, were either party stalwarts or members of the powerful exporters who were instrumental in shaping and profiting from the lumber trade. Schierbeck questioned the deeper meanings of the staples economy, and his criticisms pointed to ways in which the trade could have been operated differently.

Schierbeck was also critical of past Crown land policies that had supported the old lumber and new pulp and paper economies. In the nineteenth and early twentieth centuries, Nova Scotia lacked a clearly articulated forest policy. Most of the Crown lands were sold off cheaply to lumbermen who high-graded the forests. Crown leases were let on generous financial and cutting terms, and any forestry stipulations attached to them in the arrangement were routinely ignored. Schierbeck's reform of Crown land policies began with a review of all Crown leases already issued by the government. Here Schierbeck found that most lessees violated the forest stipulations of Crown leases. In 1929, for example, the department took a Hants County operator, William McDougall, to county court for illegal cutting of Crown wood. The defence was revealing. McDougall argued that twenty years earlier he had applied for and paid for a grant of Crown land, though he had never received the grant from the department. The department countered that twenty years earlier the law had changed and that McDougall was entitled only to a lease, which permitted him to cut wood over ten inches in diameter, a cutting stipulation that he had violated. The case was unprecedented in that a violator was taken to court, and the judgment was 'awaited with interest by many engaged in the industry.' As a result of such actions, Schierbeck collected trespass fines in most of his terms in office. In 1931 he reported that 'a number of trespassers were caught, and in most cases the trespasser was allowed to buy the wood he cut illegally by paying double stumpage.' In that year, he collected $1,115.37 in trespass fines. In 1928-9 the trespass fines amounted to $3,729.50; in 1929-30, $8,090; and in 1931-2, $701.50. In subsequent years, the department collected no trespass fines. Schierbeck's criticism of the functioning of the Crown lease system complemented his critique of sawmilling techniques and lumber marketing. Both spheres, he believed, were in dire need of reform. Lumbermen, by contrast, were doubly disturbed, believing that Schierbeck poorly understood their situation.[30]

Schierbeck also stepped up measures to reform and collect the provincial land and fire taxes that routinely went unpaid by many of the larger forest landowners. As he took office, the fire tax was being increased from one-half to three-quarters of a cent per acre for woodland owners of 200 acres or more. But collection remained exceedingly difficult. At the time, Schierbeck found that no fire tax had been collected in Antigonish County and on Cape Breton Island and 'that in most counties a great number of owners had escaped the Tax altogether.' In 1931 he reported that 'most of the Chief Rangers have been forced to have writs issued against these Tax payers and, in this way, have collected a considerable amount.' Schierbeck also used his taxation records to defend himself against the attacks of the province's large lumbermen and politicians. When Liberal member of the legislature and lumberman Alexander S. MacMillan criticized Schierbeck for providing faulty statistics and imposing too heavy taxes, stating that 'most of us [lumbermen] regard Mr. Schierbeck as more or less of a joke,' Schierbeck responded that MacMillan's critique was 'taken out of his prolific fantasy' and that 'Mr. MacMillan himself has not been seriously bothered about his Land Tax, and has not paid any for the past four years.' In spite of his efforts, Schierbeck continued to have difficulties collecting the fire and land tax for the province.[31]

Schierbeck also criticized the emerging pulp and paper economy. He tackled absentee American pulp and paper companies that had acquired a high proportion of privately owned forest lands during the 1920s. The high proportion of privately owned forest lands in Nova Scotia, the collapse of the lumber market, the degradation of the forest, and the rise of the pulp and paper industry in North America had resulted in a speculative boom in pulpwood lands in the province, where American absentee pulp and paper companies sought pulpwood reserves with the eager assistance of local speculators and brokers.[32]

Schierbeck roundly criticized the speculative boom in pulpwood lands.

Soon after taking office in 1926, he boldly wrote in the *Annual Report* of the Department of Lands and Forests about 'the alarming rate in which the American pulp and paper producers are buying Nova Scotian freehold land for the export of pulpwood. Over 2,000,000 acres [810,000 hectares] of the best timber lands of the Province are in the hands of American pulp and paper companies who have no manufacturing plans in the Province and are concerned in the export of pulpwood.'[33] Most of these companies, he argued, held their freeholds and their leases passively, as pulpwood reserves, speculative holdings, or securities for capital investments. This had detrimental effects on the forests, for without use, he argued, forests deteriorated and became susceptible to insect infestations, storm damage, and early death. In addition, he argued, forests that currently had value were being tied up for alternative uses in the future. Schierbeck recommended that the companies be required either to work their leases and freeholds or to turn them back to the Crown or sell them to other parties. He supported, as part of this concern, Barnjum's proposal that an embargo be imposed on the export of pulpwood cut from private lands in order to force the companies to build manufacturing facilities in the province. Failing that, absentee companies should sell their forest lands and move home. Schierbeck also suggested a policy requiring landowners with more than 1,000 acres to obtain a licence before felling any tree on their property. He believed that this policy would force them 'to submit inventories of their holdings to the Department, accompanied by a proper working plan to sustain yield.'[34]

Schierbeck's criticisms extended beyond absentee pulp and paper companies. Schierbeck also attacked the terms under which the provincial forests were leased to pulp and paper companies operating mills in the province. One prominent issue involved determining the forested area or the volume of wood to be leased to the companies. Although the science of forestry provided a more detailed and standardized measure of inventorying forest lands than the old practice of timber cruising (or timber looking), it was still a highly political and subjective exercise. When the Nova Scotia government celebrated the coming of the first newsprint mill to the province in the late 1920s, the question of forest inventories became a hot political issue. The Mersey Paper Company, backed by Izaak W. Killam and the Royal Securities Corporation of Montreal, typically put pressure on the government to provide long-term service and security of tenure to the company's forest supplies. Mersey, not the government, clearly stipulated its own conditions of operation. Brokered by local promoters and speculators, the company entered an agreement with the province to cut 1,000,000 cords of pulpwood from Crown lands over a thirty-year period at a stumpage rate of one dollar per cord.[35]

Schierbeck brought the relationship between a captured state and a powerful corporation to sharp light through the negotiations over the Crown forests that were to be set aside to provide the 1,000,000 cords of pulpwood for Mersey's use. In defiance of his political masters and business clients, Schierbeck challenged the forest inventories submitted by Mersey's woodlands manager. The dispute focused on the extent of Crown lands necessary to meet the 1,000,000 cords that the government promised to Mersey. Offered an extensive concession in eastern Nova Scotia, the company claimed that the seven counties – Cape Breton, Inverness, Richmond, Victoria, Antigonish, Guysborough, and Pictou – did not contain enough wood. Company officials sought to select additional areas from a complete list of Crown leases that would expire during the next thirty years. Schierbeck countered that he had not 'the slightest doubt that we shall be able to furnish the total agreed upon [amount of pulpwood] in the Island of Cape Breton.'[36]

After seven months of taking inventories on the island, Mersey concluded that there were only 536,366 cords available. In response, Schierbeck noted that 'it is only reasonable to expect that the Mersey Paper Company are desirous of obtaining as much area as possible and therefore have been very conservative in their cruise.' But his sympathies stopped there. With the provisions of the Mersey agreement behind him, Schierbeck acquired the field notes of the Mersey cruise. Here he found that, although Mersey had measured all trees from four inches and up, only those above six inches in diameter became part of the company's estimates. The company had also calculated one cord as 150.6 cubic feet instead of the regulation cord, 128 cubic feet. Finally, the company had deducted from 5 to 15 percent of all wood as 'defective' and even excluded some 'barren' but reproducing lands. These accounting practices contravened the agreement between the Mersey Paper Company and the province. Schierbeck pointed out that the volume on Cape Breton Island, when corrected, would be much higher and 'practically correspond' to the DLF figure of 893,198 cords. He recommended that no areas outside the Island of Cape Breton be allotted to the company.[37]

Despite his attempts to defend the public's interest in Crown lands, Schierbeck received little support from politicians. By January 1932, Attorney General and Minister of Lands and Forests John Doull appeared impatient with Schierbeck, urging him to settle the dispute. Political pressure focused clearly on Schierbeck to settle with the company. The recollection of Schierbeck's successor, Wilfrid Creighton, suggests that the dispute was never settled. Schierbeck himself reported that, of the 171,472 acres cruised by Mersey and the Department of Lands and Forests, Mersey selected 128,507 acres consisting of a number of blocks varying in size from approximately 30,000 acres to a few hundred acres. The volume of wood was, by agreement, tentatively fixed at 800,000 cords. Additional lands were, from time to time, to be turned over to the company until its final quota of 1,000,000 cords had been reached. By the close of the thirty-year term of the agreement, Mersey had cut fewer than 100,000 cords of Crown pulpwood. The cost of the settlement was clearly minimal to the company, though it added to the list of 'offences' that led to Schierbeck's dismissal the following year.[38]

Schierbeck's credible challenge of the Mersey forest inventories illustrates well the tenuous position of foresters who practised their profession and who adhered to a professional code of ethics. Schierbeck was punished instead of rewarded for his behaviour, while Mersey's woodlands manager stayed in the company's employ until the late 1950s, ending his career as director of the Parks Division of the Nova Scotia Department of Lands and Forests. Again, Schierbeck's recommendations and actions were unique in the Canadian context, in which foresters promoted fire protection, forest education, and forest inventories, and in which services were provided *for* industry rather than, as in Schierbeck's case, obligations were demanded *of* industry.[39]

Schierbeck also challenged the industry on forest management. Forest management along European lines did not exist in Canada during his term in office. In North America, pulp mills typically clearcut forests to supply the mills. Costs determined this practice, and the woodlands managers rather than the company foresters planned the operations.[40] Invariably, it meant cutting all wood closest to the mill, regardless of age or species, as long as it could be pulped. Schierbeck's vision of forest management differed. Schierbeck believed in sustained yield management, a system in which the goal is to maintain the forest capital by harvesting only the interest. His general advice followed the textbook principle of sustained yield: 'The time it takes for a forest to reach maturity is estimated. Let us suppose this to be sixty years. The total stand of timber on the area to be leased is estimated. The Lessee is then only allowed to cut one-sixtieth part of his total stand a year and regulations must be set down demanding that the lessee cut the oldest stands first and leave the young and growing stands to increase.'[41] He maintained that the distribution of age classes in the Nova Scotia forest fit perfectly for such management, in spite of the heavy exploitation of the past. In contrast to the other provinces, which were to a large extent covered with either large areas of overmature stands or fire barrens, Nova Scotia provided a fine laboratory for such management.[42]

Schierbeck firmly believed in sustained yield management, arguing that the forest could be 'improved' through selective thinnings that would work on all types of forest tenures, including large freeholds and Crown leases. Comparing his observations and experiences in Denmark with those in Canada and Nova Scotia, he contended that physical, climatic, and forest conditions were similar. Yet the Danish forest, through the use of several thinnings over the course of a rotation, produced four to five times as much as the Nova Scotia forest and wood of far superior size and quality. Over a full rotation, the Danish forest produced approximately two cords per acre per year, while the Nova Scotia forest produced only one-half a cord per acre per year.[43]

By thinning the forest, improving its productivity, and increasing the total volume taken out over a rotation, the forest industry would be able to save money and to reorient its woods operations by reducing acreage and building permanent infrastructure such as access roads.[44] Schierbeck also spoke enthusiastically about the prospects of thinning the pine stands in the western section of the province, where the excellent climatic and germinating conditions had made the stands too dense: 'The trees are standing nearly as thick as the hairs of a dog's back, and it cannot be too strongly recommended to thin these stands by removal of the secondary species, spruce and balsam.'[45] The harvested trees could be used as pulpwood, profiting both the woodlot owner and the remaining forest.

But Schierbeck was by no means dogmatic about selection thinning. He thought that it particularly suited Nova Scotia's pine forests; for the spruce and fir stands, he believed that clearcutting, or clean-cutting, was more appropriate. The seeds of the spruce and fir, once in the ground, remained there for as long as four or five years, and once the old forest was cut the new generation was provided with sufficient light to regenerate.[46]

On the question of forest management, Schierbeck took on the giants of his industry. In 1930, at the annual meeting of the Woodlands Section of the Canadian Pulp and Paper Association, he challenged Ellwood Wilson, the president of the Canadian Forestry Association, who contended that the profession already knew enough about silviculture and reforestation. Schierbeck countered: 'I think this is wrong. In my humble opinion we know nothing about it. It is my belief that research is needed very badly.'[47] He lamented the change in the forest environment that resulted from the growth of the pulp and paper industry. He pointed to the lack of knowledge on the silvics of spruce, the disappearance of large-dimension spruce, and the replacement of spruce by balsam fir, an 'unhealthy tree, subject to all kinds of diseases.'[48] Schierbeck favoured maintaining a mixture of species and age in the forest, both necessary for sustaining yields and promoting a diversified forest economy. Large-dimension red spruce were an integral part of that system. Schierbeck also worried about the growing areas of even-aged balsam fir stands. While these stands would someday be the prime target for the spruce budworm, he stood alone among Canadian foresters at the time in his concern about the degradation of the forest. Most Canadian foresters welcomed the spread of the fast-growing balsam fir as a ready supply for the pulp and paper industry. One prominent forester, Bernhard Fernow, pioneered research into the balsam fir as a tree species for the pulp and paper industry.[49]

Schierbeck also criticized the accumulation of slash in the pulpwood clearcuts, which, he alleged, promoted the development of heart-rot fungus not only in old but also in young and sapling growth. More than 60 percent of a stand could be attacked in this way, he pointed out to Canada's pulp and paper industry men, lecturing them that red heart rot was principally responsible for the culls and sinkers in their pulpwood. But he was by no means dogmatic in his advocacy of the removal of slash. This should occur only in the final harvest of a rotation to promote regeneration. Schierbeck recommended that during thinnings all the branches and tops be lopped off and left on the forest floor to save on costs and to protect and refurbish the organic matter of the soil. His recommendations in this area also ran afoul of the pulp and paper industry. Removing slash in clearcuts remained an expensive undertaking, and lopping off and leaving slash on the forest floor during thinnings was irrelevant to an industry that did not believe in thinnings.[50]

Schierbeck's presence at the 1930 meeting of the Woodlands Section appears to have been, perhaps not surprisingly, his first and last. Judging by the response of his colleagues, his presence might not have been all that welcome. Ellwood Wilson requested that his response not be recorded in the proceedings, and the chairman characterized the Schierbeck-Wilson exchange as 'a very fine finish,' a comment that occasioned laughter from the audience.[51]

Otto Schierbeck in the Nova Scotia woods.

For Schierbeck, the promotion of selection-cut forestry to improve the growth of a forest stand was a favourite topic, one on which he wrote frequently. If adopted, he believed, it would form the foundation of a new forest sector strategy of use and marketing for Nova Scotia. In time, however, he became more and more pessimistic. In one of his last published articles, he wrote with some 'trepidation,' stating that 'I am afraid it may prove of no great interest.' He closed with the modest hope that, 'if my article only might result in that somebody would give it a try, then the intention would be fulfilled.'[52]

Conclusion

Otto Schierbeck was a professional forester in the conservationist tradition who endeavoured to practise professional forestry when considerations of partisanship remained dominant and forest industries and the broader public were unsympathetic to many of the aspirations of the conservationist movement. His appointment as Chief Forester for the province of Nova Scotia in 1926, and the broad powers accorded to him through his personal connection with forest capitalist and conservationist Frank Barnjum, provided him with extensive powers to implement some measures of forest conservation.

The common focus of forestry during Schierbeck's career – its instrumentalist view of nature, its elitist orientation, and its isolation from the concerns of rural residents – was clearly evident in Nova Scotia. Locals ambivalently received the fire protection system, supervised by full-time salaried rangers, that replaced the previous system of part-time salaried and volunteer rangers. The forest industry welcomed and lobbied for it, while rural residents, who used fire as a tool to clear land for different purposes, often opposed it. Schierbeck's program of demonstration lots to show the merits of thinning did not have a wide impact because it failed to address the issues of price and market in wood production. Schierbeck, in spite of his concern for the forest and the Nova Scotia economy, appeared as part of an 'insular professional priesthood' with few connections to other stakeholder groups in society.[53]

But apart from pushing the elitist agenda of the conservation movement, an agenda that favoured the forest industry, Schierbeck was unique in challenging the terms on which the forest industry conducted its operations. He dared to question the Canadian staples economy. He criticized the sawmilling industry for its poor utilization of the province's hardwoods, poor techniques in sawmilling generally, and poor marketing methods employed in the British market. He promoted a stricter policing of Crown leases and the collection of taxes that were routinely ignored by many of the larger land holders. He criticized the forest land tenure system in Nova Scotia, and he attempted to stop the accumulation of forest lands in the hands of large American absentee pulp and paper companies. He became a strong supporter of a provincial export embargo on pulpwood, a measure that he thought would promote local manufacturing and improve forest management. Schierbeck also advocated a more proactive approach in negotiating the terms of Crown land cutting rights for the only newsprint mill in the province, the Mersey Paper Company in Liverpool, Queens County. Perhaps most importantly, his central recommendations focused on forest management, a sustained yield approach to thinning, and the complete utilization of forests on large holdings. Through written articles, Schierbeck promoted sustained yield and selection thinning as ways of making both the forest and forest industries more efficient through the growth of better stock and the operation of more compact and efficient forest businesses.[54]

Schierbeck presented these criticisms widely, both in the professional literature, such as *Forest and Outdoors, Pulp and Paper Magazine of Canada, Canada Lumberman,* and *Forestry Chronicle,* and in the *Annual Reports* of the Department of Lands and Forests and the province's daily newspapers. In the newspapers, Schierbeck often had a more visible profile than most politicians. Many of his initiatives pointed to constructive alternatives that could have been pursued more vigorously, even within the political and economic constraints of the time. Subsequent calls for cooperative marketing and better sawing techniques and grading, as well as the formation of the Nova Scotia Forest Products Association (1934) and the Maritime Lumber Bureau (1938), were measures that Schierbeck had recommended years earlier. His abilities, public zeal, and high profile might explain why he retained his position for seven years and why the official version of his dismissal was highly complimentary.[55]

Schierbeck's fate illustrates the defeat of one forester's struggle to promote professionalism and scientific forestry during an era of patronage and laissez-faire in Canadian politics and business. It also points to another lost initiative in Canadian forestry during the Depression. Speaking more to the uniqueness of Canada and Nova Scotia, Schierbeck's dismissal represents the emphasis put on the all-important urge to develop new forest resources in a resource-dependent economy. Canada was, and continues to be, a resource-dependent economy in which export staples fuel the national and provincial economies. In this situation, Canadian foresters, whether in public or private institutions, faced pressure to cut rather than manage trees efficiently. By 1936 this notion had such power that it occasioned Schierbeck to write that 'the trend of our profession has ... turned so decidedly toward the logging end, that it is even seriously being considered to throw the appellation of "forester" overboard, substituting it by "logging engineer," and to let our technical society sink down to become a branch of the Engineering Institute of Canada.'[56]

Appendix

Working Plan and Experimental Forest (Otto Schierbeck)[57]

When, as a young graduate, I arrived in Canada in 1905 looking for a job as a forester, I met with very little success. I visited the different provincial forest organizations, but found that practically no forestry work was being carried out. I had letters of recommendation to some of the large lumber firms and visited most of them, among others Price Brothers, Mr. MacLaren and Sharples & Company, Limited, but none of them were at all interested in forestry. The only encouragement I got was from Mr. Joly de Lotbinière, whose father was then Lieutenant Governor of British Columbia. Mr. Joly de Lotbinière was very much interested in cutting on the basis of sustained yield, and was of the opinion that this could best be done by working to a diameter limit, and I came very close to obtaining the position of Chief Forester for British Columbia.

Mr. MacLaren, the big lumber merchant in Ottawa, introduced me to the Director of Forestry in the Federal Government, Mr. Stewart, who also was very interested in sustained yield and propounded his theory for me on how this could be worked out by cutting to the diameter limit. I really believe that I had a great chance of obtaining a position with the Federal Government, due to the influence of Mr. MacLaren. Mr. Stewart was extremely pleased to get in touch with a big lumber merchant like Mr. MacLaren, and I recall that during our conversation in Mr. Stewart's office he, Mr. Stewart, expressed his pleasure in getting in touch with a lumberman on the question of forestry. He said to Mr. MacLaren, in my presence: 'This is the first time that the question of forestry has really been mentioned to me by a Canadian lumberman.'

However, I did not succeed in securing the job. I was then young, not very diplomatic, and did not speak English very fluently. I recall expressing my opinion to Mr. Stewart that cutting to the diameter limit was all more or less nonsense, which of course immediately queered all chances of a future position with the Federal Government. In Toronto, I had quite a discussion with the Chief Forester of Ontario. I have forgotten his name, but I remember that there were then two or three graduate forest engineers employed by the Department, and we had a long discussion about working plans.

The result, however, was that I started railroad surveying, went to Cuba and South America, and finally returned to Denmark, where I worked in the Danish Government Service for a number of years, returning to Canada in 1919.

I then found a great change. The Commission of Conservation had been working for a number of years and tremendous pressure in forestry here in Canada, especially in fire protection, had been undertaken. In going over the Reports of the Commission of Conservation I found that Dr. Fernow, in

1910, submitted a report to the Commission on 'Scientific forestry in Europe; its value and applicability in Canada,' in which he states: 'The first and greatest need is, however, a change, a radical change, in the attitude of our people and Governments from that of exploiters to that of managers. We should realize that existing methods of treating timber lands are bad, and that a change is imperatively needed. Only when there is doubt implanted as to the propriety of our present methods of forest management, and only when people realize the urgent need of change, will the radical reform be inaugurated that we believe necessary. When that attitude is established which demands that our forests will be managed, not merely exploited, all the rest will be comparatively easy, and it will then astonish the practical men to find how much European methods and systems are really applicable, just as it lately surprised the Americans across the line.'

Since then practically every volume of the Report of the Commission of Conservation stresses the necessity of inventory taking in our forests, classification of land fit for settlement versus land only adapted to the growing of forests, forest research, etc. In other words, forest management has been talked about, discussed, advocated and recommended all over the country. Timber shortage, national disaster and blue ruin have been in newspapers, magazines, pamphlets and official reports. We are still at it, we are doing it every day, but what is the result – nothing but talk. What is the reason? Who is to blame? I maintain that the ones chiefly to blame are the foresters, the professionally trained men who should provide leadership.

When a young forester has completed his excellent education and enters the service of a lumber firm or paper company, he considers that his first solemn duty is to throw overboard all that he has learned in College. His greatest desire is to become a practical lumberman, to be able in the most efficient and economic way, to continue destroying the forests as his forefathers did for centuries. He seems to forget that when he, on his graduation day, with his diploma in his pocket, left his Alma Mater as one of the few technically trained foresters in the country, he had donned mail, lance and sword, had received his guerdon for protecting the forest of his native land.

As an excuse, we are liable to cast the blame for the dead water and mire into which the ship of forestry has entered upon the government politicians and heads of lumber firms and pulp and paper companies. Our excuse has always been, 'we can't do anything with these people, we can't get the money, nor awaken any interest.'

I ask you, gentlemen, how is it possible for you to awaken the interest of your bosses when you are not able to demonstrate the benefit of your proposals, when you yourself do not know what you wish to do! As I said before, the hue and cry of the whole forestry profession has, for the last twenty years, been working plans, but to date not a single forest in the Dominion of Canada is operated under a working plan. What do we really know about the factors essential for working out a working plan? Is there anyone of you here able to produce a yield table applicable to a comparatively large area? Can anyone in this audience prescribe the proper cutting methods for a comparatively large area? We speak about clean cutting, shelter wood cutting, strip cutting, etc., but what do we know about the effect of the different methods here in Canada? Have they ever been demonstrated? We all agree that thinning of softwood stands, girdling of birch, etc. is essential, but what do we know about the economical side of those measures? Has it ever been carried out to any extent?

It is easy to criticize, and you will probably get mad at me for being so frank. However, no progress will be made if we keep on as we have done. Let us get away from talk, let us do something, let us get to action. We have a splendid body of trained foresters here in Canada, who are fully capable of handling the situation. Let us not be like Kipling's Bandar-Log:

Here we sit in a branchy row,
Thinking of beautiful things we know;
Dreaming of deeds that we mean to do,
All complete, in a minute or two –
Something noble and grand and good,
Won by merely wishing it could.
Now we're going to – never mind,
Brother, thy tail hangs down behind![58]

The first step in order to accomplish something, is to convince our superior officers of the benefit of proper forest management. It is not enough that we refer to European methods, that we show photographs and preach about the wonder which European foresters have accomplished. We have to demonstrate the applicability of those methods here in Canada, show their benefits to a Canadian forest. We have to demonstrate that those methods can be carried out without excessive, preventive costs, or devise methods whereby costs are not prohibitive, and the sooner we do this the better.

The only way it can be done is by establishing a proper experimental forest, which should not be a small patch of cut-over abandoned [sic] land, in which no one has any interest or cares for, but should be a forest of at least twenty-five thousand acres in extent, adjacent to a pulp and paper or lumber mill. The forest should contain mature timber, young immature timber, young growth, reproduction, and should contain the different softwoods and mixed types, pure hardwood types, and even some barren land, where artificial reforestation can be demonstrated. A working plan should be prepared for this forest, different cutting methods on large areas should be demonstrated, thinning, girdling or cutting out of undesirable species should be practised in the young stands, proper records kept of reproduction after cutting, etc. The forest should be operated as an economic proposition, that is, it should be adjacent to a pulp and paper or large sawmill where all the products of the forest could be utilized. A proper accounting system should be established, the books being kept in such a way as to make it possible to ascertain the exact cost of each operation and each experiment. All operations should be followed through a series of sample plots.

The forest should be located in a place of more or less easy access, with accommodations for visitors, and excursions and meetings of technical associations should be held there. It might be advisable for the forestry students of the universities and ranger schools to spend part of their time each year in the forest.

At least five of these experimental forests should be located in the Dominion; one in British Columbia, one in the Prairie Provinces, one in Ontario, one in Quebec, and one in the Maritime Provinces. They should all be in the charge of the best technically trained men available.

I consider the establishment of such experimental forests absolutely necessary. It should be the first step towards the establishment of forestry in the Dominion, and if contemplated by either the Federal Government or any of the Provincial Governments, I wish to issue a warning not to follow the example of the United States Department of Forestry. This Department has laid out a great number of experimental forests, and issues bulletins annually, giving a great amount of data and information based on research work undertaken in those forests. Practically all of their experimental forests are small areas, ranging from fifty to one thousand acres, and the experiments undertaken are of no commercial value as they are carried out on small one acre plots. They have a certain scientific value, but absolutely no practical [value].

The experimental forests should, in my opinion, be selected very carefully, so as to be sure that all the forest types are included. They should be dedicated as experimental forests forever and should be acquired, if necessary, by expropriation from private individuals if suitable vacant Crown Land cannot be found. If sufficiently strong recommendations are pressed by the different societies of forestry engineers, by heads of the different forestry departments in the Provinces, by the Canadian Forestry Association and by the Federal Forestry Branch, I do not think that there would be any difficulty in establishing the forests. I hereby earnestly solicit your support in putting this scheme over. To my mind it is the first step in the establishment of forestry in this Dominion.

Next to the establishment of the experimental forests, I consider it of the utmost importance to do away with the old, antiquated diameter limit regulations, which act as brakes to all proper forest management throughout the Dominion. I have already shown, in the beginning of my speech, how the diameter limit was a hindrance to the proper development of forestry in Canada twenty years ago, and even yet it is still in force in practically every Province. To the general, uninformed public it is an excuse for not carrying out proper forest management. The man in the street, who, once in a while, gets stirred up by alarming articles in the press, regarding our forests, will say to himself, 'Oh well, it can't be so bad as all that. I know that the lumbermen and pulp and paper men can't cut down trees below a certain diameter limit, so that in itself must be a proper safeguard for the forest.' Politicians and even the Ministers of the different Provincial Forest Services in all probability take a somewhat similar attitude, and are therefore reluctant in getting down to real business.

I would strongly recommend that this Society go on record condemning this antiquated method of conserving our forests [and] pointing out that it is not enforced, is technically unsound, and a hindrance to the introduction of proper forest conservation.

3
John Bigelow: Nova Scotia Nationalist and Forest Reformer

The politics, economics, and science of forestry in Nova Scotia underwent a remarkable change from the early 1920s to the late 1960s. In the early 1920s, the lumber trade still dominated the forest industry sector despite more than a century of high-grading the best stands of pine and spruce sawlogs. The trade was still part of a colonial economy in which local buyers or agents controlled the export of crude lumber to English Empire and American markets. The allocation of wood was coloured by patronage connections, in terms of both the purchase and the lease of Crown lands. The production of lumber was conducted by hundreds of small portable sawmill operators who pursued an increasingly degraded and/or distant supply of sawlogs. After the middlemen had taken their share, these sawmillers received little net return and had little incentive either to upgrade their operations or to practise forest management. Forestry as a science and profession was virtually absent or was used as a crass tool for personal emolument.[1] A Department of Lands and Forests was only formed in 1926 (relatively late, even in Canada), and any attempts by the department to regulate (rather than service) the forest industry were met with considerable resistance.[2]

By the 1960s, the situation looked very different. The province's pulp and paper industry had reached unprecedented heights. Three large transnational corporations were operating pulp and paper mills in the province: British Bowater Mersey in Queen's County, Swedish Stora Forest Industries on Cape Breton Island, and American Scott Paper in Pictou County. Almost all the province's Crown lands, at approximately 25 percent of all forest lands, were leased to the pulp and paper companies. Over half of the province's annual wood cut fuelled the pulp and paper mills. The Department of Lands and Forests had become part of a professional state bureaucracy, staffed with trained foresters and forest rangers, most of them graduates of the Faculty of Forestry at the University of New Brunswick and the Maritime Ranger School in Fredericton.

Yet the Nova Scotia transition from lumber, patronage, and laissez-faire forestry to pulp, bureaucratic control, and managed forestry was not complete. Even in the mid-1960s, the lumber industry remained important, though the colonial legacy continued to render it inefficient, and the forests were increasingly degraded. Politics continued to play a role in the era of pulp and bureaucratic and professional control. The transnational pulp and paper corporations were invited to the province through generous provisions of Crown forest lands, hydroelectric power, and other concessions. On the other hand, pulpwood marketing legislation for small woodlot owners, who owned over 50 percent of all forest lands, was repressed in order to provide a cheap source of wood fibre for the pulp mills.[3] Forestry remained a low priority for the provincial government, with the Department of Lands and Forests holding a weak position within the hierarchy of departments. After the tumultuous years under Otto Schierbeck, who strove to make the department an efficient free-standing bureaucracy, it emerged as a service department to industry, concerned primarily with suppressing forest fires, taking forest inventories, and administering leased Crown lands. The department was not interested in dealing with the economics of the lumber and pulpwood trade, nor was it accorded the mandate to negotiate the terms of entry of the pulp and paper companies to the province.

The objective of this account is to shed light on the nature of the transition from lumber to pulp, and patronage to professionalism, through the perspective of a remarkable forester and civil servant. John Bigelow served the forest industry in various capacities from the 1930s to the late 1950s. Throughout his career, he embodied a Nova Scotia nationalism fighting the colonial and imperial legacies of power and privilege associated with the lumber trade, the excessive prerogatives of international resource capitalism in the province, and the readiness of local politicians to compromise the welfare of the province in favour of transnational pulp and paper companies. This was essentially an economic nationalism, promoted by reforms aimed at generating jobs and markets at home, building local value-added industries, and emulating and adapting the best technological innovations available elsewhere to the Nova Scotia situation. Bigelow took a special interest in the social and economic inequities and political interferences in the Nova Scotia pulpwood and lumber trade. In fact, he abandoned employment as a professional forester when he became convinced that the prospects for forest management depended on broader social, political, and economic reforms. He spent over two decades working to promote the organization of small woodlot owners and to reform the backwardly mercantile character of the lumber trade. He was instrumental in bringing a new Swedish pulp and paper mill to Nova Scotia in the late 1950s, a mill that he thought would open a progressive new market for small woodlot owners' pulpwood.

All these initiatives fall into the category of what James Kenny has called 'the possibilities of the interventionist state,' the limited but still proactive measures that might have changed the conditions of resource exploitation in the Nova Scotia forest industry.[4] Bigelow's career further reflects the competing views of several government bureaucracies and the disparate paths of development that they favoured. Although a professional forester, he focused his efforts through the Departments of Agriculture and Trade and Industry, which at different stages became the champions for small woodlot owners and sawmillers and the advocates of a more diversified and competitive pulpwood and lumber market. Opposed or at best indifferent to these efforts was the Department of Lands and Forests, a forest protection service that showed little concern with the economics of the forest industry or the status of private woodlot owners.[5] Finally, Bigelow's career shows how a broad provincial industrial strategy based on corporate concessions overrode the social concerns of bureaucrats and professionals. Bigelow thus confronted formidable obstacles. Although he made some material progress and presented some innovative ideas, his efforts were often stymied by the powerful political and economic forces that he sought to challenge.

Formative Years

John Bigelow was born 16 September 1906 in New Salem, Cumberland County, Nova Scotia, a descendant of New England planters who came to Cornwallis Township, Nova Scotia, in the early nineteenth century.[6] His father was a contractor and a builder of large wooden structures, such as ships, wharves, bridges, breakwaters, and buildings.[7] In 1912 young John went to the forest for the first time, in search of wood to build a scow. There were many more trips after that. But it was reading that first aroused Bigelow's interest in the economic welfare and plight of rural people. One influence was Elbert Hubbard, the social philosopher and pacifist who started an American-oriented social movement. Hubbard published two influential and widely read magazines, to which the Bigelow household subscribed: *The Philistine: A Periodical of Protest* (1895-1915) and *The Fra: A Journal of Affirmation* (1908-15).[8] The magazines' subtitles illustrate well Hubbard's philosophy. *The Philistine* was a magazine in the tradition of muckraking journalism, 'flippant, slangy, rude and crude,' and intended to jolt readers into thought. But Hubbard was not wholly committed to this tradition. He thought that there was 'enough of the literature of depreciation and defamation,' and he thus offered *The Fra* as an antidote. *The Fra* was initially published for businessmen, celebrating the industrial development of the progressive era, but it quickly received a wider readership. Hubbard described the journal as standing 'for beauty in typography and ornament, as well as beauty in spirit,' a description that corresponds to Bigelow's memory. Hubbard's publishing ventures came to an abrupt end in

1915 when Hubbard and his wife died on the *Lusitania*, which was sunk by the Germans in the Irish Sea.

After grade school and high school in Canning from 1912 to 1923, Bigelow completed two engineering courses (standard mathematics toward a science degree and drawing) at Acadia University from 1925 to 1928. They were useful for his father's line of business, and Bigelow used these skills while working intermittently in a variety of jobs.[9] Fascinated by trees, however, he often wondered about the scientific basis of tree growth. That is why he was drawn to the study of forestry in 1931.

When Bigelow entered the Faculty of Forestry at the University of New Brunswick, Canadian forestry education (and forestry as a science and profession) were closely linked to the pulp and paper industry. Having enjoyed record growth in the decade preceding the Depression, the industry was a major sponsor of Canadian forestry schools and the major employer of its graduates. The marriage was further sealed by the concurrent annual meetings of two organizations: the Woodlands Section of the Canadian Pulp and Paper Association and the Canadian Institute of Forest Engineers.[10] Bigelow's experience in forestry school reflected this situation. Bigelow entered with the expectation of learning the practical sides of operating small woodlots and forest nurseries and conducting reforestation in Nova Scotia, but he soon discovered a clear bias in the curriculum toward large lumber and pulpwood operations. The course in forest management relied on an American textbook in which the focus was on how to manage forests for large landowners, one of the examples being from the pine stands of northern Michigan. In Nova Scotia, most of the pine stands had been cut.

There were also courses on measurement for standing timber and lumber, important for the bankers who advanced capital to the large pulp and paper companies and for the companies themselves in planning long-term fibre supplies. One course dealt with water management or how to manage rivers and lakes for log driving. This was not of much use in Nova Scotia, where rivers were short and easily handled, but was useful for larger pulpwood operations. The emphasis, then, was on large corporate wood supply systems most applicable in New Brunswick, Quebec, and Ontario. There was nothing that dealt with the economy and management of small woodlots, which accounted for about one-third of New Brunswick's forests and one-half of Nova Scotia's.

The pulp and paper industry bias permeated the student lounge in forestry school as well. There were eight students, including Bigelow, in the UNB entry class of 1931. Three were from Quebec, two from New Brunswick, two from Nova Scotia, and one from Ontario. Most discussions revolved around landing or moving to a 'big job.' As mergers and shake-ups occurred in the industry, there was always talk about who had ended up with the big job. All students looked to or dreamed about big jobs in the west. When graduating, the class was told by the dean that all good students went to work for pulp and paper companies. This was reinforced by the summer jobs of the students, which invariably meant working for a pulp and paper company or a provincial forest service. Bigelow, for example, worked as a timber cruiser for the Nova Scotia Department of Lands and Forests, the McLeod Pulp and Paper Company Limited, and the Minas Basin Pulp and Power Company Limited.

Like many other foresters during the 1930s, Bigelow had problems finding permanent employment upon graduation. He spent one summer with a senior forester, Brian Alexander, cruising timber lands for an American outfit that had taken over the forest lands of Nova Scotia forest operator Frank Barnjum.[11] Part of their job was to look up titles of the Barnjum lands, a task that took them to Pickles and Mills, Barnjum's agents in Annapolis Royal. There the caretaker, 'Old Ernie Mills,' gave Alexander and Bigelow a free hand to peruse the material. But these jobs were tenuous. Bigelow's engagement was periodic, and his place was eventually taken by the son of the American owner, Mr. Fessenden. Bigelow was also hired to inventory forest land for Minas Basin, but he quit when he was given the message that he was no longer needed.

Working in Newfoundland

Eventually, Bigelow got the chance to practise his profession on a permanent basis when Miles Gibson, the forestry dean at the University of New Brunswick, called to tell him that the Anglo-Newfoundland Development Company of Grand Falls, Newfoundland, was looking for a forester. A short while later, he received a phone call from Henry Crowe, Anglo's woodlands manager, who asked for an interview in Halifax. There Bigelow was told that 'the bosses want me to hire a professional to provide figures on inventories.' After a three-hour interview, he got the job, and in 1935 he started forest inventory work at the company's Badger Brook Division.

It was in Newfoundland that Bigelow was first confronted with some of the negative social consequences that flowed from the pulp and paper industry. It was fishermen who cut pulpwood for the company. They were outfitted by the outport merchants in the spring, fished during the summer, and sold their catch to the merchants in the fall. Invariably, they ended up cutting pulpwood for winter wages to make up for their negative balance with the merchant. The merchant system of debt was firmly in place, and the clergymen would neither touch it nor talk about it.[12]

It was also in Newfoundland that Bigelow was exposed to the inefficiencies of a backward and ill-planned colonial industry.[13] When Canadian John McGinnis, a well-known cost-cutter in the industry, was brought in to help the situation, he caused a major controversy. He was welcomed with a traditional liquor party to put everybody in a good mood. Suddenly,

McGinnis stood up and announced that, while the company was running a nice little mill with a nice little hydro development, and producing a fair quality of newsprint, it was not making any money. McGinnis then asked each of the eight division heads to stand up and state how he would cut costs, starting the next morning. When McGinnis came to Bigelow's division to examine the records, he pointed to the discrepancy between the moving costs from stump to mill between two years. On being told that the climate could be ideal in one year but disastrous in the next, he responded resolutely: 'To hell with the weather, you've got to get along with the weather if you're going to work here.' Bigelow contributed to the rationalization by suggesting that an extra tractor be ordered as a source of spare parts.

The fact that hiring and promotion turned more on Old World connections than on ability was an important contributing factor to the inefficiencies of the Newfoundland pulp and paper operations (as well as to Bigelow's growing sense of nationalism). Woodlands manager Crowe was a Nova Scotian who had faced, but overcome, such prejudice. His rival for the position was H.H. Wilding-Cole, an Englishman whom Bigelow observed would tackle everything but didn't know anything. After being bypassed in favour of Crowe, Wilding-Cole openly declared that he should have had the position, stating 'I had a better birth than Mr. Crowe ever had.' Nevertheless, during his three years in Newfoundland, Bigelow saw more and more Canadians and Newfoundlanders taking over management positions from the English.

But while the reorganization of the company might have been good for the bottom line, it did little for the health of the forest or the company's pulp cutters. After three years of service, Bigelow had grown disillusioned with his job and ambivalent about his profession. All that seemed to matter was 'getting the wood out of the forest as fast and cheap as possible.' He had made a small contribution in providing knowledge on the company's forest supply and how it would evolve over the long term. At one time, he had convinced his boss of the importance of assessing the damage caused by a major forest fire. No forest management, however, was practised. Balsam fir stands were clearcut, but regeneration was good. One thing that he would have liked to do, had the opportunity arisen, was to explore the nature and causes of spruce budworm infestations. He had certainly seen the spread of the budworm in Newfoundland. Infestations started in various spots, then spread gradually in circular patterns over a three- to four-year period, but then died down. The largest areas afflicted were up to 100 acres.

Overall, Bigelow's introduction to industrial pulp forestry was a formative experience. Over these three years, Bigelow had performed cruises, laid out logging camps, and supervised operations on all the company's divisions. At the time of his resignation, he was the assistant superintendent of the Millertown Division, where 90,000 cords of pulpwood were cut annually. He had also seen the social and economic realities of woods work from the ground up and its particularly stark impact upon coastal villagers. To Bigelow, the organization of a woodlands division involved more than camp bosses, cutters, haulers, and rivermen. It fit into a social fabric including company towns, seasonal incomes, merchant creditors, and impoverished communities. This was also part of the understanding that he took away from his time at Anglo-American.

Promoting Pulpwood Marketing for Small Woodlot Owners

When Bigelow returned to Nova Scotia in 1938, the province was in the process of coming out of the Depression. The provincial Department of Agriculture was at the time particularly active in supporting the province's farmers.[14] It was Waldo Walsh, the director of the department's Marketing Division, who provided Bigelow with a chance to come home. When Bigelow was in forestry school, the department had alerted the school to its interest in woodlands owned by farmers. Before moving to Newfoundland, Bigelow had talked to Walsh about a position, though there were no prospects at the time. But he had a good hearing and the backing of the minister of agriculture, John MacDonald, who was the MLA for the constituency where he went to school. A few years later, while he was in Newfoundland, he received a formal letter from Walsh indicating that there was a position available in the woods-marketing section of his division.

The Department of Agriculture was at the time developing programs to improve marketing practices for all Nova Scotia forest producers and exporters, especially small woodlot producers.[15] This represented a continuation of previous work in other commodities.[16] Walsh had already worked with farmers, organizing marketing and quality control in poultry and a number of other agricultural products. When travelling in Cape Breton, he noticed the importance of wood products in the local economy. At that time, local and German buyers and agents were travelling the island in search of pulpwood. The demand and prices for pulp, paper, and newsprint were high. Yet woodlot owners, cutting and selling pulpwood at roadside or shipping points, were at the complete mercy of buyers. Walsh thought it appropriate to get producer and marketing cooperatives established, and he called on Bigelow for assistance.[17]

Although there was always resistance by industrial processors to the collective marketing of farm commodities, the problem was particularly acute in the case of wood. Unlike the food industries, the wood industries tended to be strongly integrated backward into primary production, owning private timber limits and holding Crown leases as sources of logs. Although these firms were eager to buy fibre from the farm and non-farm woodlots that accounted for half of Nova Scotia's forests, this was most valuable when the price of woodlot timber fell below that on company woodlands. Furthermore, the pulp and paper industry was in a sustained slump during the Depression, after a booming expansion in the 1920s. Any effort that would have the effect of increasing production costs would be stoutly resisted, regardless of the plight of the small woodlot owners.

There were several segments in the Nova Scotia pulp and paper industry in 1938. The largest integrated operator was the Mersey Paper Company in Liverpool, Queen's County. Nova Scotia native Izaak Walton Killam was a Montreal financier who sponsored the project in the late 1920s. He found his general manager, Colonel C.H.L. Jones, among the 'cross-section of better people' in the exclusive men's clubs of Montreal.[18] Retired from the British army, Jones had no work or living experience in Nova Scotia. However, he landed in Liverpool as 'the Lord of creation.' He was an autocrat and insisted on getting his own way with a compliant province, municipality, and community.[19]

Since the mill opened in the teeth of the Depression, its employment and revenue generation tended to command high political attention. As the predominant pulp purchaser across western Nova Scotia, Mersey's woods division was a force to be reckoned with. The company's extensive areas of freehold lands and Crown timber leases served not only as securities for investments and pulpwood sources but also as a leverage in setting the price of pulpwood from private suppliers. The provincial state offered several concessions. Electricity was sold at concessionary rates. Fire protection ranked among the best in North America, with the government building and staffing lookout towers and raising local crews for fire fighting. Although Mersey paid the usual fire tax, it was quite modest, and 'there wouldn't be a summer that they couldn't see anything good from it. The company never had to budget a cent for fire protection.'

After Mersey came the smaller groundwood pulp companies. Largely American-owned, they had arrived in Nova Scotia after the First World War. They capitalized on the wave of bankruptcies that hit the sawmill industry, purchasing extensive timber limits during the liquidation period. Small groundwood mills were built near the mouths of most major river valleys, whence the logs were conveyed in spring runoff. Many of these American subsidiaries aimed not to operate their lands and manufacture pulp but to hold timber limits for speculative gain. Bigelow had worked earlier for one such agent, Austin Parker, who handled the Barnjum lands while serving at the same time as an accountant with Mersey (which eventually bought these lands). Another leading speculator was the Hollingsworth and Whitney Company of Boston, the largest land holder in Nova Scotia. This firm had three overriding concerns: minimizing its tax bill, keeping fires out of the woods (at government expense), and extracting wood as cheaply as possible. While Hollingsworth cut some pulpwood for export and sold some stumpage to sawmillers, it was primarily interested in holding the lands for speculative purposes. In 1955 the Hollingsworth lands were sold to Scott Paper Company of Philadelphia.[20]

Bigelow's plans to organize woodlot cooperatives and marketing networks ran squarely into the commercial purchasing networks run by the leading companies. These networks were of several types. European mills shipped loads of Maritime pulpwood across the Atlantic, employing export-buying agents. There were never more than three or four such agents in Nova Scotia in a given year. They were all outsiders from New Brunswick or Quebec, but they knew how to handle people in rural communities. They visited all the small communities in search of producers, offering contracts for pulpwood supply. As competitors, they had little interest in talking to Bigelow, and he only came to know one in any depth.

A somewhat broader network serviced the local pulp and paper companies.[21] They employed networks of 'buyers' to procure private wood. Mersey had up to half a dozen such buyers. They tended to be influential men in regions of potential pulpwood supply. Often they were local politicians, such as town councillors or members of the legislature. Others were merchants with a desire to 'muck about the trade.' Such people became exclusive buyers in designated areas. They began with contracts from firms such as Mersey, calling for the supply of designated volumes of wood at stipulated prices. In return, they were granted sole buying rights for their areas, so that producers approaching Mersey independently were refused and referred back to a buyer. These buyers tended to be quite isolated from one another, as Mersey dealt separately with each and was not averse to playing them off each other. Although the company would plan subregional allocations of its total annual requirement, this would never be shared with the buyers, who remained ignorant of the big picture.

The buyers would approach the larger pulpwood producers (i.e., woods operators) with the suggestion 'I will run a bill for you if you supply me with pulpwood.' Once this was agreed, prices, volumes, and delivery times were settled. Ideally, the buyer wanted the pulpwood delivered to siding or port, ready for loading onto a railcar or one of the Mersey ships that plied the Nova Scotia coast. However, if the pulpwood was delivered to roadside, the buyer had to arrange onward truck transport. The producers would then buy supplies, arrange crews, and cut the wood. Farmers could receive subcontracts from larger producers as part of this system but no financing. For financing, the farmer had to deal with the local bank, with his prospects depending on his reputation and his credit record. In those days, a small loan was less than $1,000, while a large one did not exceed $10,000. The large producers also bought wood from individual farmers. The small volumes, coupled with the extreme poverty of the times, normally forced farmers to accept rock-bottom prices. A farmer offering five cords of wood might be met with the retort 'Hell, I'll burn that in my woodstove this winter.'

In Bigelow's experience, this pulpwood-purchasing system was highly exploitative. Like the Newfoundland fishermen, Nova Scotia farmers benefited little from the pulpwood trade, though the buyers and larger producers were reasonably well off. The elaborate system of intermediaries, positioned between small producers and the mills, relegated farmers and other woodlot owners to a price-taking position. There were additional burdens in the form of cheating at the scale and unreasonable terms of delivery. Everything considered, Bigelow believed that the pulp companies could obtain their wood twice as cheaply from farmers as on their own lands, which were normally held in reserve. Bigelow recalls one small pulpwood operator working on a tract of 1,500 to 2,000 acres. His was a very inefficient operation, contracted to a buying agent known for his practice of short-measuring carloads of pulpwood. When the operator trespassed inadvertently on adjacent Mersey land, the company insisted that he pay compensation in cash. The operator suggested that the stumpage be taken off his deliveries, but this was not deemed sufficient. When threatened with a lawsuit, the man replied: 'Sue me. I have a backload of debt and a house full of kids. You can have either.'

This commercial pulp-purchasing network was one of the factors prompting the Department of Agriculture to tackle forest product marketing in the late 1930s.[22] Here the department approached pulpwood in the same way that it would treat any other farm commodity. The first initiative centred on pulpwood-marketing cooperatives for export sales, and some progress was made in the eastern part of the province. It was here that the cooperative movement was strong, and some of the groundwork had already been laid by the Department of Extension of St. Francis Xavier University.[23] Bigelow developed cordial relations with this office, which extended over the balance of his career.

The first cooperative sale of pulpwood for export occurred in Richmond County on Cape Breton Island.[24] The woodlot owners there were desperately poor, but they possessed relatively large lots of up to seventy-five acres. Most of the participating suppliers lived in Louisdale, D'Ecousse, and Arichat. Other areas on the shores of the Bras d'Or lakes included Grand Anse and Whycogomagh, which was the central shipping point. Not all communities responded positively, however. Despite a plentiful supply of good wood, there was little interest in East Bay, where – despite the many small farms – most men were employed elsewhere. Similarly in Baddeck, home to the Oxford Paper Company (the largest Crown leaseholder in the province), there was little support for cooperative pulpwood marketing. Oxford manager George Harvey, a major local employer, certainly offered no encouragement to potential competitors.[25] Where a pulpwood cooperative was formed, the majority of the farmers tended to join. Those who declined usually had strong alternative contacts, political connections, or little need for supplementary income.

Bigelow also visited several communities in the western and northern parts of the province, though never with much success. In Annapolis Royal, where farmers cultivated dyke lands and owned woodlands on the sides of the valley, he found very little interest. Here the farm woodlots were treated as sources of fuelwood or incidental income from occasional sales of logs. The buyers' prices were never questioned. The same was true in a community on the Northumberland Strait. Here there was a strong logging tradition, and in places the forest had been cut back five or ten miles from the coast. Yet there was little interest in forest management or marketing. The forest was seen as a depletable resource, to be cut until it was gone. In areas where farms were marginal, the leading supplements were hens or pigs, and wood production languished at the bottom of the list. Trying to reach these people was frustrating, and Bigelow concluded that in such circumstances cooperative work had to be seen as a long-term prospect linked to education.

Consequently, his pulpwood cooperative program centred on the export market, where the prospect of success seemed to be greater than in the closed commercial system of domestic mills. This proved to be a sound step. The export agents were more than willing to deal with reputable producer associations, with the wood cut and loaded largely by the cooperatives themselves. One of the reasons for success here was the stress placed on the quality of the wood shipped. Europeans wanted prime pulpwood, consisting of at least 60 percent spruce, peeled or strip-peeled, and a minimum of four inches in diameter. However, if the overall quality was high, the cooperatives would slip in extra balsam fir and call it spruce.

Where they took root, the pulpwood export cooperatives made a difference in the returns to small woodlot owners. The prices improved over a number of years, and there was a small increase in the domestic purchase price, as awareness of the terms of the export contracts spread.[26] Unfortunately, the offshore market dried up with the outbreak of the Second World War. However, an alternative market arose when the British government sent over officials to buy pit props for its coal mines (to replace Scandinavian supplies cut off by the war). With their new knowledge of markets, prices, and shipping techniques, the cooperatives sold a significant volume of pit props over the years that followed.

Yet when overseas markets failed to recover after the war, the marketing cooperatives atrophied. The local buyer network again claimed firm control of the pulpwood market, and Mersey continued to be a particularly hard-nosed buyer. When some overseas contracts collapsed in the early war period, the Department of Agriculture approached the company to buy the wood. It refused at first, on the grounds that the wood was already peeled and would not fit the system of operations in Liverpool. Later, however, Mersey bought the wood, though at a heavy discount.[27] In general, neither Mersey nor Minas Basin proved overly concerned with the woods cooperatives either before or during the war. An attitude of indifference best describes their stance, especially while the cooperatives were confined to eastern Nova Scotia, well away from the companies' prime catchment areas.

The pulp companies were careful to keep their distance from farmers. When the department discussed its marketing plans with industrial woodland managers, they would invariably declare 'We don't want to get mixed up with those farmers.' What this really meant was that farmers would have to fit into the existing purchasing networks. More generally, there seemed to Bigelow to be a cultural bias against small farm and non-farm woodlots. At times, this verged on the common prejudice of businessmen against a backward peasant class. Mersey's forester Ralph Johnson used to ask 'Why do they stay on their land? Why don't they get out of here, the way I did?' At other times, it was rationalized by dismissing small operators as marginal to the industry, with short-term expectations of quick cash but little long-term commitment. This could be reinforced by the bias of timbermen and foresters, who saw landowners as ignorant of forest-tending techniques and prone to destructive practices.

Bigelow considered this both misguided and ironic. Neither Mersey nor its contractors showed any higher commitment to forest management. The goal was always 'to cut its yearly volume as fast and cheaply as possible.' Seaborne, the woodlands manager, continually proclaimed that forest management was not his 'headache,' and Johnson was never allowed to get mixed up in woods operations.[28] Faced with criticism of its forestry methods, the company tended to blame the local tax system, arguing that it forced landowners to clearcut their timber.[29]

Reforming the Forest Trade

The second strand in the Department of Agriculture's forest product reform program involved the sawmill sector and lumber marketing in particular. This situation was far from satisfactory during the Depression years. Maritime lumber was suffering severely for the lack of quality control and grading standards. Nova Scotia sawmills turned out a crude staple product, which occupied the bottom end of the export market. The local market was limited, often informal, and filled with rejects from the export trade. Overall, it was a sector in disarray.

In structural terms, Nova Scotia sawmillers were smaller in scale and more geographically diffuse than those in other provinces. The system of primitive equipment, degraded forest lands, and predatory middlemen seemed to lock lumbermen into the production of low-grade and low-value commodities. Nova Scotia lumbermen seemed to look to New Brunswick with awe, observing that larger and more sophisticated mills made it possible to bypass various layers of brokers and realize better terms. Bigelow remembered when a new agent with an interest in opening a new business visited Nova Scotia in 1938. Given a list of some 700 producers compiled by the Department of Lands and Forests, he soon returned to announce that he had visited everybody on the list. He was struck by the fluidity and disorganization of the trade, and he left the province without pursuing a business relationship.

The export segment of the lumber business was oriented around a number of British buyers who sat atop the trading pyramid. The larger of them, such as Pharoh Gane and Brandts, served as buyers for various lumber-consuming enterprises in the United Kingdom. To this end, they bought Nova Scotia lumber under an all-inclusive 'merchantable' grade and delivered it in lots to British customers. Part of the process involved culling poor pieces as the loads were landed. Although a partial cull is inevitable, the absence of domestic grading standards opened the way for wide variations in quality as well as exaggerated losses and rejects. So long as the ill-defined 'merchantable' category persisted, Maritime producers and exporters alike were at a serious disadvantage. Returns were depressed, value was lost, and forests were continually degraded.

As with pulp, the Nova Scotia lumber business was beset with intermediaries. Exporters generally followed two strategies in procurement. The larger ones employed salaried fieldmen who travelled the region in search of producers. A single exporter might have up to twenty sawmills supplying lumber, and, while a fieldman usually maintained a stable core, contracts could shift according to past performance and available terms. Exporters could also help to finance the sawmillers, either directly or indirectly. Since banks tended to regard small lumbermen as poor risks, affiliation with an exporter could offer a form of security. Exporters had less difficulty in obtaining bank finance for their advances. (Bigelow remembered one lumberman who was turned down by one bank manager only to be approved by another. Returning to the former, the lumberman offered him a job, which the manager politely declined, stating that he was perhaps not qualified to operate a complicated lumber business. The lumberman replied that this was not the position he had in mind. He was offering a job pushing a wheelbarrow to dump sawdust outside the mill – a duty usually assigned to young boys!)

The relationship between lumber buyer and sawmiller worked as follows. The exporter gave a contract to the lumberman for a volume (perhaps 1,000,000 board feet) of boards sawn to designated dimensions. Using either his own or a leased portable sawmill (capitalized at around $25,000), the lumberman located a piece of land or stumpage rights covering perhaps

200 acres. He would then recruit his crew and put them to work. This sort of operation varied in size from 500,000 to 4,000,000 or 5,000,000 board feet per season. The price paid to such operators was invariably low, and so was the quality. As the crew worked through a stand of trees, the sawmiller cut only what the contract called for, leaving all other trees (including the slash and the sawdust that posed such severe fire hazards) on the lot. In the process, the combination of high-grading and wastage invariably left the forest in a deteriorated state. In the vernacular, such operations were styled 'woodpecker mills.' They seldom showed a profit on a consistent basis. Financed by the export agents and dependent upon them for contracts, the portable sawmills worked from year to year, changing hands as circumstances required but unable to break out of a cycle of marginality. Some operators tried to do business independently of the agent-exporter system, but this proved difficult. G.W. McLelan of Oxford, Cumberland County, did not want to deal with the British buyers. He set up a planing mill and sawed specifically for the American market.[30] However, even lumbermen who turned out good-quality products were trapped by the agent-exporter system, which did not reward higher value. D.A. Huston, a producer from Upper Musquodoboit in Halifax County, was meticulous about quality but was paid no more than anybody else.

There was some variation across the lumber sector as well. Some lumbermen were farmers operating on their own lands, while others cut more widely and intensively. Many evolved over time, from small home operations to working neighbouring lands to eventually buying stumpage from large foreign landowners such as Hollingsworth and Whitney. Many of the larger lumbermen and exporters made money in spite of the low-quality/price trap. Although they maintained a low profile, there were half a dozen prominent export firms in the Maritimes. Names such as Mullins, MacCullough, and Wilber were common in the Halifax business, as were MacKay, McKean, and Colter in Saint John. Bigelow recalled the fieldman who handled all of McKean's buying for more than twenty-five years. Travelling constantly, he knew all the lumber producers and operated on a secretive basis. In New Brunswick, the Fraser Company was in a different league, as a large-scale shipper of industrial wood to the American market. Halifax exporter Charles MacCullough accumulated enough wealth to pump $100,000 of earnings into the South American utility Brazilian Traction and went on to make a fortune in stock trading during the Second World War. After that he was more of a financier than a lumber trader. Outside Halifax and Saint John, Roy Jodrey from Minas Basin was one of the largest exporters.

The Nova Scotia government was not unaware of the problems of the lumber industry, but it was cautious in the face of entrenched business power and uncertain how to proceed. One significant initiative to reform the lumber industry was launched by a group of growth-oriented sawmillers with progressive business ideas. In 1934 the Nova Scotia Forest Products Association (NSFPA) was formed as a lumbermen's organization. Bent on improving production practices and product quality, it set about organizing this diverse and isolated business. Many of its early proposals involved government action on matters as diverse as lumber grading and workman's compensation.

In 1939 Bigelow started to work on reforming the Nova Scotia lumber trade. It was Walsh's opinion that 'nothing can be done without a formal organization.' Consequently, they agreed on the importance of supporting the NSFPA. On a voluntary basis, Bigelow served as its secretary from 1939 to 1943. The major concern of the organization at the time was to reduce the workman's compensation claims in the industry. During this time, he came to know most of the leading figures within the industry and became thoroughly conversant with their concerns. He also used the opportunity to promote the Marketing Department's pursuit of uniform grades for eastern spruce lumber.

This expertise soon drew Bigelow in an unexpected direction. From 1943 to 1945, he worked in Ottawa in the Office of the Timber Controller, one of the supervisory agencies directing wartime production. It controlled all exports of wood during the war and granted permits to lumbermen to log, cut, and export various products.[31] Since the Maritime industry was so disorganized, the government of Nova Scotia was asked to supply a knowledgeable local man for the position. Bigelow was approached, and, after initial hesitation based on considerations of his young family, he accepted. Part of his job involved screening applications for export permits to determine whether the firms were established and able or fly-by-night profiteers. He also had to deal with irate lumbermen protesting the retention of certain shipments in Canada.[32]

After the war, Bigelow returned to Nova Scotia. Not long after, he left the civil service to join the Maritime Lumber Bureau (MLB) as its full-time secretary-manager. Formed in 1939, the MLB was a regional organization which aimed to install uniform grading standards for eastern Canadian lumber. It sought to do for marketing what the FPA sought to do for lumber production.

Progress came the hard way, as the agent-exporter system still militated against collective actions and technological change. Bigelow made some gains in providing better price and market information for producers, collecting wholesale prices for lumber in Boston and quoting prices from various publications in Scandinavia. These figures contributed to the growing public knowledge in the industry about comparative price movements, a subject about which exporters remained tight-lipped. In response to queries, their answers would invariably be 'Now why would you want to know that? You know we don't give out that kind of information.' Any answers offered would either be distorted or limited to partial dealings. However, the long-run effect was clear. As a result of determined and sustained effort, the chronic secrecy and general ignorance of market trends that characterized the traditional corporate buying networks gave way by the 1950s, though slowly and not completely, to a commercial clearing house and independently graded products.

The low price of Nova Scotia lumber remained an obstacle to technological change and improvement in the industry. Producers would not invest in new machinery, blaming their decisions on 'big bouts with their bankers.' And while the Second World War constituted an excellent time to reform industries, with plenty of federal government money available for industrial restructuring, nothing happened in the Nova Scotia sawmilling industry. This was in stark contrast to British Columbia, where the industry was transformed both in technology and in productive capacity. When Bigelow found out that a specialty saw, costing ten dollars more than the cheapest saws then used, could overcome the problem of the ugly ends of Nova Scotia lumber (which contrasted to the beautiful ends of Scandinavian lumber), he was surprised to discover how few sawmillers adopted it.[33] 'Why should I spend money on a planer saw?' was the general response. Those who switched did so reluctantly and complained bitterly afterward. By the time Bigelow quit the bureau in 1950, about 75 percent of the lumbermen had adopted the saw, but it was a hard battle all the way. There was no incentive for the producers to change their operations since the agents and exporters never told the producers when (and indeed if) they passed on a better price for a better-quality product.

During Bigelow's time with the Maritime Lumber Bureau, grading services were developed for Maritime spruce lumber, training schools were established for graders, and a certification system for export lumber was put in place. The grading and shipping of certified lumber were accepted by all Maritime exporters, as well as those on the Gaspé coast of Quebec. Overall, however, progress among the sawmillers was still slow, and by the time Bigelow left in 1950 only a minority of lumber producers had been affected.[34]

In 1951 Bigelow embarked on a couple of ventures to improve the marketing of lumber on a smaller scale for private ventures. He first worked a short stint as the manager of Scotia Sales Limited, a joint export arrangement between three Maritime lumber exporters: Lockhart from Moncton, Hawkins from Riversdale in Nova Scotia, and Hickman from New Brunswick. They were all discontent with the overseas market, the agent-exporter system, and the frequent disputes over quality and price. In addition, they all wanted to explore the American market and the possibility of improving quality. Bigelow was charged with this duty. He soon found out that the model to follow was the Fraser Companies in New Brunswick. All the large buyers whom Bigelow approached in the United States stated that they wanted quality equal to Fraser's. The firm had, over a period of two to three generations, obtained better prices in the United States than any other producer.[35] When Bigelow looked into its practices, he found that when Fraser shipped its lumber there was always a man at the boxcar (or other shipping point) who picked out all bad pieces and lumber that was dirty or had excess knots. He picked out perhaps only two or three pieces per boxcar, but doing so made all the difference.[36] In the end, the initiative failed, however, because it was believed that the market was too scattered and the centralization of the three operations would be too costly. Consequently, they stuck with their old markets. After Bigelow left Scotia Sales Limited in 1952, he took up a brief appointment with a Mr. Libman, an internationalist who shipped pulpwood and lumber overseas, including to Soviet Russia. Bigelow arranged a couple of shipments for him from Saint John, one of them to Italy. Libman died suddenly, however, and so did his organization.

Fighting Political Patronage

These various efforts to reform the sawmill industry took place in the context of political pressures both for and against them. The provincial state was extremely sensitive to business pressures, and the bureaucracy was both subservient and fractured. Political support for reform came from the faction that Bigelow described as 'broad-minded and sincere' men who could see beyond their immediate personal interests. They understood the connection between the agent-exporter nexus and the crude exploitation of low-quality products. Furthermore, they recognized that improved milling techniques and graded products would yield higher-priced lumber and enhanced forest management.[37]

Many of these broad-minded lumbermen were also politically active. They included Liberal Senators W. McL. Robertson (appointed in 1943) and Charles G. Hawkins (appointed in 1950). Hawkins had links, through his sawmill business, to the whole of rural Nova Scotia. He was also a close friend and confidante of J.H. MacQuarrie, the minister of lands and forests from 1933 to 1947.[38] Another such figure was Elmer Bragg, a farmer, sawmiller, and Liberal politician from Collingwood Corner in Cumberland County.

On the other hand, there were powerful structural forces acting against reform. Although broad-minded, Robertson and Hawkins were part of the hierarchical system of exporters, other middlemen, and small woodlot owners. They strove to improve rather than change this system. Those positioned at the lower rungs of the hierarchical ladder were reluctant to accept or adopt improvements, having no assurance that they would benefit from them.

The partisan legacies of the province created additional impediments. Political office and political connections had been important factors in advancing the position of individual lumbermen in the past. After a period of over thirty years in power (1882-1925), Liberal-affiliated lumbermen were by far the most prominent in the industry. The fate of government bureaucracies was also closely connected to political affiliation. Following the changes of government in 1925 and 1933, the provincial civil service was largely restaffed by people of the 'right' political party. Interestingly, the deals struck with the major pulp and paper companies occurred during Conservative reigns.

Many lumbermen were thus still old-style politicians with partisan connections who held the view that their power and knowledge stood above the emerging state bureaucracies and industry-wide institutions. Bigelow characterized most politicians at the time as men with superior egos who referred to their constituents as 'my people' and their supporters as 'the good people.' They were not well read, and, if pressed on the nature of their profession beyond politics, they had no answer. Winning an election in Nova Scotia gave politicians the idea that they were superior to all others. While in reality local Members of the Legislative Assembly (MLAs) were little more than dispensers of petty patronage, this counted for much in poor areas. It also served to set these politicians apart from both the public and the bureaucracy. Bigelow was once asked to advise a Cabinet minister. Beginning his sentence with 'If I were you ...,' he was quickly interrupted by the indignant politician, who declared 'You are not me. To be me you have to be elected.' On another occasion, he advised an MLA to consider a particular individual for a senior civil service appointment but was told that 'He would never win an election.' This apparently disqualified the prospective candidate.

Some of the lumbermen-politicians illustrated the existence of old-style politics as well as the persistent politics of self-interest that militated against any reform in the industry. Bigelow remembered one group of politicians as merely advancing their own political careers rather than being interested in the reform of the industry. A.S. MacMillan, the Liberal premier of Nova Scotia from 1941 to 1945, was sympathetic to the efforts of reform but always filtered it through his chances for reelection.[39]

Some politicians used their political positions to advance their own business interests. C.W. Anderson was a Liberal MLA who cut wood illegally from Crown lands in Guysborough County and figured prominently in the 'Woodpecker Election' of 1937.[40] Anderson literally ran his lumber business from his government office in the legislature. He worked consistently against the Maritime Lumber Bureau and opposed any government intervention in the industry. Premier Angus L. MacDonald gained Bigelow's respect by acting personally to stop Anderson's activities.

Albert Parsons was another local 'big man,' a Conservative MLA and onetime speaker of the legislature who ran lumber operations along with a gypsum quarry at Walton in Hants County. During the 'Woodpecker Election' campaign, it was revealed that he too had cut illegally from Crown lands.[41] Once during his time with the Maritime Lumber Bureau, Bigelow invited Parsons to deliver the opening address at a demonstration of lumber grading in his community. Lumbermen congregated from far and wide. Upon arrival, Parsons approached Bigelow and declared: 'I only want to know two things from you, Mr. Bigelow. Who is paying you and what are your politics?' He then departed, leaving Bigelow to perform the honours himself. After Parsons's death, a tax audit revealed extensive unpaid taxes that were assigned to his estate.

The Department of Lands and Forests was clearly shaped by the strong political influence of the forest industry, exercised by both the lumbermen and later the emerging pulpmen. The department was primarily a protection agency, concerned with fire protection and ancillary services such as wildlife protection and the licensing of sports fishers and hunters. The department also had responsibilities for land surveying and a legacy of colonial land policies that left property and survey lines in a chaotic state. In fact, this left the department ignorant of the full extent of its own holdings.

Departmental officials were generally better connected and more sympathetic to the old established lumbermen than to the pulpmen.[42] But they were very reluctant to address questions about the production, marketing, and pricing of wood products. Forest management was only promoted within this narrow context, which meant concentrating on the 25 percent of forests in Crown hands. In the mid-1930s, Wilfrid Creighton was the only forester employed, and he had no specific duties.[43] F.A. Harrison, the head of the Department of Crown Lands under Premier Murray until 1925, took charge of the Department of Lands and Forests again when the Liberals regained power in 1933. His training was as a land surveyor, which was reflected in his concerns with survey and inventory work within the department. Harrison was suspicious of the new pulp era, looking down his nose on Hollingsworth and Whitney, the American pulp and paper company that was the largest landowner in the province. He wanted to keep strangers out of the province. Harrison was born and raised in Harrison Settlement and cherished the methods of the old days, when oxen were used in the woods and six watermills operated in the area. He rightfully criticized (in Bigelow's view) the era of steam and portability, which was so destructive to the forest.

On all important matters and many minor ones, the Department of Lands and Forests was controlled tightly by its minister. Bigelow often asked departmental officials to tackle questions of prices and markets, but he ran into an administrative blind. The response was always 'I wouldn't dare touch that. The minister would be right at my neck.' When Bigelow tried to advance these concerns, he received a cold response. Junior people in the department were even fearful of being seen with lumbermen and exporters. And when Bigelow talked to the forest rangers in the field, they made it clear that the lumber trade 'wasn't their interest.'

One major departmental initiative that did not seem to threaten the private forest trade was the Crown land buy-back program introduced in the 1930s. Invariably degraded by commercial cut-overs, these lands could be purchased from private owners because no one else valued them.[44] Choice timber lands were more expensive and were generally left to commercial buyers. On at least one occasion, for example, Bigelow was charged with negotiating the purchase of a large tract of fine timber, known as the Tusket lands, in the southern part of the province. Several other parties, however, were interested in the lands, and Bigelow was told by the minister to stop his efforts, and the lands were sold to Mersey.[45]

Following the war, Bigelow's career was periodically enmeshed with the Department of Lands and Forests. In 1945 the Woods Marketing Division was transferred from the Department of Agriculture to the Department of Lands and Forests, though this transfer was reversed not long after. A different sort of coordination became possible in principle during periods when the two departments were headed by the same minister. This occurred under Liberals Art Mackenzie (1947-53) and Colin Chisholm (1953-6). However, the departmental rivalries remained deeply rooted. In 1947 Waldo Walsh was the deputy at the Department of Agriculture, and he put a novel proposal to his minister, Art Mackenzie. Walsh suggested that the deputy responsibilities at Lands and Forests be divided for a year, with Bigelow appointed director of marketing and Provincial Forester Wilfrid Creighton appointed director of production. In itself this implied a basic redefinition of the department's mandate, elevating forest-marketing policy to unprecedented heights. However, Walsh seemed to be advising as much on personnel. He suggested to Mackenzie that he 'leave it that way for a year and receive their reports, see how each man did, and then at the end of the year make a decision who would be Deputy.' Walsh added that

> I knew very well that if they followed this practice who would be the Deputy – the man who had some knowledge of something beside silviculture, [who knew] something to do with the economics, the business of it, and John Bigelow knew it.
>
> Art Mackenzie agreed to this ... I think he fully agreed with me – ... but, much to my surprise and complete disgust, he came in about a month later and said the government had appointed Wilf Creighton, and, as much as I like Wilf – I respect him, he is honourable and he is decent, he has done a

> lot of good things for Nova Scotia – but I just couldn't help feeling that he [Creighton] was one of those who had to work with the big companies, and that is the trouble here. My Godfrey, could you or anyone else understand the statement that Waldo Walsh or the Department of Agriculture was sitting in the offices and in the hands of Canada Packers and CIL or any of those other big companies that [are] mucking around Agriculture and have done so much harm to the progress of Agriculture in this country?[46]

The Department of Lands and Forests thus remained aloof to the price and marketing issues for the indefinite future. In 1948 Bigelow wrote to the minister to suggest the formation of a new Economic and Marketing Branch, but the questions had already been settled.[47] Eight years later, the two departments were once again assigned to separate ministers, and the possibilities for cross-fertilization receded further.

Bigelow speculated that an added factor in the hostility toward the organizing of woodlot owners and lumbermen was the overall fear of socialism. He often spoke up in favour of the small woodlot owners and charged that the price of pulpwood was far too low. For this he was often warned. At one time, a senior Cabinet minister stepped into his office to inquire whether he had made statements to the effect that the price of pulpwood was too low. When told yes, the minister warned: 'That's a very serious charge, Mr. Bigelow.' On another occasion, the minister of agriculture cautioned him not to get too close to the St. Francis Xavier people who promoted the cooperative idea. In 1952 Bigelow was taken to task by a member of the legislature, D.D. Sutton, for criticizing the low price of pulpwood, but Premier MacDonald came to Bigelow's defence.[48]

The impact of the Department of Lands and Forests on forest management was equally abysmal. Most initiatives for forest management came from elsewhere. Bigelow was part of the team that wrote the report of the Nova Scotia members of the Canadian Society of Forest Engineers in 1944.[49] Its main author was Mersey forester Ralph Johnson, who was a hard worker at this type of thing, though he did not appreciate the economics of the resource. Although this report was presented to the government of Nova Scotia, little of it was implemented.

The significant piece of legislation at the time was the Small Tree Act of 1942, which was more an initiative of the politicians than of the Department of Lands and Forests.[50] To Bigelow, the continuing flaw in Nova Scotia forest policies, including the Small Tree Act, was the failure to harness forest management to forest economics. This was quite different from acceding to the short-term business interests of one or another constituent interest. Political interventions tended to reflect the reigning interests of the day, while the Department of Lands and Forests bureaucracy remained economically illiterate by choice. As a result, Bigelow considered most of the forest policies aimed at sawmillers and woodlot owners in his day to have been misdirected. From his perspective, working markets and positive financial returns were the keys to forest management, not vice versa.

Bringing a Pulp Mill to Nova Scotia

The 1950s were a time of unprecedented economic growth and increasing (though limited) opportunities for Canadian provinces and their bureaucracies to influence the nature and conditions of the path of development. Opportunism and deference toward a growing segment of transnational capital operated cheek by jowl. Local conditions and personalities and the competitive status of various industries and industry sectors determined the outcome.[51] The pulp and paper industry was growing at a fast rate, but at the same time there was a lot of competition for investment capital. In Nova Scotia, sawlogs were getting scarcer, and their quality was deteriorating. The larger sawmillers were scrambling for sawlogs and forest lands to buy. It was not uncommon for them to meet each other in the woods, looking to buy the same wood stands, so rare were good sawlogs. Given the constraints in the industry, many sawmillers were looking to get out of the business, and many politicians were turning to the pulp and paper industry.[52]

Bigelow's career took a new turn in this period when he took up a position with the Nova Scotia Department of Trade and Industry. After more than a decade of experience in forest product marketing, including stints serving trade associations, grading authorities, and the wartime administration in Ottawa, this was hardly a surprising step. It did, however, alter his policy responsibilities in important respects. Although Bigelow remained involved in the forestry sector, his new frame of reference was enterprise development and large-scale pulp and paper industry development in particular. After years of living in Amherst, he liked the idea of moving to Halifax, and the bureaucratic culture at Trade and Industry was certainly a contrast with that at Agriculture. The minister, Wilfrid Dauphinée, was a former salesman with a penchant for creating headlines. In campaigns to promote tourism, he favoured celebrity endorsements: 'Let's get Babe Ruth down.'[53]

Bigelow's initial appointment in 1952 was to a research and consulting position with the department, and his chief project became the feasibility of additional pulp and paper capacity for Nova Scotia.[54] This assignment arose while he was still in Amherst. Premier MacDonald requested him to travel to Port Hawkesbury for discussions on prospective development projects for Cape Breton Island. There was widespread anxiety in the Strait of Canso area given the provincial decision to replace the rail and car ferries with a causeway connecting Cape Breton to the mainland. The premier had dumped the ceremonial first load of rock in the spring, only to be met by a citizens' delegation fearful about the extensive loss of local jobs. Bigelow's initial mandate was broad. The premier mentioned many possibilities, including gypsum, coal, minerals, tourism, and sheep farming. However, nothing was said of the forest industry.

There was an important reason for this omission. It involved the Oxford Lease, or Big Lease, a ninety-nine-year option on 620,000 acres of Cape Breton Crown forest that was dealt to the Oxford Paper Company in 1899.[55] When Bigelow queried the premier about adding forestry projects, he cringed and exclaimed, 'Do you have to bring that up? If you do, bring it up last.' Although the lease contained various manufacturing stipulations, no local pulp mill had ever been built. The lease seemed largely to be inoperative by the 1950s, to the growing frustration of local people. For the provincial political establishment, however, the Big Lease had become something of a sacred cow. Frank Smith, a lawyer, briefed Bigelow on the background to the lease and the problems that it presented. Every time it came up for renewal, company lawyers descended on ministers to lobby for the status quo. 'Those Americans were very shrewd,' Smith recalled. They used to approach the top local law firm and then lay down the law for what they would and would not do. The precedent established by the firm of Burchill, Smith, Parker, and Fogo served them well, and over time the politicians had grown afraid to touch the lease.[56]

Bigelow recalled attending one of those meetings with the Department of Lands and Forests minister, the Department of Trade and Industry minister, Burchill (representing the Oxford Paper Company), and two lawyers from Maine. By then, Burchill was elderly, and he had lost some of his forcefulness. Nevertheless, he browbeat the two ministers, preaching the sanctity of contracts and the dire consequences of reneging, and he pointed out the importance of his law firm's contributions to Liberal election campaign funds. According to Bigelow, the Lands and Forests minister was actually trembling after one of these meetings, concluding 'I guess we can't touch this issue.' The lease was renewed in 1955 for another twenty years.

Bigelow nevertheless looked into the terms of the Big Lease, and eventually he suggested that the contract with the Oxford Paper Company could be broken. He secured an opinion from former Lands and Forests minister J.H. MacQuarrie, a lawyer who stated that the lease could be expropriated. In fact, this approach had already been followed for part of the lease when the federal government made it a condition for the establishment of Cape Breton Highlands National Park in 1935. Armed with this insight and a determination to take action, Bigelow successfully convinced a hesitant Dauphinee to travel to New York to see Hugh Chisholm, the president of Oxford. When they inquired into Chisholm's interest in selling the lease, his response was 'Well sure, I am a businessman.' This came as both a relief and a surprise, since Mersey had repeatedly told the government that

Oxford had refused to sell the lease to it. The conclusion seemed to be inescapable: 'We realized that Mersey had lied to us.'[57] It was a relieved minister who returned to Halifax to report that Oxford was prepared to sell.[58] The way was now clear for serious negotiations with prospective investors in an eastern Nova Scotia pulp and paper complex. Bigelow played an increasingly central role in this process. In 1954 he was appointed to the post of deputy minister of the Department of Trade and Industry.

Over a period of five years, Bigelow attended pulp and paper industry conventions and came to know the industry quite well. Inevitably, there were false starts in the pursuit of a major complex to anchor the primary forest industry. Once, in Toronto, he was asked to contact Quebec businessman Elliot Little, president of the Anglo-Canadian Company. This led to a high-level meeting in Halifax, including the premier, Dauphinée, Bigelow, Little, and his assistant. It got off to a poor start when Little greeted MacDonald with the question 'Well, Mr. Premier, what do you have to give us?' The government provided a variety of information but never received a firm response from Quebec. In any event, authoritative data on timber volumes was lacking. The first modern provincial forest inventory commenced in Cape Breton in 1953, and the analysis took three years to complete.

Bigelow later conferred with a Montreal-based consultant to the pulp and paper industry, John Bates. He was able to refer Bigelow to a New York businessman who was looking for a site for a high-grade bleached sulphite pulp mill. This was Karl Clauson, the sales representative for Stora Kopparberg AB of Sweden. Clauson presented Stora as 'a very old and well-established company, with a large market in eastern North America.' After Cabinet consideration in Halifax, Dauphinee told Bigelow that the government's initial response was favourable. Stora then sent several people over from Sweden to join the ambitious Clauson.[59]

The election of Robert Stanfield's Conservative government in 1956 raised some uncertainty in the course of negotiations. The project appeared to be in jeopardy, as the new minister of trade and industry, Edward A. Manson, had little interest in promoting a 'Grit' project. This placed Bigelow in a difficult position. However, after deciding that the Stora project was more important than his allegiance to his minister, Bigelow approached Ike Smith, the minister of highways and a leading member of the Stanfield Cabinet. Clauson was then sitting in his hotel room, contemplating whether to return home. Stanfield, Smith, and Manson invited him in for a meeting and explored the Stora project. Negotiations were revived, and in the end a deal was struck.

When the Stora agreement was close to being sealed, Bigelow was asked to draw up the formal terms of the Crown lease agreement. Although somewhat surprised by the request, he was elated to do so. Seeking a model contract, he contacted a friend who was the deputy minister of lands and forests in Alberta, which had recently negotiated a similar agreement with a large pulp and paper venture. Bigelow presented his draft agreement to the Cabinet, which endorsed all the provisions save one. He had included a clause for a stumpage payment of $4.40 per cord on the Crown lease of 1.2 million acres.[60] At Cabinet, this stumpage rate was reduced to one dollar per cord. Bigelow recalls a meeting with Premier Stanfield and two other ministers in which the $4.40 figure was questioned. It was Bigelow's belief that this figure reflected the state of pulp stumpage in the industry and that the province should obtain as high a stumpage as the market would bear. However, the premier thought that terms similar to the Mersey agreement (struck thirty years earlier) were appropriate. Under that arrangement, Mersey had been a good corporate citizen, and the government did not want to upset Stora by treating it differently.

Clauson, who was prepared to pay more, was delighted. According to Bigelow, Clauson 'had never expected something like that.' When Bigelow mentioned the revision to Manson, his reaction was 'Why would the premier have done that?' Bigelow urged him to raise the matter at the next Cabinet meeting, though this was never done. Bigelow considered this revision to be a disservice to the people of Nova Scotia. Throughout the search and the negotiations, he had hoped that the farm and woodlot producers would gain an essential new outlet for their pulpwood and thereby bring forest management techniques to their properties. He feared that cheap pulpwood from Crown lands would depress the price of private pulpwood and eliminate the incentive for improved forestry.[61]

Small woodlot owners were well aware of the consequences of low Crown stumpage for the price of private pulpwood. It was one of several issues (including enhanced forest management, improved logging practices, and policy advocacy to government) that led to a new campaign to organize a Woodlot Owners Association in 1959. Bigelow was involved in and supportive of these efforts. He clearly did not trust Stora to pay a premium price for private pulpwood, writing in 1959 that it was not certain 'how active they will be in deterring private landowners from destroying their growing stock.'[62] Consistent with his earlier work with Walsh, he had long supported the establishment of a woodlot owners' organization to promote fair economic returns and collective management by woodlot owners. In 1959 he wrote that 'This organization of woodlot owners has, I am happy to say, now been formed in this area on a voluntary and informal basis. It has the wholehearted support of the Department of Trade and Industry and will be aided by this Department in every way possible to develop into a formal organization with a membership of the majority of the woodlot owners of the area, authorized to do business for its members and to acquire knowledge and business experience as rapidly as possible.'[63]

Bigelow also supported the Department of Extension of St. Francis Xavier

John Bigelow with other invited speakers at the Rural and Industrial Conference, St. Francis Xavier University, 1960. From the left, John Bigelow, S. Bates, A. Laidlaw, and Alan MacEachen.

University, which was heavily involved in promoting the formation of the new organization. At one point, the director, Reverend J.A. Gillis, approached Bigelow to join his staff, and later he approached the minister of agriculture to make him a top-level liaison officer between the woodlot owner and the pulp company.[64] It was disappointing to Bigelow that the prospects of the new Woodlot Owners Association were frustrated by the continuing opposition of the Nova Scotia government and the Department of Lands and Forests.[65]

On the issue of technical aspects of forestry and forest management, Bigelow had more faith in Stora's abilities. Stora sent over people who examined everything. One of these men, Mats Chartau, told Bigelow, 'One thing you don't have to worry about is the supply of pulpwood.' The Swedes also recognized the value of balsam fir, since tests in Sweden showed that the fir yielded 5 percent more pulp than the spruce. The spruce budworm did not worry the company, the feeling being that it would die a natural death or be eradicated by insecticides. Bigelow believed that the company would buy several species of hardwood, including poplar and white birch.[66] He was generally impressed with Swedish methods of forest management, quoting at one point the development in Sweden of seed orchards, clones, and plus trees to improve the productivity of the forest.[67] For all of these Stora preparations, there was very little contact with the Department of Lands and Forests, and Bigelow maintained a suspicious view of the department. The Swedes, so Bigelow claimed, were equally unimpressed with the department. The feeling was that, 'if Bulmer [one of the foresters working on inventory] will live long enough, they [Lands and Forests] might have something that is useful.'[68] Bigelow had nothing to do with subsequent negotiations to bring additional pulp and paper industry capacity to Nova Scotia. Some years later, in 1965, Scott Paper Company of Philadelphia built a sulphate pulp mill in the central part of the province. Scott was the largest corporate landowner in the province, as well as the holder of a Crown lease of 250,000 acres.

Ending a Career

In 1960 Bigelow was offered the position of Director of Research for the Department of Provincial Treasury. His responsibilities were to continuously review and advise on the research programs carried out by the various government departments and agencies, as well as to monitor the federal-provincial arrangements and undertakings with respect to research.[69] When he asked what this entailed, he was told to find out. After a little digging, he found out that one department of the Nova Scotia government was involved in one research project. As part of his duties as Director of Research, in 1962 Bigelow was asked to accompany Ike Smith, the minister of finance, and Lorne Goodfellow, the recently appointed deputy minister, to Britain and France to study European approaches to economic planning. The common market was a big interest of economists at the time, as was the concept of large-scale economic planning. The Stanfield government was also enamoured of the success of big businessmen such as Frank Sobey and Roy Jodrey, as well as various Upper Canadian experts and businessmen whom they sought to bring into the process.[70] Goodfellow, for example, had been employed on the recommendation of an Upper Canadian executive-finding agency.

Bigelow accepted the offer to go to Europe. The group visited several government offices in Britain, where the Canadian Consul had set up various appointments. Most of the talks went over their heads, but Bigelow took detailed notes. They had one meeting with one government department in France. The latter was headed by Giscard d'Estaing, but he was not able to see them personally. The trip to France was a general disappointment because everybody was on holiday. Bigelow nevertheless took extensive notes again. They came in handy later when Smith asked him to write a report. This turned out to be quite voluminous, and Bigelow was personally congratulated by Smith for his effort. It directly influenced the design of a novel business-labour consultative mechanism known as voluntary economic planning (VEP).[71]

Stanfield was committed to attracting new manufacturing industry to Nova Scotia, and Ike Smith offered the French planning system as a model for economic consultation. Bigelow did not personally support this scheme, however, and openly advised Premier Stanfield to shelve the project. Bigelow's feeling was that the government might as well get an economics professor to advise it. The French equivalent was staffed by a small army of civil servants and bureaucrats who were knowledgeable on planning issues. Nova Scotia had no people to fill such functions. Bigelow believed that his assessment at the time was borne out by the record of VEP. 'You'll have a hard time coming up with anything useful they've done.' Their existence was justified by their own public relations efforts.

In the 1960s, Bigelow's relationship with the politicians gradually soured. He worked closely with the Nova Scotia Research Foundation. This support extended to recommending the closure of his own research directorate in favour of an expansion of the foundation. He also defended the foundation's effort to remain independent from a merger with its New Brunswick counterpart. The foundation was sponsoring the development of important projects at the time, among them the sonar buoy. Bigelow advanced other points of view that were not popular with government. He advised the government against backing the heavy-water plant at the Strait of Canso. In doing so, he referred to the experience of a pulp mill at Bear River in the 1920s. Water was needed for the venture, so saltwater was piped to the plant from the ocean. After a few months, the pipeline and machinery had corroded, and the plant failed. The heavy-water plant venture at the strait was based on the use of saltwater. It failed for the same reason. The Nova Scotia government chose to listen to the promoter and speculator of the project, a Mr. Spevack, who made off with $2 million of Nova Scotia taxpayers' money.[72]

Overall, Bigelow became discouraged and disillusioned with the government strategy of attracting industry at any cost in the 1960s and the associated failures, such as the water plant at Port Hawkesbury and the Clairtone factory. When asked about a final assessment of his career, Bigelow believed that he had worked hard in the interests of the people of Nova Scotia. He had pushed certain initiatives to their limits. If and when they had not worked, he had moved in other directions. He had no regrets.

Conclusion

John Bigelow's career serving the Nova Scotia forest industry for close to three decades offers telling insights into a provincial political economy undergoing the change from sawmilling and patronage to pulp, paper, and bureaucratic control. Bigelow was a Nova Scotia nationalist insofar as he consistently supported measures to maximize the welfare of provincial residents and institutions. In this, he disregarded class, status, and sectional differences, and he showed a sharp perception of the public interest. Early in his career, he had an opportunity to pursue these goals through government

John Bigelow in retirement, Halifax, 1995.

service, and he continued to see the state as a powerful potential agent for the public good. At the same time, this did not stop him from acknowledging the backward and prejudicial character of partisanship and patronage. Bigelow was also an economic nationalist, working to promote home and value-added industries and fair returns for the province's primary producers. He understood the regressive consequences of rural mercantilism while recognizing the progressive possibilities of dynamic and organized markets in increasing efficiency and income returns. He was fascinated by trees and forests early in life, and this fascination is what propelled him to go to forestry school and then to work as a professional forester with a large pulp and paper mill in Newfoundland in the 1930s. It was there that he became sceptical of the social benefits of pulpwood cutting and disillusioned with the practice of forestry. He also developed an insight into and aversion to the colonial ventures and legacies in the forest industry; this included businesses being run for the parent operation rather than the colony and the staffing of positions based on personal connections rather than ability.

In Nova Scotia, Bigelow took up an appointment with the Woods Marketing Division of the Department of Agriculture, where he worked to reform the social inequities in the pulpwood and lumber trade. In 1954 he summarized the situation in the Nova Scotia forest industry as follows:

> We have portable sawmills setting up for a time in an area, cutting available saw logs, employing a few people, buying a few farmers' logs, then moving

> to another area and repeating the cycle – always dependent on what some agent in Halifax or Saint John or Moncton chooses to tell them of market conditions – no standard of quality – no stability. For thirty years there has been a market for some pulpwood ... Tremendous variations in price and an almost complete lack of competition between buyers have been outstanding features of this pulpwood market from the beginning. In the past twenty years, Christmas trees have moved from this area in some volume. Again, the American buyer, or the agent for a buyer, has told the forest owner what his trees are worth, how many and what quality he can sell. Pitprops have been cut and sold to the local mines for many years and in times of war, or other disturbances of world markets, some export demand will develop. The reasons for changing demand and prices for this product are not usually known to the forest owner. That is the story of the utilization of the product from this two million acre forest in Eastern Nova Scotia.[73]

The economic obstacles were complemented by political ones. Partisan politics were connected closely to the forest industry. Lumbermen were close to the Liberal Party and had plenty of influence in, and sympathies from, the Department of Lands and Forests. The larger lumbermen found no challenges from the department but mere services in the shape of fire protection, land surveying, and the Small Tree Act (which was conceived outside the departmental bureaucracy but then administered by it).[74]

It was the Department of Agriculture that took up the fight for small producers in the lumber and pulpwood trade. Bigelow was very much part of that effort. The department's marketing division basically argued that the promotion of good forest management and forest industry vitality was dependent on reforms in the pulpwood and lumber trade rather than only the provision of technical forest practices advice. The division thus focused its efforts on 'a consistent policy of quality production, standard grade, and organized sales.'[75] This constituted an alternative to the dominant policy at the time, illustrating what Kenny has called the 'possibilities of the interventionist state.'[76]

This alternative policy was further refined in the 1950s when Bigelow became part of an effort to expand the pulp and paper industry in the province, an expansion that he believed would help to promote a 'better utilization' of the Nova Scotia forest.[77] In the mid- to late 1950s, as deputy minister of trade and industry, he was instrumental in bringing the Swedish Stora Kopparberg pulp mill to Port Hawkesbury on Cape Breton Island. But he did not foresee this as a sellout to a foreign corporation. Bigelow drove a hard bargain on the conditions of the company's entry into Nova Scotia, and he did not trust the company's policy on pulpwood buying.[78] He expected fair stumpage payments for Crown wood, and he continued to actively support the organizational efforts of woodlot owners, believing that both initiatives would serve to improve forest management in the province.

In the end, however, politics interfered with the realization of Bigelow's pulp and paper development scheme, and the Conservative government's political agenda of attracting industry on concessionary grounds won the day.

Appendix

The Utilization and Marketing Problem of Farm Woodlot Products in Nova Scotia (John Bigelow)[79]

Decline of Forest Industries

It is evident to the most casual student of the problem that our farm woodlots are in a particularly favourable situation to contribute an important part of our timber supply, and at the same time to present an opportunity for forest management. Yet we know that these same farm woodlands provide a far from satisfactory source of supply for even local requirements, and that proper management and utilization practices on them are practically unknown. The very fact that they are accessible and privately owned, factors which should increase the value, seem to have worked against them.

Unrestricted and improper cutting by cycles of high prices for certain products, or periods of depression and lack of other sources of income for the owner, have depleted farm woodland timber supplies to the point where many farmers are burning coal or oil for fuel. On these same farms there is probably being abandoned more and more submarginal agricultural land, thus throwing more emphasis on forest land use. Yet this change from agricultural to forest land is being accompanied by a steady and alarming decline in forest industries extending over several decades. The abandonment of farms has probably been hastened in many communities by the decline and loss of the forest industries, which provided a number of opportunities for employment and markets for farmers' products.

Ownership and Reforestation

We hear today much talk and many opinions expressed of the necessity for reforestation, and the necessity of an increase in the area of publicly-owned land. This would seem to seriously question the ability of private ownership to redeem its responsibilities in the use of land. We feel that before any effort to restore submarginal lands to economic productivity can be successful, whether it be by public ownership or other means of reforestation, satisfactory outlets for the products or substantial forest industries must be established. Otherwise the value of such reforested lands will be restricted to recreation and aesthetic possibilities or watershed protection.

We have then, in this Province, a situation in which our forest industries are declining while the area of well-located, accessible forest land is increasing. The reason for this obvious maladjustment can be found in the present condition of the forest resources and the lack of management in the past.

Present Utilization Practices that Result in Forest Depletion

The existing lack of control of our forest resources has made no provision for the maintenance of enough growing stock or merchantable size to afford a continuous supply of the kind needed by the industry. The basis of profitable industry is bound to be destroyed if the best trees are continuously removed, leaving only the poor and small trees after each operation. Growing stock cannot be maintained if trees of small size are cut indiscriminately. The productive capacity of the forest must be realized on trees of merchantable size and quality, and not dissipated on bush and sapling growth. Restoration of our forest and farm woodland growing stock depend largely on the kind of available markets and on the manufacturing conditions used in preparing the material for these markets. Under existing conditions farmers have little opportunity to dispose of their timber of sawn material size except by selling to portable mills or industrial operators.

Portable mills are usually wasteful and inefficient. They must then obtain their stumpage at a low figure, and their operations will have no regard for the future productivity of the forest. Whether the sale is made on the basis of a log scale or by a lump sum the farmer is at a disadvantage because he probably does not understand log scales and their application as well as the mill man, nor is he so well qualified to estimate the quantity and quality of wood on the stump together with the cost of getting it out. Returns to woodlot owners from sales in this market are therefore usually considerably below the real market value of the product.

Portable mills have not proven to be a stabilizing factor in the industry. Usually operated by the owner with little or no financial backing, he is absolutely dependent on a broker or exporter for disposal of his product, and is unable to take advantage of favourable markets or to avoid stress sales at times of low prices. He has then no particular incentive to properly saw and season his lumber or to eliminate waste. His major objective is more likely to be the next setting to which he must move his mill. Obviously under these conditions the market will remain in a haphazard and uncertain state, with the buyer of the stumpage and of the lumber dictating the terms, with neither the farmer nor the millman getting value for his products.

This system of handling the product is also a serious hindrance in the way of any efforts that the rest of the industries may put forward toward orderly marketing. It is not quite fair to lay all the blame on portable mills. There are many small, permanent mills which continue to function, but their situation in regard to quality of manufacture and methods of sale is no different from that of the portable [mills]. The mills are probably obsolete and uneconomic in arrangement and location. They operate for only part of the year, but are convenient for the custom sawing requirements of local people. They do make a certain use of local timber, but few, if any, of them are prepared to compete in any measure in the general market. In some communities where wood-using factories or small industries are established there is a certain market for special material for special uses, such as hardwoods for handles, dowels and furniture stock. Individual farmers are very likely to have only limited quantities of these special materials for sale, and if they can reach the market at all probably cannot get value for their product.

There is also a third class of market available to the farmer with which we are all quite familiar, that is the pulpwood, pit prop and fuel market. Where there is a steady demand for this class of material it should be an asset to practical management. What is happening in Nova Scotia, however, is that when the prices for these products reach attractive levels, production of as large a quantity as possible for the present year becomes the major objective of the majority of woodlot owners. This lack of restraint on the part of the owners results in further destruction of the basis for other industries requiring larger material, and at the same time further reduces the economic productivity of the forest. We believe that markets for this small material should be used to facilitate management and improvement of the woodlot and to utilize those parts of the forest not suited for more valuable products. The experience and study of European countries have conclusively shown that the yield of this sort of material is just as great under some plan of integrated production as when all of the trees are cut at small scale.

We realize that intensive methods commonly practised in Europe are not practical here because of our lack of markets for wood materials of small size and poor quality.

Management and Marketing

We feel that existing marketing and manufacturing practices will not lead to the satisfactory development of the farmers' woodlot of this Province. Local wood-using establishments or reliable export markets are needed which will provide an outlet each year for the disposal of timber in small quantities, which would result from proper management of farmers' woodlots.

If wood-using industries are established they must be well located and properly organized so that they can successfully compete in the various markets that would be available. It would appear, therefore, that some form of centralized management of the woodlands available to these industries would be highly desirable, if not absolutely necessary, in order to insure the continuous supply of raw material without which they could not exist.

Without some form of centralized management the probability is that owners of woodlots will continue as in the past to concentrate their cuts on the small remaining amounts of the best timber, which is going to result in eventual complete exhaustion of supply. We realize that any success in the establishment of local wood-using industries based on proper management practices will be dependent on the ability of these industries to utilise the large quantities of inferior-quality material which covers so large a proportion of the depleted and mismanaged woodlands in the majority of our communities. Improvement of these depleted forests will require disposing of much poor-quality, small-size material, and a conserving of practically all the better trees for future growth. Any form of profitable operation is going to require an integration of use, almost impossible to obtain through existing facilities, or by individual owners working alone.

Some of the outstanding faults in our lumber industry as a whole, and they all apply with equal force to farm woodlot production, have been summed up as follows:

(1) High percentage of low[-quality] material.
(2) Relatively small amounts produced by individual owners.
(3) Inadequate methods of distribution incident to small operations with all attendant drawbacks from the buyer's standpoint.
(4) The lack of co-operation in distribution on the part of the majority of small producers.
(5) The absence of information on quality in the various trade requirements.

In developing a program and formulating a policy of assistance to woodlot owners, the Nova Scotia Department of Agriculture and Marketing felt that with growth and ownership as they exist in this Province, and with a limited budget available, the most good would be accomplished by attempting to correct the last three of these unsatisfactory conditions. To correct these conditions we are working on a policy which we generally divide into three principal parts:

(1) Evolve a form of organization for the primary producers that is applicable to the local conditions of each community.
(2) Develop a system of forest management that will enable the rundown and depleted woodlot to recover from past exploitation, and will enable them to realize the full growth potentialities of the soil.
(3) Build up a utilization system so that the products of the organizations may readily find markets.

We consider the first part of this policy the most important at the present time. The successful operation of the other two parts depends on its satisfactory completion. We must have an organization that can fix responsibilities, be flexible and efficient in operation, and at the same time be under the ultimate control of the farm woodlot owners. Ownership and control must remain with the producer, and woodlot owners must be admitted without too great a cash outlay, or without being submitted to uncertain, personal liabilities. Continuity of operation, mutual benefit, and future expansion must be guiding principles of this organization.

We have in Nova Scotia the necessary legislation to organize co-operative associations, and we have a growing number of organizations of farmers who have made a study of, and thoroughly understand, the basic principles of co-operation. These groups are succeeding in the marketing of their farm produce and expanding into other fields, one of which is the marketing of forest products. With continued loyalty to their organizations, these farm woodlot owners are bound to succeed in increasing the income from their woodlots by putting improved-quality products on the market in a regulated and orderly manner. Once this increase in income has been attained, each individual then sees the necessity of the second part of our policy, sustained yield management and conservation. Through an organization the necessary instruction and advice can be given to inaugurate this part of our policy.

It should be noted that with this programme, the demand for the instruction and advice comes from the producer or woodlot owner. It is not forced on him by legislation or other means. Also included in our progress are lectures at our Agricultural College, and the various short courses and farmers' meetings throughout the province. While it is much too soon to predict the ultimate value of the information thus distributed, it is already noticeable that the individuals of the organized groups lose no time in putting their education and information to practical use.

Once a community is organized on the correct basis the advantages of co-operation are so apparent on all sides that it is not a difficult problem to direct their efforts into the proper channels. Information as to quality and trade requirements can be passed on to and discussed with organized groups to the advantage of the whole producing area, or community. To attempt to do the same and obtain any results with an unorganized community of rugged individualists is almost a hopeless task.

When a sufficient number of communities are organized to warrant it, our aim is a central organization for the whole Province or perhaps the Maritimes. With this end in view we are directing our most intensive efforts toward providing the necessary and proper information and education to unorganized and uninformed communities, and directing the efforts and expansion of those already organized into the proper channels.

We realize that this is not a programme that will show spectacular or large-scale results immediately, but we do feel that we are laying a sound foundation for proper farm woodlot management and conservation; one that will endure periods of prosperity and depression and be capable of gradual growth and expansion until the farm woodlots of this Province reach a point of maximum production and contribution to the wealth of the Province.

We feel that in the co-operative movement we have a very logical means at hand for solving our farm woodlot problem, and indeed our whole forest products marketing problem. With five co-operatives actually engaged in the production and marketing of their forest products in this year, we feel that we have some small but effective results to show for our one-year effort. By the example of these organized associations operating successfully in localities far from ideal for the purpose, we hope to spread the knowledge of management and conservation to all parts of the Province.

Reproduced courtesy of the Nova Scotia Department of Agriculture.

4
Lloyd Hawboldt: Bringing Science to Forest Policy

Canadian forestry after the Second World War has typically been described as industry-focused, instrumentalist, and utilitarian.[1] Provincial departments of lands and forests have negotiated long-term Crown leases with large integrated pulp and paper companies, where the concerns for local communities and forest environments have been secondary to industrial wood production. Foresters have typically been described as allied closely to industry interests, using industrial means to protect and boost the wood fibre supply.[2] It is only relatively recently that environmentalists have pushed foresters to take account of whole forest ecosystems, though we are still not certain about the seriousness of this commitment.[3]

There is some credence to this picture, but it also ignores the subtleties and dissensions within forestry and among foresters historically. This chapter explores the career of forest entomologist and self-taught forester Lloyd Hawboldt, who served the Nova Scotia Department of Lands and Forests in various capacities from 1944 to 1977. During that time, Hawboldt drew on forest entomology and European sustained yield forestry to challenge the increasingly dominant objective of industrial forestry: the extreme manipulation of forest environments by mechanical and chemical means to protect and boost the production of wood fibre (and where other functions of the forests appear as constraints to this basic objective). This challenge offers insights into how the forest environment, forest community, and forest economy could have been served differently. It also suggests that the alternative forestry regime proposed by Hawboldt failed for two basic reasons: first, because of its insensitivity toward the needs of the small woodlot owners' community, and second, because of the overwhelming political support given to the pulp and paper industry and the particular brand of forestry that it favoured.

From a more general perspective, Hawboldt's career illustrates the tension between what Aldo Leopold has called type A and type B forestry. Type

A foresters, Leopold wrote, are 'quite content to grow trees like cabbages, with cellulose as the basic forest commodity. [They] feel no inhibition against violence; [their] ideology is agronomic.'[4] They also consider forest 'problems' (e.g., insects, weeds, and fire) as isolated variables, amenable to isolated treatments. Productivity gains are sought through the use of fertilizers, genetically improved tree stocks, synthetic pesticides, herbicides, clearcuts, and artificial reforestation with trees of the same age and species (often exotics). The forest is here actively manipulated and changed to accommodate industrial needs. Typically, the adoption of the various management options has been linked to the demands of the pulp and paper industry, which came to be dominant in the Nova Scotia forest economy during Hawboldt's tenure with the Department of Lands and Forests.

Type B foresters, by contrast, see forestry as fundamentally different from agronomy. They employ natural species and manage the received environment rather than create an artificial one. They prefer natural reproduction and worry on biotic as well as economic grounds about the loss of certain tree species. They are also concerned about the forest functions sometimes labelled 'secondary' (in commercial terms): as wildlife habitats, recreation grounds, watersheds, and wilderness areas. To Leopold, type B foresters feel 'the stirrings of an ecological conscience.'[5]

Hawboldt's career illustrates the tensions, permutations, and the potential of marrying types A and B forestry, as well as their collective shortcomings. Hawboldt drew from European sustained yield practices to inventory the Nova Scotia forest and to suggest methods by which to make the forest more productive for fibre production. This clearly puts him in the type A category, though his approach was more balanced and benign in its manipulation of the forest ecosystem. By contrast, Hawboldt's rejection of insecticides for controlling forest 'pests,' in favour of biological and silvicultural controls, clearly puts him in the type B forestry camp. In the present discourse of ecosystem management, his forestry 'hybrid' may be criticized for being utilitarian, where words such as 'desirable species,' 'pests,' 'weed species,' and 'sustained yield' figure prominently, and for not taking into account the importance of preserving 'nature for nature's sake.' If seen in the context of the times, however, when resource development was all-important, antimodernist sentiments were in decline, and modern environmentalism had yet to be born, Hawboldt displayed a radical synthesis. In holding these views, he drew on alternative paradigms and ideas that were widely respected at the time (and which have recently been revived).[6]

A less positive aspect of Hawboldt's forest management approach was its failure to connect with the broader society. Although Hawboldt was well versed in the debate over forest management strategies, he seldom acknowledged the concerns of small woodlot owners and sawmillers in a woods market increasingly dominated by the pulp and paper companies.

These concerns were seen to lie beyond his sphere of expertise and beyond the perceived mandate of the Department of Lands and Forests.[7]

Here we explore the interplay of the above themes through three phases in the career of Hawboldt. In the first phase, from 1938 to 1952, his activities were dedicated exclusively to forest biology (with a focus on entomology), and he had free reign to pursue his own research agenda, based on university studies and the extension of scientific inquiry into various field situations. This research was locally based, yielded insightful results that occasioned international attention, and had important implications for a less agronomic forestry. In the second phase, from 1952 to 1969, Hawboldt's research skills were constrained and funnelled into forest management and policy questions. Here Hawboldt assimilated some important European methods of forest mensuration to the Nova Scotia situation. They were then combined with his knowledge in entomology to advance a comprehensive forest management plan for the province. This period held the potential of marrying types A and B forestry into something new. However, in the third period, from 1969 to 1980, Hawboldt's alternative forestry paradigm was eclipsed by broader type A currents, though not before significant contestation.

Early Years

Lloyd Hawboldt spent his childhood years in Halifax. His family was well known in marine circles in the Chester area, where a mechanically minded uncle had developed the one-cylinder Hawboldt marine engine. Hawboldt's father was a machinist who, on the urging of a family friend, pushed his son away from engineering and toward entomology instead. During the depths of the Depression, Hawboldt entered the Nova Scotia Agricultural College in Truro, where he studied for two years before going on to MacDonald College (the agricultural faculty of McGill University) in Montreal. There he completed a BSc in agriculture in 1938 and an MSc in forest entomology in 1946. Throughout his studies, Hawboldt focused on the study of entomology, taking courses in economic and systematic entomology, insect ecology and zoology, and writing a thesis on *bessa selecta*, a parasite associated with the European spruce sawfly.[8] The Department of Entomology was particularly strong. Its prominent and influential teachers included W.H. Brittain, the dean of the faculty who had served as Nova Scotia's Provincial Entomologist from 1915 to 1925, and Melville Duporte, an insect morphologist and strong theoretician.[9]

The focus of entomology at the time was on the life histories and habits of insects, together with the parasites that attacked them. This opened the way for 'biological treatments' against problem insects by introducing the appropriate predatory antidote. Hawboldt's understanding of entomology and biological control was widened by his work for the Nova Scotia

Department of Agriculture during the summer of 1937. As a student assistant studying apple orchard insects and agricultural pests in the Annapolis Valley, he was brought into contact with Allison D. Pickett, an authority of international fame in the field of biological controls.[10] Pickett was at the time conducting experimental spray tests to identify insecticides that would control orchard pests with a minimal degree of damage and a minimum of repeat applications. This was a particular concern given that many orchardists were applying a dozen or more treatments each year. Following Pickett, Hawboldt was suspicious of the increasing use of chemicals in agriculture and forestry. From 1938 to 1944, Hawboldt's entomological knowledge was further strengthened by his work as an Agricultural Assistant with the federal Department of Agriculture in Fredericton, New Brunswick. Here his responsibilities centred on the Eastern Section of the Canadian Forest Insect Survey, covering the province of Quebec south of the St. Lawrence River, together with the three Maritime provinces. The survey data were gathered manually by sampling field locations, then growing organisms in the headquarter's insectory.

Exploratory Beginnings

In 1944 Hawboldt took up the position of Provincial Forest Entomologist with the Nova Scotia Department of Lands and Forests. The department was then still in a formative period, and the influence of the forestry profession was just beginning to be felt. The senior bureaucrat, F.A. Harrison, was a long-serving member of the land survey staff who lacked professional credentials, but he was open-minded about bringing in the necessary expertise. The department employed only three foresters at the time: the Provincial Forester, Wilfrid Creighton (who succeeded Chief Forester Otto Schierbeck in 1933); an Extension Forester, Dave Dyer; and Dave Hudson. In this fluid environment, Hawboldt was able to carve out a preferred position oriented to research. While working in Fredericton, he often visited his home in Nova Scotia and dropped by the offices of the department to talk to Harrison.

The presence of biological problems helped Hawboldt's cause. One prominent concern, and a pet project of Harrison's, was a disease affecting the province's declining moose population.[11] Another problem was the extensive dieback of yellow birch, which was affecting the whole of eastern Canada. This was of major concern to both company and government forest managers, since the yellow birch was a valuable source of veneer logs for the plywood industry.[12] Other insect problems included the European spruce sawfly, the larch sawfly, the eastern spruce barkbeetle, the forest tent caterpillar, the balsam woolly aphid, the spruce budworm, the white-marked tusked caterpillar, the European winter moth, and the balsam needle midge (on Christmas trees). From the standpoint of forest damage, the top threats were the spruce sawfly, the spruce budworm, and the balsam woolly aphid. It was this plethora of 'insect problems' that prompted Harrison to hire Hawboldt as the first and only forest entomologist in a provincial Department of Lands and Forests. Hawboldt was then one of the few university-trained personnel at the DLF, which offered him the unique opportunity to pursue his research field in a practical setting.

The Division of Forest Biology, 1947-52

In 1947 Hawboldt formed a new Division of Forest Biology to tackle the birch dieback phenomenon. The division contained a powerful team. Hawboldt was the director. Ken Greenidge, with a strong background in physics and mathematics, was the botanist and tree physiologist. He later went on to Dalhousie University and later still to the Department of Biology of St. Francis Xavier University in Antigonish. There he was able to develop a course on forestry that was equivalent to the first two years of credit at the University of New Brunswick forestry school. Another member of the team was Doug Redmond, a forest pathologist who later went on to become chairman of the Canadian Institute of Forestry.[13] Then there was Simon Kostjukovits, a White Russian forester with extensive experience in Europe.[14] Forester David Dwyer, discussed in detail in a later chapter, was part of this research group, as was wildlife biologist Dennis Benson. The group had facilities at the agricultural college in Truro and at the military barracks at Debert, which became available after the Second World War.

The group in the Division of Forest Biology made great headway in finding the cause of the birch dieback. At the CFS research station in Fredericton, where work was already under way, it was generally believed that the cause lay with the bronze birch borer. However, the Hawboldt group took a different tack. They observed that the disease was less advanced in Nova Scotia than in New Brunswick, and it was easy to conclude that the bronze birch borer was not the main culprit, for there were many diseased birch trees that showed no sign of borer presence. The Nova Scotians distinguished two phases of dieback: one in undisturbed stands, and one in stands that had already been cut through for softwood logs. Beyond this, the DLF people kept open minds. As Hawboldt recalled it, 'We tried everything.' In the end, they were driven to look underground, to examine the birch root structure. At field sites in Cape Breton at Lake O'Laws, they installed water pumps and spent three summers washing yellow birch roots in order to classify living and dead roots. This was an exhausting job that involved excavating the root structure of dying trees by washing them with high-pressure hoses. Significantly, they found early symptoms in the feeding roots. The evidence increasingly pointed to root damage as the prime cause of injury to the trees, with subsequent injury to the crowns of trees, and the birch borer then arriving as a secondary effect. As for the cause of root damage, there was considerable support for a hypothesis that climatic change (a fall in mean precipitation and an increase in temperature, especially between May and August) was the primary cause, with tree weakness, undercutting, and crude sawing as contributing factors. Eventually, Fredericton also came to this view.

Hawboldt's research approach, with many disciplines represented, served his team well. Their work was at the cutting edge in their field, and results were published in the most prestigious academic journals.[15] Most of their intellectual exchange took place with the research community in the United States. They had visits from Dr. Ray Hansborough and Dr. Harvey MacAloney, as well as other senior researchers of the US Forest Service. Professor Diamond of Yale University and Dr. Hugh M. Raup, director of the Harvard Forest, also showed interest or visited their research sites. Diamond and Raup asked Hawboldt to write a monograph on hardwoods, but he declined after losing all of the data on the birch dieback in an office fire. Hansborough later wrote an article that appeared in the *Toronto Star Weekly* on the extensive death of hardwoods in the eastern United States and Canada.

Building on their own and others' experience, Hawboldt's team was planning more investigations of the maple dieback. Quite unexpectedly, however, Hawboldt's biological research mandate came to an end in 1952, when the provincial and federal governments announced that henceforth all biological research would be centred with the federal Canadian Forest Service. Formally, the decision was explained by the need to avoid duplication, since the CFS could mount an integrated research program more efficiently than separate agencies could. However, an investigatory monopoly also had its drawbacks, as illustrated by Hawboldt's own experience with birch dieback. In fact, institutional and professional rivalries may have been part of the problem. Although the DLF group had made progress with its Nova Scotia research on the dieback, the findings conflicted with both the problem definition and the findings of the CFS Division of Forest Biology. Reginald E. Balch, the man in charge in Fredericton, was visibly irritated with Hawboldt. When Hawboldt visited Fredericton and conveyed his findings on root damage, Balch openly chastised the person in charge of the federal birch dieback studies, Frank Morris, for not pursuing this line of inquiry, in spite of the fact that he had never been instructed to do so. Balch was often in Halifax consulting with more senior DLF personnel, particularly DLF minister A.W. Mackenzie (with whom he had attended the Ontario Agricultural College).

It was at these levels that the ultimate decision was reached. For Hawboldt, a great opportunity was lost. The research group drifted apart, as Redmond went to Ottawa and Greenidge to academics. Overall, the Canadian Forest Service had never supported the Nova Scotia Division of Forest Biology. Then, as later, its researchers had to justify their existence by publishing, and this often had little relevance for the practical and local problems confronting the Nova Scotia forest community. In order to get relevant assistance for his dieback studies, Hawboldt had to push the service in Ottawa to order one of its botanists (Arthur Skolko), a specialist in tree pathology, to join him. Once the study was completed, both Skolko and Ottawa wanted senior authorship.

The abrupt closure to Hawboldt's research efforts on the birch (and pending maple) dieback illustrates not only the centralization of research efforts in Canadian forestry but also the favouring of research on softwood species. Research on the dieback of hardwoods was becoming less relevant to industrial needs of the Canadian staples economy, at the time increasingly dependent on softwoods for the pulp and paper industry. While important during the war, hardwoods were otherwise seen as dispensable by the forest industry. Foresters internalized this bias over time, through their thinking on the 'hardwood problem.' This referred to the natural growth advantage of hardwoods when competing with the more commercially valuable softwoods in mixed stands. Many pulpmen welcomed the coming of the dieback because it provided more room for softwood regeneration. In Canada, the wider research agenda in forest biology was increasingly driven largely by insect problems threatening softwood species such as spruce and balsam fir.

The Division of Extension, 1952-71

The end of the Division of Forest Biology dealt a serious blow to Hawboldt's alternative forestry vision, but it did not preclude Hawboldt from pursuing scientific questions in different ways. It simply meant that research had to be harnessed more tightly to forest management needs rather than driven by the urge to fill gaps in the knowledge base of forest biology. In the language of the day, the role of the Department of Lands and Forests was to extend the results of forestry research to the various actors in the industry, to concentrate, in other words, on the 'extension' function. For Hawboldt, this brought many opportunities. As he put it in a 1955 article, 'Although the Department of Lands and Forests does not have a research programme as such, frequent opportunities arise for personnel to conduct a certain amount of experimentation.'[16] Hawboldt pursued these opportunities from his new assignment as Director of Forest Biology and Extension (1952-9) and Director of Extension (1959-69).

Over the next twenty years, there were innumerable questions that called for trained scientific input on matters of extension and forest biology, and Hawboldt was called upon to contribute more often than not. Major questions included the promotion of forest management among small woodlot owners; mensuration (measurement) issues associated with ongoing forest inventory surveys; responses to spruce budworm infestations; and the design of effective silvicultural treatment practices, as natural (wild) regeneration gave way to more intensive cultivation. The approaches taken in these various areas exhibited a blend of approaches and biases, ranging from utilitarian (type A forestry) to ecological (type B forestry) notions (stemming from European sustained yield forestry and entomology). In his application of European sustained yield forestry, his mensuration studies lay at the cutting edge of Canadian forestry (though they also served the pulp and paper industry development agenda). When exploring biological issues, Hawboldt remained true to his training and insights in forest biology and entomology. These convictions came increasingly into conflict with the agronomic forestry paradigm, which centred on the emerging pulp and paper industry and was increasingly advocated by the Canadian Forest Service, as well as taught at the forestry school in Fredericton, where most Maritime foresters were trained.

In his constrained position as Director of Forest Biology and Extension, Hawboldt made contributions in four major fields: extension services, tree mensuration, spruce budworm treatments, and forest management. We begin by exploring the technical and educational bias of extension forestry in the 1950s, which later formed an obstacle to the building of a broader alliance in support of an alternative forestry paradigm. The technical and educational bias in extension forestry had a long tradition in Nova Scotia when Hawboldt took up his position. As far back as the late 1920s, Otto Schierbeck had advanced demonstration thinning programs, visited Boy Scouts, and introduced annual Arbour Days. David Dyer was designated the Extension Forester on staff in the 1940s. When Hawboldt joined the DLF, extension forestry was still centred on public education. Through school visits, children were encouraged to join the Green Forest League, whose credo called upon members to 'be a true friend of the trees of Nova Scotia and the Folk of the Forest by learning and never breaking [a number of] rules.' The department supported forest management courses at the agricultural college and the Nova Scotia teachers' college in Truro. It also collaborated with the Nova Scotia Forestry Association (the provincial affiliate to the Canadian Forestry Association) on public awareness campaigns. This association sprang into high gear for the annual Forest Week in May and ran the Smokey the Bear program.[17]

The department's first large extension project under Hawboldt's direction concerned blueberries. In those days, extensive barren lands were being burnt by local people to promote blueberry growth. These burns were invariably informally organized, and the barrens were considered open-access areas. Harvesting and marketing the blueberries were also done informally, generating substantial incomes and a source of subsistence for rural residents. The burning was done as a means of pruning low-lying growth by killing old shoots and enabling new ones to grow out. The new shoots grew vigorously during the first year as new fruit buds appeared. A heavy fruit yield was anticipated in the second year. This practice was especially common in the major blueberry counties. There were massive barrens at the backs of Digby, Yarmouth, Shelburne, and Queens Counties, as well as in parts of Cumberland, Pictou, and Guysborough.

The blueberry economy was part of the informal sector, loosely organized by local communities and untouched by bureaucratic control and formal commerce. The Department of Lands and Forests was suspicious of such resource management under popular control. It considered burning the barrens a danger to the surrounding forests, as evidenced by many recorded fires, and a threat that might lead to soil exhaustion. Already in 1927, the Department of Lands and Forests had tried to cope with this 'problem.' On the initiative of Otto Schierbeck, a statute was passed to facilitate blueberry cooperatives on barren Crown lands. In the early 1950s, the deputy minister of lands and forests, Wilfrid Creighton, wanted to go even further by replanting much of the barrens in pine and redirecting the wildberry industry onto abandoned farmland, where it could be under tighter government control. From the department's perspective, the barrens should be reforested.[18] Creighton referred to them as the ragtag and bobtail lands of Nova Scotia, the 'unwanted' lands. The province bought back many of these barren tracts for twenty-five cents per acre, and many of these lands have since become forested.

To local communities, the barrens were never 'barren.' They provided a rich source of subsistence and tradeable goods for the household or informal economy. Local blueberry growers thus protested against Schierbeck's cooperative scheme and Creighton's pine plantations. Several plantings were actually burnt.[19] This confrontation was part of the larger process of commerce and bureaucracy displacing the informal economic activities and popular resource management schemes of local communities.[20]

The treatment of the blueberry barrens by the DLF was clearly insensitive to the demands and needs of many rural residents. Hawboldt's blueberry manual was similarly indifferent to rural residents, as it was largely a guide on how to grow blueberries commercially.[21] In fact, when it was published, 'all hell broke loose' when it landed in the middle of a bureaucratic turf war. Several government departments at the time were vying to oversee the commercialization of the blueberry sector. The federal and provincial departments of agriculture had begun the battle. When Nova Scotia wanted to promote old-field blueberries, Ottawa resisted and mounted a particular defence of the high-bush cultivated blueberry areas in the Annapolis Valley around Kentville. This left the provincial agricultural department with the wild segment but little interest in promoting it. In fact, the horticultural specialist was told to steer clear of it. Yet with the appearance of Hawboldt's bulletin, the Nova Scotia Department of Agriculture believed that the

Department of Lands and Forests had encroached on its domain. The outcome was that Agriculture took over the more obvious 'farm' aspects of the blueberry sector, while Lands and Forests carried on with its foothold in Crown lands.

Hawboldt's blueberry guide, however, was never used by the DLF Division of Extension. Its focus was on producing a maximum yield using the agronomic techniques of the times, including controlled burns, fertilizers, herbicides (including 2, 4-D and 2, 4, 5-T), pesticides (such as DDT), and the most up-to-date mechanical devices for picking. Absent too was any information on how to organize production into either private or cooperative enterprise or how to deal with industry in gaining a better price in the marketing of the product.

Not long after, the Christmas tree industry was also handed to the Division of Extension specialists, with similar results. Since the 1930s, Nova Scotia had been shipping trees to the northeastern US market. After the war, the level of exports rose rapidly, doubling and then tripling to a level exceeding 3,000,000 trees by 1952. As Hawboldt learned, this was a complex market, involving growers (landowners), cutters, buyers (dealers), and wholesale and retail handlers. The best tree stock was balsam fir, which eventually became the staple of the export trade. Although tree sales brought more than $1 million to woodlot owners and workers, this market was inherently unstable. Product quality varied tremendously, as good trees vied with poor culls that threatened to drive the entire price structure toward the lowest common denominator. This suggested the need for a grading scheme. Moreover, Christmas trees were widely regarded as a wild resource to be cut and shipped as found rather than cultivated for maximum value. There were occasional difficulties with pests (e.g., the balsam fir midge) and with poor cutting and handling practices. The Department of Extension sought to change much of this, arguing that 'Christmas trees are a crop.'[22]

Hawboldt visited the more active Christmas tree growers to develop materials for this program. The Murray Feltmate family of Guysborough County was experimenting with stump culture, which allowed several trees to be grown from a single stump. The Feltmates exemplified many of the more innovative practices in growing Christmas trees. In 1953 Hawboldt and DLF forester Gordon Maybee travelled to New York and Boston to examine the wholesale and retail ends of the Christmas tree trade. Their report explained the several phases of the industry and the factors that determined the final conditions of sale. They stressed that the label 'Nova Scotia balsam' was recognized as one of quality but that the delivered product often fell short and therefore jeopardized future sales. The message was clear: 'Nova Scotia has the potential for quantity production of quality Christmas trees, if woodlot owners will give attention to their crops, and if shippers will discourage the cutters from gathering "trash."'[23]

Despite the circulation of these bulletins, progress was slow. In 1957 the size of the export shipment peaked at about 3.8 million trees. A decade later, it was less than half this size. A number of factors were involved in this decline. Despite Hawboldt's warnings, the quality of Nova Scotia stock had fallen, while American suppliers were beginning to get into plantation growth, which yielded more attractive trees. In addition, the artificial tree began to make inroads in the US market in the late 1950s, rising later to account for as much as half of all sales. The policy recommendations advanced in the 1950s by Hawboldt and Maybee were taken up slowly.

What is remarkable about the Christmas tree trade is that the Department of Extension advice was limited to the technical aspects of growing and marketing the trees. This included the use of insecticides and herbicides, which Hawboldt believed were used with a calculated risk under commercial pressure.[24] Little effort was made to assist woodlot owners in gaining bargaining strength in a trade rife with predatory middlemen. It is small wonder, then, that the efforts to establish standardized grades failed, because most small producers thought that they would not benefit from producing a better product.[25] Additional bulletins were issued over the years, explaining the advantages of shearing and fertilizing the trees, but the suspicion lingers that the Extension materials influenced the more active and well-capitalized growers while failing to reach the more marginal participants. And while some county-level growers' groups were formed, it was not until 1975 that the Nova Scotia Christmas Tree Council was formed.

It is thus interesting to note one striking omission from the roster of Department of Extension initiatives. This is the relative absence of projects targeted at the organization and protection of private woodlot owners in the blueberry, Christmas tree, sawlog, and pulpwood markets. There were no efforts to emulate the activities of John Bigelow, the representative of the Forest Products Marketing Division of the Department of Agriculture in the 1930s, who was instrumental in organizing several cooperative ventures in Guysborough and Inverness Counties before the Second World War.[26] In fact, there was no Extension field structure when Hawboldt arrived. This contrasted sharply with the situation in the Department of Agriculture, where the district 'ag reps' filled a key liaison role between scientific research stations and farmers. In the Department of Lands and Forests, the rangers were the key contacts with the public, but they were more equipped and inclined to promote public education than to support organizational efforts among woodlot owners.

Forest Research Activities

Hawboldt's contribution to Nova Scotia forestry proved more innovative in fields other than extension. In the early 1950s, the federal government gave a major boost to the provincial management mandate by financing the first modern forest inventory to be conducted along scientific principles. This was a pivotal event for the Department of Lands and Forests, as no credible estimates were available hitherto on the forest stock of Nova Scotia. With Ottawa's financial support, this glaring gap could now be remedied.

The forest inventory was needed for two principal reasons: to meet the needs of the pulp and paper companies vying for mill sites in the province, and to serve as a basis for a provincial forest management plan. However, there was a serious question about where to start, since the forest estate was seriously depleted and management practices were still primitive on the whole. The Small Tree Act (STA) of 1942 restricted the cutting of certain species of trees below a diameter of ten inches at breast height. This was at a time when the depleted softwood forests were under pressure from an army of small portable sawmillers who would run any stem through their blades to make a rough two-by-four. But despite the commendable objective, Hawboldt had many reasons for thinking that 'the worst kind of forest management is a diameter limit cut, and the worst kind of diameter limit cut was Nova Scotia's Small Tree Act.' It often promoted high-grading because it assumed that smaller trees were younger, while in fact they might have been 'over-mature.' There were too few foresters to enforce the act, and those who did were often half-hearted. Finally, when loggers applied for exemptions to clearcut, the licence was given almost automatically. Over time, this had the effect of discrediting the act, since it came to be seen as virtually unenforceable and full of loopholes. But if global rules were defective, one of the first steps in formulating an alternative forest policy was to develop a credible forest inventory.

The plan was to combine the analysis of aerial photographic data with cruise data compiled from sample plots on the ground. In three consecutive years, the photographic work covered Cape Breton, then the northeast mainland, and finally the southwest mainland. In each case, the second year was devoted to the ground cruises, the fieldwork being contracted to the Quebec firm of Bélanger and Bourget. The raw data from the inventory studies were voluminous. They included 22,000 photographs, 550 forest-type maps, and twelve ledgers of cruise data. When a decision was taken to publish a summary of the results, together with a discussion of their economic and public policy significance, the job went to R.M. (Dick) Bulmer, the forester in charge of inventories, and Lloyd Hawboldt, the director of the biology division. Bulmer provided the calculations, and Hawboldt wrote the text of this 171-page book.[27]

It was in this study that Hawboldt developed new measurement techniques for Canadian forest inventory work based on European concepts and methods. Although they were widely used, the conventional form-class-volume tables (which estimated volumes by tree height and diameter) were being questioned in some quarters as very cumbersome for measurement.

Lloyd Hawboldt and Simon Kostjukovits (white hat) on the Antrim woodlot, Halifax County, 1951.

Hawboldt directed his staff to develop new volume tables for the province-wide survey. The key figure in this new initiative was Simon Kostjukovits, the expatriate Russian forester who joined the department in the fall of 1949. He brought a central European forestry perspective distinct from the North American training being offered at the time. Born in Russia in 1888, he had learned his trade at the Polotsk Forest Technical School and the St. Petersburg Forest Institute, and he later attended the Kiev Military College. After fighting with the White Russian Army, Kostjukovits lived in Estonia from 1919 to 1944 and was displaced to Austria until 1949, when he emigrated to Canada. Thus, he was well into his sixties when he joined the Department of Lands and Forests, where he continued to work until retirement at eighty.

For Kostjukovits, the basic concept for site measurement was the 'normal stand,' in which the crowns of trees would just touch one another, leaving neither overlap nor open space. When he first joined the department, he spoke very little English. However, by 1951 he was able to explain his ideas more clearly. The prevailing approach in Canada was the strip cruise, by which the diameter and height of each tree along a designated line were measured to yield the volume. The Dominion (later Canadian) Forest Service developed forest volume tables, and local tables were also prepared for particular settings. These form-class-volume tables could be used to derive timber volumes for ten-inch diameters at fifty-foot heights. The process was very similar to a log-scaling exercise.

By contrast, Kostjukovits advanced a site-quality, normal yield approach. Here the key site-quality indicator was the tree height-to-age relationship. The variables of site class, height, and stand density could then be correlated in the normal yield tables. This offered both conceptual advances and advantages in field application. For the next two years, Hawboldt, Kostjukovits, and their assistant, David Dwyer, prepared experimental tables by measuring 'normal' stands around the province. Once eligible stands had been located, they staked out quarter-acre plots (two chains by two chains) and measured the height of perhaps a dozen dominant and co-dominant trees, which were then aged by increment borings. The tree diameters were also measured, and in each plot a 'model tree' (of average diameter and height) was selected and felled. At two-foot intervals along the trunk, they cut cross-sectional disks, from which they could calculate curves on the growth rates of the trees. By the end of the field plot studies, the group had compiled the set of normal yield tables used in the Nova Scotia forest inventory.[28]

It was a measure of the novelty and theoretical originality of this approach that it triggered considerable scepticism within the department. One forest inventory specialist, Dick Bulmer, refused to accept the site-quality, normal yield approach, but Hawboldt supported it strongly. As a test case, he located an area scheduled for clearcutting and did a cruise by the site-quality method. After the logs from the cut were sawn out, the cruise fell within 2 percent of actual volume. After this was repeated several more times, it was accepted even by the sceptics. Today it is the standard cruise method.[29]

While the measurement techniques that Hawboldt and his team pioneered in Canada were not threatening to type A foresters, the results that flowed from them certainly were. The analysis was blunt. It established that sawlog material was being harvested at levels far beyond the annual allowable cut, while a substantial surplus existed in hardwood pulp material. Both trends were framed by the striking deterioration in forest quality, demonstrated by comparing the current inventory to Fernow's 1910 figures. This was attributed in large part to poor harvesting practices. In effect, the practice of high-grading sawlogs served to degrade the forest structure and reduce the yields per acre. The increase in balsam fir at the expense of other species was a case in point. Balsam fir is an aggressive species in competition with other softwoods such as red spruce, hemlock, or white pine. If the stands are allowed to persist to maturity, the latter will outlast the balsam, and the site will be renewed in sawlog material. However, if the spruce and pine are logged before maturity, then the balsam will reassert itself in higher proportions when the site is renewed. Bulmer and Hawboldt also drew attention to the age structures, particularly in the balsam fir and spruce stands of Cape Breton Island, which were highly skewed toward mature and over-mature stock. This was a serious impediment to the goal of steady harvests from mature stands.

The 1958 report on the Nova Scotia forest inventory, titled *Forest Resources of Nova Scotia,* closed with a set of twelve recommendations, together with a list of sound silvicultural rules for landowners. The authors (Bulmer and

Hawboldt) pointed out the need to achieve a surplus rather than a deficit of growing stock each year; to reduce the sawlog harvest by at least 50 percent per year; to adopt cutting restrictions to achieve better utilization of timber; and to increase the use of hardwood. Overall, it amounted to a charter of future forest policy. Looking back, even Hawboldt was a little surprised that such a forthright document was approved for publication. Perhaps it had something to do with the fact that the presentation was grounded in themes with deep roots in Nova Scotia forestry, in Fernow's classic report of 1910, as well as the DLF Annual Reports of the late 1930s.

During the same time that the Nova Scotia forest inventory was being prepared and released, the government of Nova Scotia under Premier Robert Stanfield was committed to a new strategy of economic diversification. Even before Stanfield's election, municipal politicians and community leaders were promoting new pulp ventures for both the Canso Strait and the Eastern Shore. The Bélanger-Bourget inventory provided a critical input to this process, as far as the DLF was concerned, by establishing the wood supply parameters on a sound basis. The Bulmer-Hawboldt study suggested that Cape Breton was increasingly becoming a pulpwood forest. At the same time, the report was cautious about adding new capacity and thereby increasing pressure on the resource at such a vulnerable moment. More generally, for the province as a whole, Bulmer and Hawboldt had the following advice: 'Any expansion of the pulpwood industry must be with the use of hardwoods.' Furthermore, 'if the pulp industry expanded on the basis of [the] total inventory of pulpwood resources, there could be no future sawlog industry.'[30]

There was, however, some room for pulp and paper industry expansion, and this was realized by Nova Scotia Pulp Limited (NSPL), a subsidiary of the Swedish firm Stora Kopparberg, which built a pulp mill to utilize Cape Breton balsam fir and spruce. This fit rather easily with the DLF appraisal, and there was little debate from the foresters. In fact, Hawboldt came to be friends with NSPL's woodlands manager, Hans Lindgren, after extensive discussions on the state of the sizeable Crown limits on the Highland Plateau. In addition to Lindgren and his assistant Jan Weslien, Stora sent in Mats Chartau, its global forestry consultant, whom Hawboldt remembered visiting Truro to grill the Crown foresters on inventory issues. Ultimately, Stora found the inventories adequate. The sulphite pulp mill was built at Port Hawkesbury and came into production in 1962. Stora even offered Hawboldt a job. On the strength of his rapport with the Swedish managers, he was tempted. However, he ultimately declined, fearing that life in industry might become too stressful. At the time, he had no doubt that there was room for Stora, from both the timber supply and the silvicultural standpoints.

The spruce budworm figured prominently in Hawboldt's forest management framework. Hawboldt had received reports of small outbreaks of spruce budworm on the west coast of Cape Breton Island, near Cheticamp, Strathlorne, and Inverness. It seemed that the insects had come from New Brunswick in great moth flights. Hawboldt articulated his approach to the spruce budworm infestation in a 1955 departmental memorandum.[31] In a memorable phrase, he described the budworm as 'nature's harvester of mature and over-mature balsam stands.' The most effective short-run response was to let the budworm eat itself out of stands, salvaging as much of the dead and dying wood as possible. The long-run solution was to apply silvicultural tools to establish a more balanced age structure and more diverse range of species to make the forest budworm resistant. Environmental solutions were pursued that were grounded in Hawboldt's training in entomology and his understanding of the ecological dynamics of the spruce budworm and the spruce-balsam fir forests.

Hawboldt had seen at this point the practical use of biological controls in Nova Scotia, and he was convinced that most insects could be treated with biological or environmental means. Indeed, no insect seemed to gain ascendence before it was 'knocked down' by some parasite. The release of two parasites in southwestern Nova Scotia wiped out the European winter moth. The most spectacular example, however, related to the European spruce sawfly, which was eradicated not by a parasite, strictly, but by a virus. A wasplike parasite (*Microplectrum*) was gathered in what was then Czechoslovakia and released in huge numbers on the Gaspé Peninsula and in the Maritimes. The spruce sawfly was killed, but when surveys came in they showed that a virus rather than the parasite had killed it.[32]

Hawboldt's conceptualization of the spruce budworm and other insect problems was very different from that of the foresters in neighbouring New Brunswick, where DDT spraying of the budworm started in the early 1950s. Indeed, as the DLF's expert in entomology, Hawboldt was under considerable pressure from the Canadian Forest Service to support the introduction of a spray program in Nova Scotia at the same time. However, he resisted, with the support of his deputy minister, Wilfrid Creighton. Hawboldt argued that the problem with chemical sprays was that they slowed or stopped the advance of the budworm before its food supply had been exhausted. Not only did this preserve the conditions for its return, but it also sped up the cycle of return. Consequently, Hawboldt advised against spraying, in 1955 and again in 1975.

His reasoning had crucial implications for the Cape Breton highlands portion of the substantial Crown lease (1.2 million acres) granted to Stora in 1959. The Bélanger-Bourget inventory had established that the predominant species was mature to over-mature balsam fir, the prime host for the budworm. Hawboldt remembered flying over the highlands with Stora's Hans Lindberg. They were convinced that an extensive road-building scheme and a well-planned cutting schedule of the over-mature stands could minimize the future danger. Both shared the view that the spruce budworm was nature's harvester and that spraying would only prolong an outbreak. Hawboldt recalled exploring various healthy trees and finding traces of the budworm embedded in the growth rings of the trees. The implication here was that young healthy stands could survive epidemics and continue to grow.

Hawboldt was also instrumental in developing new forest legislation in the early 1960s. By the late 1950s, there was a growing consensus that the Small Tree Act should be replaced with some kind of silvicultural statute. This was a favourite subject of discussion at the annual meetings of the Nova Scotia Section of the Canadian Institute of Forestry (CIF) and at the Nova Scotia Forest Products Association (NSFPA). Hawboldt was charged with drafting a silvicultural bill that reflected the collective concerns of the key political and administrative people. He shut himself into a motel room for a week and wrote the bill. He then passed it on to Ed Haliburton, the minister of lands and forests, and Ike Smith, the finance minister, who came up with the 1965 version of the Forest Improvement Act.

Although the act was not as comprehensive as the one first produced by Hawboldt, it provided for the establishment of District Forest Practices Improvement Boards and (through an amendment in 1968) a central Provincial Forest Practices Improvement Board. The boards were composed of a wide array of forest sector interests, including the pulp and paper and sawmill industries, foresters, small woodlot owners, wildlife groups, and government officials. Through consultation and education, the boards were to formulate and enforce rules pertaining to the timing and methods of harvesting and managing the woodlands of the province. No immature trees were to be cut without the permission of the boards, which could also request that all commercial harvesting operations be conducted in accordance with stipulated practices. The final substantial provision enabled the boards to prescribe that commercial operators use every effort to harvest all possible saleable wood of commercial value. These terms were clearly written from the standpoint of protecting the integrity of the forest environment. The boards had the right to prevent young forests from being cut, to prohibit clearcuts and conversions of mixed forests to monocultures, and to stipulate that 'over-mature' and 'budworm-prone' stands of white spruce and balsam fir be harvested.

The DLF Extension Foresters were ready to serve on these boards. Ralph Hale, for example, served as the first secretary for the provincial board. Another veteran of the Division of Extension, David Dwyer, also served as secretary in later years.[33] There was room for DLF foresters on the district boards as well.

Several factors, however, undermined Hawboldt's forest management scheme for the NSPL/Stora lease. The departure of Stora's first foresters

(Hans Lindgren in 1963 and Jan Weslien in 1966) constituted one blow. Under their direction, sample plots were established on the Cape Breton highlands, frequent consultations took place with Hawboldt, and a long-term forest management plan was being developed for the Crown lease. When they left, this work was abandoned, and Stora's forest management plans turned out to be short-term operational or harvesting plans. Although they left many plans behind, the NSPL strategy began to change. Under subsequent foresters and woodlands managers, harvesting priorities shifted to the more accessible and younger stands, where roads were cheaper and easier to build. The District Foresters in Cape Breton and the eastern mainland, Allister Fraser and Angus MacMillan, who dealt closely with the company on an ongoing basis, approved these plans. At the senior level, the minister and deputy minister approved the plans too. Although the NSPL act called for the annual submission and approval of forest management plans, ministers were never willing to invoke these powers to force adjustment or compliance. On the one hand, so Hawboldt argued cynically, it may have been good that the company plans were quickly approved, for the District Foresters did not have the ability and resources to develop alternative plans. On the other hand, however, Hawboldt and Bulmer were available to assist the District Foresters and the company in developing long-term forest management plans, had the political will been available.

The second factor that undermined the Hawboldt management strategy was the coming of a third major pulp complex to the province in 1965. After Stora signed its Crown land deal and commenced production, Scott Paper (which had considered but abandoned plans to build a mill in the province in the mid-1950s) renewed its interest in Nova Scotia. Bulmer and Hawboldt were asked to prepare a paper on the potential wood reserves to supply a third pulp mill. In effect, they weighed the Scott proposal against the available inventory figures. Finding insufficient reserves on the mainland, including Crown lands, they advised against the project. This advice proved unacceptable to the politicians. G.I. (Ike) Smith, then minister of finance and deputy premier, convened an extraordinary meeting at the Hotel Nova Scotia in Halifax. Virtually the entire senior staff was summoned from the Department of Lands and Forests to defend the report: Creighton, Burgess, Bulmer, and Hawboldt, as well as all the District Foresters. Also in attendance was Minister of Lands and Forests Ed Haliburton. Travelling down from Truro, the staff brought cartons of data from the inventory and later sent back for more. The meeting began at 9:00 a.m. on a Monday morning, and sessions continued day and evening for the entire week.

Hawboldt remembered this meeting with ambivalent feelings. It began with a presentation by Hawboldt and Bulmer, who argued that there was not enough pulpwood available in the province to supply an additional mill. All of their assertions, however, were closely challenged and scrutinized. Ike

Smith, a renowned trial lawyer by profession, acted as the 'Grand Inquisitor.' Systematically, he challenged the assumptions of the inventory, on yields, growth rates and age classes, hardwood and softwood volumes, and annual allowable cuts. While Smith showed a keen intelligence, he could also be belligerent and nasty, recalled Hawboldt. 'He proceeded to tear down the assumptions of our case and build up his own.' Dick Bulmer, the forester in charge of inventory, 'got an awful going over' and proved not to be as firm as Hawboldt in his views. But in the end, even Hawboldt conceded to the political *force majeure*. It was clear that the politicians wanted to recalculate the reserves and permissible cut levels to accommodate the proposed mill as a fulcrum for economic growth. At one point, Hawboldt's attention strayed, and he lost touch with the discussion for a while. During this time, he was asked a question, and his response was 'It doesn't matter if I didn't hear the question, nobody will listen to my answer anyway.' Putting it delicately, Hawboldt believed that the foresters were 'stretched beyond what our small "c" conservative minds would have accepted on their own' (i.e., without pressure).[34]

Leaving the meeting, Hawboldt felt an old impression confirmed: that the Department of Lands and Forests had never fully been accepted and trusted by Cabinet and the politicians. The conference ended on Friday afternoon, with Ike Smith presenting a report to Robert Stanfield. The premier later took it verbatim to the Cabinet, where Scott's Crown lease request was approved. Five years later, the Crown reserves were stretched again. As part of a deal that added a newsprint mill and extra pulp capacity, Stora's lease was expanded from 1.2 to 1.5 million acres.[35]

Hawboldt was critical of the large Crown leases offered exclusively to Stora and Scott. There was tragedy in the fact that many people, government and lumbermen alike, had 'put their lives into the Crown lands, and then they were gone overnight.' When Wilfrid Creighton started the buy-back program in the 1930s, to convert private lands into Crown title, he never imagined that these lands would be given away so cheaply and completely. Creighton was a sawmill man, as far as Crown land management was concerned. These were also the areas where Hawboldt had begun to apply sustained yield management based on the mensuration data from the 1958 inventory.

Inevitably, these concessions of Crown lands to attract pulp and paper companies in the late 1950s and 1960s put increased pressure on the provincial wood supply. Hawboldt's forest management strategies were abandoned, and the stipulations of the Forest Improvement Act were violated. Immature trees were cut, and none of the pulp and paper companies honoured the principle, advocated by Hawboldt and Bulmer (and earlier by Fernow) that, if one does not take account of young growth, the forest will be depleted. Harvesting practices by the three large pulp and paper companies were reduced to clearcuts. And Stora launched its 'New Forest' strategy based on monocultural plantations.[36] Finally, the provision that commercial operators 'use every effort to harvest all possible saleable wood' turned into a scramble for the cheapest and most accessible wood. Scott hit its own lands hard in the 1960s, cutting the most accessible trees, whether large or small. Stora did the same. As a consequence, many sawlogs were ground up for pulp, in spite of tree-length logging providing an excellent opportunity to sort logs into pulplogs and sawlogs. In many instances, ironically, the alleged presence of pests rationalized the clearcutting of sawlogs for pulp, while over-mature and budworm-infested balsam fir and white spruce were left standing.

As in the case of Division of Extension services offered in the 1950s, lack of regard for the socioeconomic positions of small woodlot owners and rural residents also contributed to the failure of Hawboldt's management plan. The basis of that failure is well expressed in one of his descriptions of the Divisions of Biology and Extension. The Extension program, Hawboldt wrote, 'includes the various aspects of an information and educational service, as well as individual and public relations activities.' Forest biology, on the other hand, had to 'do with insect and disease problems as well as silvics – or the nature of forests and trees and how they grow. Silvics also involves measurements of growth and yield.'[37] Nowhere in these descriptions did Hawboldt touch upon the social economics of wood supply and particularly on the position of small woodlot owners in the pulpwood market, as forest managers or as an organizational force. It was perhaps small wonder, then, that, as educational efforts were having a slow impact, he began to look to legislation to promote forest management. In 1963 he wrote that 'It might be expecting far too much to consider that education alone would ever do more than a partial job. Possibly a hand-in-hand employment of education and legislation may be required.'[38] Still missing in these options was the social support necessary for small woodlot owners to 'accept' the legislation.[39] Hawboldt never thought it appropriate for the DLF to address these issues, though he did take part at various meetings as a forest technical advisor. He also believed that organizational issues should be left to the Department of Agriculture. There was always, Hawboldt recalled, friction between the departments. On one occasion in the early 1960s, when he took a group of foresters and 'ag reps' to MacDonald College to sit in on a week-long course in woodlot management, it upset the deputy minister of agriculture. Some senior DLF officials were openly hostile to the attempts of small woodlot owners to organize themselves.[40]

In summary, an opportunity was lost in the early 1960s when the forest inventory, forest bureaucracy, forest management and mensuration system, and forest legislation offered the constituents for an embryonic ecological management regime. An emerging organization of small woodlot owners was also in place, but this group was given little encouragement by the Nova Scotia government and the DLF bureaucracy. Already by the end of the 1960s, industrial forestry was well on its way to winning the day, and Hawboldt was to see many of his innovations and initiatives challenged, abandoned, or subverted over the next twenty years.

In Senior Administration, 1968-77

The late 1960s saw the opening of a new era in the Department of Lands and Forests and in the career path of Lloyd Hawboldt. The retirement of Wilfrid Creighton in 1969 triggered a more general realignment in senior forestry ranks. R.H. Burgess took over the deputy minister's job. He had risen from Provincial Forester to assistant to the deputy minister, with special responsibilities for industry relations. Hawboldt, who had already taken over the field budgets in Truro, was appointed to Burgess's former job of administrative assistant. Today this position would more accurately be described as that of assistant deputy minister, since it was held by senior personnel with extensive influence rather than junior assistants. The two men had distinct but complementary interests, and in Hawboldt's view they made a great team. Together they tackled the new policy challenges of the 1970s, including the emergence of a new budworm infestation and the rise of the federal-provincial forestry agreements. These were cost-sharing programs promoting forest management. They poured millions of dollars into forest management and boosted the employment of foresters of the DLF and the monies available for various forestry programs. The implication of the programs will become more obvious in subsequent chapters, which cover foresters whose careers were closely bound up with the federal-provincial agreements.

Both the budworm infestation and the federal-provincial agreements reflected the mammoth steps taken by industrial forestry in consolidating a hold on the forest management regime of the province. The federal forest service clearly favoured the industrial option to forest management. Hawboldt was thus never favourably disposed to the federal-provincial agreements. There was always a sense that the federal government enjoyed dominance in all negotiations over the agreements. This applied with even greater force in the 1970s, when the new generation of joint forestry agreements was being framed. 'We were subservient to them [the federal government],' Hawboldt recalled. During one negotiation over whether road building, so crucial for the effective management of the budworm, should be cost-shared, the federal government said no, while the province said yes. It happened that both federal and provincial teams picked up their papers and left the negotiating room. The province won the fight over cost-shared roads, but it was by no means an easy fight. Hawboldt's view was that federal forestry monies should be given unconditionally. The federal support system was also very tenuous. Hawboldt remembered the ARDA project that funnelled money into forestry but constantly threatened to dry up and leave everybody hanging.

The industrial forestry bias of the federal-provincial agreements became clear with the proclamation of the Forest Improvement Act in 1976. In Nova Scotia, the first Forest Resource Development Agreement (1976-81) was conditional (at federal insistence) on the proclamation of the Forest Improvement Act. However, the agreement provided no funds for the implementation of the act. Instead, it concentrated on the delivery of forest management services (plans and treatments) on the ground by covering the costs of engaging professional foresters to assist woodlot owners. The management philosophy here was more industrial, intensive, and geared toward increasing the fibre supply for the pulp and paper industry.[41]

The return of the budworm in the 1970s triggered the most controversial and stressful period of Hawboldt's career and forced some hard decisions on the Nova Scotia government. The pressure on the wood supply was intense. With the CFS and New Brunswick now in the third decade of chemical sprays, this position was fast becoming a professional forestry orthodoxy in eastern Canada. With the departure of Hans Lindberg and Jan Weslien, a new generation of Canadians and more spray-oriented Swedes had taken over at Stora. Soon the non-spray, or 'silvicultural,' option came under increasing attack. Hawboldt, however, remained true to his convictions. Throughout this period, the DLF remained opposed to chemical spraying. Hawboldt took part in public meetings against spraying and even appeared in an anti-spray film (*Budworks*) by environmentalists Neil Livingston and Andy McNeil. Although Cabinet approved Stora's spray application in February 1977, the decision was quickly reversed under public concern about health dangers. Even the threat by Stora president Erik Sundblad to close the mill in five years unless spraying was allowed proved fruitless.[42]

At the same time, Cabinet was still unwilling to wield aggressively its powers of approval over Stora's annual Crown land cutting plans to enforce the Hawboldt plan. It would not venture beyond persuasion. Hawboldt remembered a meeting in Baddeck with Department of Lands and Forests Minister Maurice DeLory and Stora officials. Hawboldt continued to advocate a strategy of harvesting the worst-infected stands first, leaving the least-affected areas for limited selection cutting or strip cutting. The Bélanger and Bourget 1955 inventory of Cape Breton Island could have guided this process.[43] Stora, by contrast, wanted to 'salvage' the healthier and more accessible stands before they were infected by the budworm and then spray the worst-affected areas. DeLory and Hawboldt 'held their own the whole night,' but in the end they could not prevail. Several hours later, Hawboldt experienced a serious heart attack and was airlifted back to Halifax for medical treatment.

Stora proceeded as planned. Hawboldt did not consider the so-called salvage to have been a salvage operation at all. In addition, rather than selection cutting or strip cutting of limited tracts in the healthier stands, wholesale clearcuts were the order of the day. In the process, more and more areas fell into age class 1 (i.e., under twenty years), adding to a predicted starvation period of wood some time in the future. The dispute was about different management philosophies. Hawboldt focused on biology and ecology, while the company foresters focused on short-term profit. The company prevailed.

These stressful times took their toll physically and emotionally on senior officials of the DLF. Both Hawboldt and Burgess took early retirement at age sixty. Burgess retired in 1976 but later had second thoughts and considered returning to work. For Hawboldt, the heart problems were a key factor. Moreover, 'the politics were getting unbearable' by that time. Hawboldt retired the next year but experienced no similar regrets.

Hawboldt remained active in one capacity after he retired in 1977. It constituted a last-ditch effort to save some remnants of a more ecological type of forest management. Itself politically embattled, the Provincial Forest Practices Improvement Board commissioned a manual for forest practices under the provisions of the Forest Improvement Act. Drafted by a young Quebec consultant, the finished product was judged inadequate by the Steering Committee, which found it ambiguous and rambling. Far back in 1966, Hawboldt and Gary Saunders had written the short but tightly focused bulletin titled *A Guide to Forest Practices*.[44] It was intended to be an interim field guide, pending the preparation of the full manual. With the

Lloyd Hawboldt speaking at a public meeting on the spruce budworm spray issue, October 1976.

entire project in jeopardy in 1977, the board approached Hawboldt to help out, and, as he recalled, 'like a damn fool I was interested.' It required virtually a complete rewrite. All sections were drafted and debated by the board, and all recommendations were approved unanimously. Ultimately, the manual was published in 1980 as the 200-page volume *The Trees around Us*.[45] Written in accessible language, it offered planning and operational tools for the ordinary woodlot owner. However, it was heavily criticized by the pulp and paper companies, and it was never put to general use as an extension of the act. Despite his work on the final product, Hawboldt did not see much hope for the Provincial Forest Practices Improvement Board. There was never anything wrong with the act that a few amendments wouldn't fix. But there had never been a political or administrative consensus after the passage of the legislation, and the timing of the forest manual was either too late (to support the statute) or too early (for its time).[46]

Conclusion

Reflecting on his thirty-year career with the Department of Lands and Forests, Hawboldt invoked a common statement of professional foresters: the problem with forestry is politics. The dictates of politics seldom coincide with those of forest management. In the first place, they are based in such different time frames. As elected politicians, ministers plan on a four- to five-year cycle, since the end of their mandate is their reckoning time. By contrast, good forest managers plan according to the rotation age of their forests, which may vary from forty to 100 years. Consequently, Hawboldt believed that it is a mistake to mix forestry and politics.

The political cycle often determines the fate of government programs, particularly whether they expand or contract. This is particularly harmful to forestry programs, which are extremely vulnerable to short-term pressures. This befell the Division of Biology in 1952 and the Division of Extension programs in the 1950s and 1960s. The same held true for firefighting. If there was a big fire, or a series of fires, the allotment of funds grew. But then, after a few years without major fires, the funds dried up. At least until the next fire.

Hawboldt also thought that business pressures compromised forest management. Sawlog and pulpwood forestry, for example, are guided by two entirely different philosophies, with the former operating on a cycle almost twice the duration of the latter. However, the key political decisions of the 1960s, on Crown land allocation, forest practice regulation, and industrial development, ran directly counter to the dictates of classical forestry.

Hawboldt always tried to stay out of politics and was convinced that, if politics is left out of forestry, foresters would agree on most points. At their annual professional meetings under the Canadian Institute of Forestry, they all seemed to think the same way.

There is, though, something more to the story. Forestry is not an objective but a subjective science, and Hawboldt's concept of forestry was very different from that advocated by the emerging pulp and paper industry. His concepts of what constitutes a forest ecosystem and forest management were *distinctive before their time,* stemming from his training in entomology and European sustained yield forestry in the 1930s, 1940s, and 1950s. As Hawboldt stated when assessing his pioneering work in Nova Scotia, 'I had an advantage, I was not a forester, I was an entomologist.' The assumptions of his outlook can perhaps best be gauged by quoting one of his mentors, A.D. Pickett, who may have paved the way for Hawboldt's commitment in resisting the spray option.

> I want to say that we believe the proper approach to the orchard insect control problem, or any control problem, whether plants or animals, is through a complete ecological study of the problems involved. If insecticides are necessary to keep injurious species below the economic level, we must take the trouble to determine the over-all, long-range effect of the chemicals on all the factors in the environment. We are of the opinion that the use of chemicals should be considered supplementary to environmental resistance. When we reach the point of using DDT ... or other new insecticides as routine control measures in the same way as we have used inorganic chemicals in the past, entomologists interested in biological controls might as well throw in the sponge.[47]

Hawboldt's forestry perspective was also expressed in the recommendations that flowed from the provincial forest inventory in 1957 and in the forest management philosophy of *The Trees around Us* in 1980. Hawboldt thus believed in nurturing, building, and restoring the Acadian forest to suit the demands of a diversified forest economy. According to these principles, he believed that the spruce- and pine-dominated forests of central and western Nova Scotia lent themselves particularly well to sawlog management. It was a sacrilege to practise pulpwood forestry in these areas. By contrast, pulpwood forestry was particularly well suited for the white spruce- and balsam fir-dominated forest of eastern Nova Scotia and Cape Breton Island. In this country, it was inadvisable to manage the forest for sawlogs.[48] Industrial foresters, however, had no faith in the ecological management philosophy of the likes of Pickett and Hawboldt. Ralph Johnson, the forester at Bowater Mersey, called Hawboldt an 'impractical idealist.'[49] The spray-focused foresters in New Brunswick also operated under a different set of assumptions. In 1953, Barney Flieger professed that 'There is no control known in nature to stop a [spruce budworm] epidemic.' The integrated workings of various components in an ecosystem were irrelevant here, and the problem components were treated separately. As Flieger put it, 'The outcome, then, unless some new trend appears, seems to depend mainly upon the relative persistence of the insect in epidemic form on the one hand and of the forest sprayers on the other.'[50]

Hawboldt's legacy in Nova Scotia forestry was to demonstrate and advocate scientifically credible alternatives to the increasingly dominant industrial forestry paradigm.[51] The failure of his alternative approach to take hold involved several factors. Partly, it flowed from a conceptual limitation within his analytical approach: the separation of ecological and social dimensions, and the virtual dismissal of the latter.[52] Fundamentally, however, his approach proved politically innocent, depending upon his prestige in elite policy circles and his proximity to executive decision makers. Technocratic in inspiration and lacking a coalition of supportive interests, the Hawboldt approach could not surmount opposition from the pulp and paper industry, its political supporters, and the professional group of industrial foresters who supported them.

Appendix

Toward 'Budworm-Proofing' the Forests of Nova Scotia (Lloyd Hawboldt)[53]

Spruce budworm is regarded as 'nature's harvester' of mature and overmature fir-spruce forests. Large continuous areas of mature pure fir forests are especially susceptible and vulnerable. Given the right forest conditions and ideal weather the budworm periodically goes on the rampage until brought under control by natural means, including eating itself out of house and home.

Because the forest growth cycle and the budworm outbreak cycle are so closely related epidemics appear about every thirty years if left to proceed naturally. This may vary in time and over specific areas from twenty to fifty years depending on the age class structure of a forested area.

Thus in 1922, following the outbreak of 1910-1921, J.D. Tothill observed, 'It is plain also that the next outbreak may be expected when the existing fir reproduction now being released ... become tall enough to pass through the crown of the forest so as to form an immense food supply for the insects. On the basis of average annual growth, the next general outbreak may be expected at any time after the lapse of about thirty years'; and, '... the next outbreak is likely to be more severe than the present one has been and to extend over greater areas.'[54]

Just thirty years after the above statement we were into another outbreak. It was 1952 that the first aerial spraying was carried out in New Brunswick. In this epidemic of the 1950s pressure to spray in Nova Scotia was successfully resisted, and the outbreak which was most severe on the Cape Breton highlands collapsed through natural causes in 1957, after five years of activity. An estimated 100,000 cords of fir were killed in the highlands, of which 60,000 cords were salvaged.

Currently the spruce budworm is again a topic of much concern throughout northeastern North America. It threatens fir and spruce forests over an estimated area of 150 million acres. We in Nova Scotia are sharing this concern particularly in the fir forests of the Cape Breton highlands, but also to a lesser degree in northern counties of the mainland.

Between 1970 and 1973, spruce budworm attacks reached severe proportions in Kings, Annapolis and Cumberland counties mainly on white and red spruce. This outbreak is believed to have arisen through moth flights from New Brunswick. By 1974 the Annapolis Valley infestation had collapsed naturally but persisted in Cumberland County and appeared quite dramatically in parts of Antigonish, Inverness, Victoria and Richmond counties. Severe defoliation occurred over parts of these counties in 1975. The Canadian Forestry Service expects this to continue in 1976 based on egg and larval surveys in late 1975.

Aside from letting outbreaks develop and collapse naturally, which in the past has taken about five years in Nova Scotia, the only effective short-term means of reducing budworm to levels at which trees are kept alive is through aerial spraying with insecticides. In 1955 this writer speculated on the possible outcome of spraying should it be undertaken in Nova Scotia. These views were based on budworm literature, observations on past and the then current activities on Cape Breton as well as experiences in orchard insect control with sprays. The following is quoted from an article of May, 1955:

> It is clear that spruce budworm epidemics arise as a result of the forest conditions. Spraying to kill a portion of the current budworm population does not alter the susceptible nature of the forest, nor can it be expected to eliminate the pest population ... Since epidemics built up before under those circumstances they may be expected again regardless of any temporary relief from sprays.
>
> Thus one can visualize repeated applications of spray in what is likely to be a vain attempt to offset a natural trend.
>
> In the case of a valuable stand of timber which cannot be harvested within the next few years, the pest population may be kept down enough by spraying to protect the stand until operations are under way. Such action has merit. However, in the case of mature and over-mature stands which cannot be harvested within another twenty or thirty years, what possible assurance could there be for success? Within that time spruce budworm could be expected to return to such a susceptible forest, perhaps very quickly.
>
> Should it be possible to keep mature and over-mature forests alive by

spraying, what is their future? Fir should be harvested by the time it reaches an economic age of sixty years. Beyond that the decrement of age, the invasion of rots, plus the continued susceptibility to 'frustrated,' and possibly repeated budworm attacks will within a twenty- or thirty-year period greatly reduce the value of the stands. Past experience has shown that uninhibited budworm epidemics, which are part of the 'natural climax of the spruce balsam stand,' result in dense new stands. A very real danger in holding old stands, lacking in virility, lies in the sacrifice of good regeneration for the future.

On the other hand, should a spray program be undertaken and then prove unsuccessful in preserving such stands, the investment would be jeopardized. If the wood is required it would have to be salvaged very quickly. Then would arise the necessity for more investment for roads, and no doubt consequent stumpage concessions.

Mechanically an aerial spray operation may be successful for any given year, but as has been pointed out 'the long-term effect will be measured in terms of how much change there has been in the population trend.'

The presence of an insect like the budworm in epidemic form is evidence of neglect. Again it is emphasized that, if the wood is required, then a management programme is essential. Any spray programme considered necessary should be supplementary only to the more permanent one of prevention. It should be considered as a temporary measure and not an end in itself.

The budworm epidemic is present because of the large quantity of susceptible forest. Either need of the wood introduces 'the wise use of the axe' to correct the situation, or we would be wise to let nature take its course and ensure a well-stocked, harvestable stand within the next 60 years.

Tothill (1922) suggested the forests could be made 'budworm-proof' for the future. He recognized it 'would be wholly impracticable were it not for the fact that there is a period of about thirty years in which to bring it about, and for the fact that fir has recently come to have a value that enables it to be cut at a profit.' 'It can scarcely be emphasized too strongly that no remedy is likely to be applied unless a beginning is made while public opinion is focussed upon the subject.'

By 1952 the thirty-year grace had passed. The budworm returned. No programme was initiated to 'budworm-proof' the forest. Industry not only operated but expanded since 1922, while large acreages of fir in eastern Canada were permitted to become mature and over-mature. The conclusion is that the wood was not needed by the forest land holders. Now that this old neglected wood is being felled naturally (to make way for a new crop) there is a great hue and cry against budworm.

If we are being sincere and if the wood from these acreages will be needed, before recently cut areas become harvestable, then now (1955) is the time to start 'budworm-proofing' the forest for the future.[55]

Someone has said words to the effect that 'if we learn anything from history it is that we learn nothing from history.' Thus in 1975, twenty years after the above quotation and fifty years after Tothill's views, we are in precisely the same situation with respect to the growth cycles of fir and budworm on the highlands of Cape Breton Island. Other prominent forest entomologists and silviculturalists since Tothill have advanced the same views. Had these recommendations been acted upon we would now be well down the road toward 'budworm-proof' forests.

Two distinct problems exist in Nova Scotia. One is typified by the practically pure and highly susceptible stands of fir on the Cape Breton highlands. It is unlikely that any appreciable change in the type of this forest is practicable or possible other than what may occur through natural rotation or succession. White spruce when mature is an equally susceptible species in the lowlands of Cape Breton Island along with fir. This association is more amenable to manipulation.

Fir is not so natural a species on the mainland of Nova Scotia as on the Cape Breton highlands. Selective cutting practices in the past and the dying of beech and birch have contributed to an abnormally high fir population. The presence of so much fir increases budworm hazard to the more naturally occurring spruce. Recent short-rotation cutting practice of spruce-fir types tends to increase fir content and therefore budworm hazard.

Two essentials are indicated in forest management to 'budworm-proof' the forests. One is to develop mixed forests and reduce fir content especially on the mainland and the lowlands of Cape Breton. The other is to regulate harvesting so as to develop a broken distribution of age classes of softwoods and particularly in the fir forest of the Cape Breton highlands. The rigid utilization of fir within its biological rotation age is essential but not at the expense of longer-rotation red spruce.

A basic requirement to forest management as related to spruce budworm is a land classification to show susceptible and vulnerable acreages in a forested area – that is, a hazard map. This involves identification of age classes of stands and if possible by site quality. Priorities for cutting are decided by best judgement based on this information. Roads are built to meet these priorities. Some of the considerations in the decision-making process are as follows:

(1) Spruce budworm is part of the aging process in the natural climax of the fir-spruce stand – the budworm is nature's harvester.
(2) Large areas of pure, mature and over-mature fir are the most susceptible and vulnerable to budworm epidemics. Similar areas of white spruce are high-hazard forests.
(3) Generally three to five years of heavy budworm feeding in a stand will result in tree mortality. An additional two or three years are available to

salvage trees before serious break-down of wood. On the highlands of Cape Breton this may be extended to as much as five years.

Despite this generality, old, poor site-quality fir have been killed following the severe feeding of a heavy population which flared up in one year. Aside from this extreme there is a period of five to ten years from the first year of general heavy feeding in which to conduct presalvage and salvage operations.

(4) The objective cannot be strictly one of presalvage and salvage cutting. The primary objective should be to regulate cutting so as to reduce future hazard. Crash harvesting undertaken to liquidate an area without consideration for age and site quality over a period of five or ten years will not accomplish this.

Instead of the partial distribution of age classes now present such action will create more of a one age-class forest. Not only is this biologically unsound, but it defeats the principle of sustained yield by creating a future shortfall of wood until young growth matures.

Added to this acreage crash of harvesting will be the older, poorer site, more vulnerable stands which die from decadence, budworm attack and both.

(5) Inherent in these considerations is of course the principle of annual or periodic allowable cut being controlled more by acreage, age and site than by volume. Area allowable cut presents the only means of developing a broken-aged forest of even-aged stands. It also gives encouragement to increase volume yields by whatever silvicultural means possible.

(6) In the current crisis situation with budworm it is not possible to predict what will happen and thus determine what course must be taken with respect to cutting. Historically outbreaks have lasted about five years and then collapsed naturally. The more decadent stands suffered the greatest mortality and quicker. The younger and better site stands showed more resistance and less loss. Highest cutting priority should be directed toward recovering the fibre from the former rather than the less vulnerable. In following this course of action less acreage will be reverted to the same age class. Also, the volume taken will more nearly approximate the permissible yield than would be the case if this type were sacrificed in favour of cutting higher volume, less decadent, better site stands.

(7) Priorities for cutting should be based on all considerations with the primary objective as the deciding factor, without unnecessarily overloading the wood supply, without reverting any more acreage than necessary to age class 1 and by utilizing as much of the decadent and killed fibres as possible.

(8) Should the situation develop to such a degree that there is an overload

of wood supply trade arrangements should be made with other in-province users. Only as a last resort should the province lose value added by exporting raw wood.

(9) Mortality due to budworm or decadence or both if salvaged is no loss. The acreage involved and put back into the regeneration stage must be greater than the area allowable cut. This imbalance is a temporary unfortunate set-back for which adjustments can be made in future with intensive management. The absolute balanced age class distribution ideal is probably unattainable. But the closer the manager works toward it the more vigorous and 'budworm-proof' will be his forest and the greater the yields on a sustained basis.

(10) An absolute essential to these considerations and the primary objective is an adequate network of permanent roads. Normally these might be planned and developed in an orderly way over a longer period of time. With the current situation a crash program of road building is essential.

Roads must be planned to service the highest priorities of cutting areas and not to gain access to cut high-volume, less vulnerable stands. Investment in roads has a far greater potential for gain than to put it into spray to keep old trees alive to get older.

Spraying to keep trees green until these can be harvested is not an acceptable alternative to forest management. 'Ecologists have been aware for many years that the use of pesticides prolongs the need to continue their use.' 'The annual application of insecticides to preserve large forest areas from the spruce budworm prolongs outbreaks and presents a serious hazard to environmental quality.' Such observations by prominent scientists of today are not to be ignored in the light of experience over the past twenty-five years with chemical warfare against budworm.

Reproduced courtesy of the Nova Scotia Department of Natural Resources.

5
Donald Eldridge: Advocate for the Industrial Forest Interest

During the period from 1950 to 1990, the Nova Scotia forest economy underwent a remarkable change. The erstwhile dominant sawmilling sector experienced a slow but prolonged economic decline, occasioned by a fall in the availability of sawlogs, changing markets in private timber lands, and the reassignment of great tracts of Crown land from sawmill to pulp mill licensees. In primary wood production, the volume of pulpwood surpassed that of sawlogs for the first time in 1963. As with all industrial shakeouts, there were winners as well as losers. Many of the larger sawmill companies increased their capacity and extended their woodland base, while many smaller outfits sold out and closed. For many of the growing survivors, the key lay in 'integrated use,' the conversion of sawmill wastes into woodchips for sale to pulp processors. By the early 1970s, pulp and paper was established firmly as the leading element in the provincial forest industry, while also enjoying the support of the provincial government and the professional forester community. At the same time, a dynamic woodlot owners' movement was challenging this new order. Finally, the fibre squeeze occasioned by new Crown timber arrangements with Nova Scotia Pulp and Scott Maritimes, together with the precipitous decline in private woodlot output, resulted in a dramatic expansion of the provincial forest service in 1977, to oversee the new era of intensive (industrial) forest management. It was the bureaucracy's role to administer the coordinated federal and provincial strategy designed for this purpose.

Donald Eldridge rode the crest of prosperity of this transition from beginning to end, holding positions in both the private and public sectors. Frustrated by his youthful experience as a government forester, he spent most of his career in private employment. From 1951 to 1968, he worked for a large sawmilling establishment, the George Eddy Company. From 1968 to 1979, he was the executive director for the most influential forest industry lobby group in the province, the Nova Scotia Forest Products Association. He then served as deputy minister of the Nova Scotia Department of Lands and

Forests from 1979 to 1987. Eldridge closed his career in the position of Commissioner of Forest Enhancement before retiring in 1990.[1]

Eldridge's long and close association with private business strongly influenced his outlook on forestry matters and his approach and attitudes to work in the public sphere. Eldridge remained an industry advocate and a strong proponent of business-driven forest management. He was also committed to consolidating the shared interests of sawmills, pulp mills, and loggers into a single industry block, with the Forest Products Association at its centre. He was thus a natural exponent of the instrumentalist and productivist agenda of postwar forestry, a program based on the premise that trees were grown to be harvested by commercial fibre processors. This continued to serve him well in midcareer when directing the affairs of the NSFPA, but it confused and frustrated him in the public sector, where as deputy minister he had to deal with a large and expanding bureaucracy and non-industrial constituencies advancing forest values that he thought were in conflict with business prerogatives. Eldridge was particularly antagonistic to the small woodlot owners' movement, which sought higher prices and wider markets for its products and expanded public assistance for forest management. Later he reacted in a similar vein to the environmental movement, which challenged chemical sprays and clearcut logging in favour of more ecological methods of forest harvesting.

On only one major issue, and then only late in his career, did Eldridge diverge significantly from mainstream corporate thinking. He was concerned that many industrial forest methods were destructive of wildlife (or, as the traditionalists put it, 'game') habitats. This criticism owed much to his personal background as a hunter and the implications of the fact (which he often repeated) that hunting and fishing licences generated more revenue to the province than forest stumpage payments from Crown lands. He favoured guidelines and regulations (as a measure of last resort) to correct this situation. At the same time, Eldridge had little sympathy for the emerging environmental movement in Nova Scotia. He routinely dismissed environmental activists as ill-informed extremists. His commitment to conservation was a harvester's perspective, centred on sustaining resource stocks for future use. He did not see the natural world as an ecosystem balancing the needs of diverse species.

Eldridge embodied in many ways the contradictions associated with the Nova Scotia transition from a sawlog to pulpwood economy. This is likely what catapulted him into (and later out of) his term as deputy minister of lands and forests. He shared with the old sawmillers a distaste for government intervention, taxation, and public spending. He also shared with them a belief in the efficacy of personal political contacts to advance business interests. These sentiments were in many ways the product of an earlier era, when rural constituencies of frugal farmers and small businessmen held pronounced influence with provincial governments. Such beliefs nevertheless served him well and gave him the high profile in the forestry community that justified his appointment as deputy minister.

Once in the position of deputy minister, however, Eldridge's views clashed with the values of a new era. While Eldridge supported the forest management regime of the pulp and paper industry (a regime that in several ways undermined the fibre supply of the sawmillers), he opposed the growth of bureaucracies and the impersonal organizational networks that linked governments and society. This led him to resist new policy instruments for forest management, particularly those targeted toward other forest constituencies. But such assistance was of course in accordance with the growth of urban areas and large businesses, both of which were happening at the time and demanded larger and more intricate spending and taxation arrangements. The federal and provincial funds made available for forest management were also part of the growth of federal spending in Atlantic Canada to boost regional prosperity.[2] Eldridge failed to appreciate these changes. He therefore became an increasing anachronism. The tensions that followed between Eldridge the public servant and his staff and various clients were only resolved when he was appointed Commissioner of Forest Enhancement, a post of limited powers that shifted him to the margins of the forest policy community.

The Early Years, 1945-51

With his surveyor's licence in hand, Donald Eldridge began his career in forestry in 1945 by entering the forestry school at the University of New Brunswick. This followed a summer job working with the Nova Scotia Department of Lands and Forests. The surveyor's skill proved useful in landing summer employment. During the summer of 1948, Eldridge cruised timber stands for the largest corporate owner of Nova Scotia forest lands, the Hollingsworth and Whitney Pulp and Paper Company of Maine. After graduating in the spring of 1950, he spent a summer surveying for the DLF in Guysborough County. Eldridge recalled a strong sense of inadequacy when he was handed his forestry degree. Despite multiple offers from private companies in Maine, he went to work full time for the department in the fall. He was one of five UNB graduates to join the public service that year as it began the transition toward a more professional basis. The new deputy minister, Wilfrid Creighton, had just completed his first full year on the job, and the process of recruiting District Foresters had just begun. The rural rangers still ruled their local empires, and there was very little comprehensive data on forest conditions. Fire protection was a central part of the DLF mandate, and the forest inventory would not begin until 1953.

In less than a year, however, Eldridge was feeling restless in the civil service and exasperated by the petty politics in Nova Scotia. By this point, he had met C.W. Anderson, the legislative member for Guysborough, who had cut Crown wood illegally in the West River area of the county. Anderson was a colourful man with red hair who went by the name 'The Woodpecker.'[3] In spite of being convicted for his activities, he was publicly unrepentant, stating that he would steal wood to feed local Guysborough people every time. In another petty but annoying case of partisanship, Eldridge remembered surveying for the Crown in 1946. Anderson's son Murray was also a member of that crew. At one stage, they moved operations from Larry's River to New Harbour, a distance of less than ten miles. As the crew approached New Harbour, the local MLA declared that if they worked in New Harbour they had to use New Harbour men. Eldridge contacted his superiors in Halifax on what action to take, and he was ordered to move the crew back just far enough to remain in the Larry's River area.

The Eddy Years, 1951-68

In 1951 Donald Eldridge took a job in industry as the surveyor, Chief Forester, and woodlands manager for the George Eddy Company. The head office was at Bathurst, New Brunswick, where a huge planing mill operated. From there, the company shipped lumber to markets in Montreal, Toronto, and New England. At the time, the business was run by George Eddy's two sons, Bob and Ches. Eldridge's responsibility was to procure timber and run the company's sawmills in Nova Scotia. Eldridge was also in charge of buying forest lands for the company. Clearly the right man for the job, he retained this position until 1968.

Eddy Lumber owned about 29,000 acres of forest land when Eldridge joined the company. Nova Scotia sawmillers were riding unprecedented prosperity at this time. The reconstruction of Europe sparked tremendous demand for building materials, and the Nova Scotia industry hit record production in 1947, when more than 800 mills sawed more than 450,000,000 FBM. However, the lumber trade is notoriously cyclical, and by 1955 the boom had passed. The total output of sawn lumber was cut in half by 1967. Like the other lumber companies, Eddy's business depended on an increasingly tight supply of sawlogs. This was met by a continuous search for new forest lands. Bob Eddy was the son who had vision ahead of the family to recognize and react to these trends. His policy was not to sell a single acre but to buy as much land as the company could afford, and Eldridge was given a timber-buying budget for this purpose.

Most of the purchases were small properties, though Eddy did buy some large tracts. Eldridge's first large transaction involved 29,000 acres from the

King Brothers in Oxford, Cumberland County. Another big forest tract amounting to 25,000 acres was bought from the Scotia Lumber Company in 1952. Later Eddy bought a 20,000-acre property without even commissioning a cruise, since the company knew the quality of the people with whom it was dealing.

For the most part, however, Eldridge purchased small forest properties with mature trees. It was a judgment call every time a lot was bought. Most of the central Nova Scotia sawmills at the time grew in the same way, and they competed intensely for timber. Eldridge was constantly bumping into his counterparts from Riversdale, Bragg, McClelland, Elmsdale, and other Truro-area mills. They bought small 100-acre lots on estate sales. In some cases, municipal assessors drove people to sell by setting the assessment too high for owners to afford. Eddy was affected by this as well, and every year Eldridge spent time in court appealing various assessments.[4]

There were also many local middlemen who took a profit by brokering land deals between local owners and timber buyers such as Eldridge. Some of the active brokers were Les Bond; Clarence Mason; Rayburn, Rayworth, and Giddens; and the Ward Brothers in Oxford.[5] They often employed unusual methods in their land dealings. One individual threw down pages of an old Simpson's catalogue while being shown around a property to mark the trail so that he would not be shown it again. Once a prospective piece of land had been identified, the broker consulted old forest land index sheets to see who was listed. Then the broker offered $500 to the owners to sign over their lands. He then started a quitclaim deed, and checked to verify that there were no heirs, before turning the property over (i.e., selling it) again.[6] These procedures added to the mountain of title deficiencies that plagued Nova Scotia. In dealing forest land, Murray Anderson sometimes gave four or five options to purchase, at $1,000 each. Eldridge recalled one case in which he wanted $100,000 for a 25,000-acre tract. One option holder, Rayworth and Giddens, showed it to Eddy, asking $150,000. Had they closed the deal, Rayworth would have netted $50,000 for 'flipping' the land.

Eldridge devised his own system, which generally worked effectively. When he got an option on a piece of land, he gave it to Lorne Clarke, a young articling student in Eddy's law firm (and later a Nova Scotia Supreme Court judge). While Clarke looked up the title, Eldridge sent cruisers into the woods. As soon as he heard a reasonable price from the owners (e.g., $10,000 for 200 acres), he pulled the cruisers out and went directly to check with Clarke. If the title was 'good,' then he bought it right away. Speed was essential because he knew that Riversdale and the others would be right on his heels. Often Eldridge had a contractor putting in a road within a week. The Crown was never a major competitor for prime timber. It had to follow a laborious four- or five-stage approval process, by which time any valuable property was long gone. While Eldridge would close a land deal in a week, it could take the government up to two years. There was often a difference in the quality of the properties as well. Creighton describes the forty-four Crown purchases of 1950, totalling 122,219 acres, as follows: 'This was the largest area added to the public holdings in a single year, bringing the area purchased up to 389,585 acres [in total], or more than one-third of the target of 1,000,000 acres. While these lands were cut-over or burned and sometimes both, they had once been highly productive. With time and care they would support good forests again.'[7]

Eddy's sawmill operations involved air-drying lumber (for six months to one year, depending on drying conditions) before it was shipped. Green lumber weighed too much. At the peak, the company cut 40,000,000 feet per year, mostly with portable ('woodpecker') mills. The largest of these mills cut 5,000,000 feet, but most cut between 1,000,000 and 2,000,000. Many of the mills were manufactured by the Oxford Company in Nova Scotia. This was a firm with a solid reputation. If one of its mills broke down, an overnight crew was sent in to make repairs. Setups of this type were found across Nova Scotia and southern New Brunswick. At one point, Eddy alone had fifteen portables operating in Nova Scotia, from Shelburne and Lunenburg Counties in the west to Guysborough and Cumberland Counties in the east. All of these sawmills employed local labour and reinforced village life. For example, the Archibald family in New Town, Guysborough County, ran an operation including a water mill, their own timber limits, a local labour force, and wood purchases from local farmers. They were still operating in 1994.

The Eddy Company financed these portable sawmills, and Eldridge had to watch them carefully. In a sawmill operation, two groups of workers were especially touchy, and they were the keys to the operation: the sawyers, and the deal carriers. If either one stopped work, the operations shut down. Eldridge also had to monitor the yard transactions. Eddy used to buy logs in the yard, scaling on the truck, with cheques issued before the seller left. This had to be done carefully, though some sawmillers were totally trustworthy. Aubrey Gilroy was one such man whom Eldridge gladly financed.

On a typical 100-acre tract, Eldridge put a contractor in, roughed out a road, set the saw in a central place, and hauled logs on trails from the corners of the lot to the mill. A horse or tractor and a sled were used. When all the logs were cut, the crew hauled out both lumber and saw and moved on to the next tract. Both spruce and balsam fir were cut, though the latter was unpredictable as sawlog material. One could sometimes cut a fir that was fine at the butt and top but rotten in the middle. But once cut, spruce and fir were worth the same in the market.

Reflecting on his early years with Eddy, Eldridge noted that there were only marginal improvements made in the sawmilling sector: 'Over time we got to know how to build roads. They lasted better, and could be used longer.' It became common for the sawmill men to have trucks to haul wood to the mill. Some did log contracting as well, either with their own crews or with subcontractors. But in the end, Eldridge conceded, 'Nobody ever got rich off of sawmilling. They made a living.' The portable philosophy was to cut a woodlot and then let the land go as a tax sale. The main reason that the woodpecker mills and the forests could survive such an exploitative system, in Eldridge's view, was that the companies had the flexibility to enter and exit the market: 'They were in a position that they could roll with the ups and downs of the market. [When] markets got bad, they just stopped cutting and stopped getting into trouble.'[8] By contrast, for Eddy and some of the other larger sawmillers, the philosophy was to cut and then to hold on to the lands.

There was little forest 'management' in Nova Scotia during Eldridge's time with Eddy. A lot of high-grading went on, spruce and fir in the east, pine and hemlock in the west. There was nevertheless some attempt to regulate the cutting of certain species. The Small Tree Conservation Act of 1946 provided a diameter restriction of ten inches on the felling of three targeted species: hemlock, pine, and spruce. While the simplicity of this limit could be seen as crude and mechanical, it was compatible with a system of selection logging in which lands were cut partially on a repeated basis, while the quality of the forest improved. This contrasted with the 'take the best and leave the rest' practice of many independent portables. A more serious limitation, however, was the exemption of small, privately owned forest lands, which sharply restricted the act's application. In retrospect, the Small Tree Act was an uneasy compromise. It fit the needs of the lumber sector. It also had the stamp of approval of the deputy minister of lands and forests, Wilfrid Creighton, whom Eldridge characterized as 'a tremendously good person for Nova Scotia.'

Eldridge recalled some stories on the implementation and consequences of the Small Tree Act. Additional rangers were hired in the late 1940s to enforce the act. They inspected the various cutting operations and could tell from the log piles in yarding areas what was being cut. When they found undersized wood, they put a yellow tag on the log brow, which meant that the wood could not move. In effect, it had been seized, and unless the operator could explain why he had made the cut he would lose the wood.

These regulatory powers had the potential to sow bitter conflicts, and there were certainly times when rangers were confronted with enraged woodlot owners or contractors. Eldridge recalled a case in Hilden in which a ranger was thrown out of a landowner's house during a dispute over the small-tree regulations.[9] Eddy's policy was to follow the stipulations of the act, setting the minimum cutting limit at an eight- to ten-inch diameter.

This way the company could go back every ten years to cut trees again. On one ten-acre tract near Camden, it made three cuts during Eldridge's time with the firm.

Some woodlot owners insisted that the provisions of the act were not sufficient. When Eldridge was with Eddy, he once bought a stand at Glenray that belonged to the deputy minister of lands and forests, Wilfrid Creighton. Creighton wouldn't let the contractor, Aubrey Gilroy, cut trees under twelve inches in diameter. Eldridge shared many lumbermen's ambivalence about the act. For them, the regrowth advantage of a diameter limit was always set against the possible weakening of the stand through thinning. On one 400-acre lot in Hants County, the contractor cut red spruce to the ten-inch limit, and the result was a forest that looked just like a park. But then Hurricane Edna arrived in September 1954. It rained for the first two or three days, and then the winds came. The Hants lot went over like candlepins since it had been opened up too much. It was salvaged for pulpwood, as were many of the Mersey lands, where millions of feet were lost. Whether or not a 'century storm' was the proper measure of diameter cutting, the debate continued.

When the pulp and paper companies began buying woodchips in the 1960s, the sawmilling industry was transformed. In the past, the sawmills used to burn slabs and edgings, which went directly from the green chain to the fire. When Bowater Mersey started buying chips, Brookfield Box was the first to join in, and Eddy was second. At first, portable chippers and debarkers followed the sawmills around. They were on trailers that could be hooked up to the mills. The first machines were Morebarks, and later came the Norman eight-knife chippers. Sometimes in the portable system, slabs were hauled from two or three mills to a central location for debarking and chipping.

Eddy encouraged all of its contractors to shift to chips, but many were reluctant. When one outfit wouldn't go into chips, Eldridge bought a barker and chipper and set them up as a separate company on the basis that the contractor and Eddy split the profits fifty-fifty. Two years later, the contractor bought Eddy out of the project. Once sawmillers had made the switch to chipping, there were more investments to be made. Mersey insisted that the chips be 92 percent bark free. This was strict even by today's standards, as Scott still receives a 'whack of bark' in its loads. But it forced sawmills to acquire ring debarkers. Furthermore, many of the mills became permanent as they adopted debarking and chipping technologies.

It was sawmillers' increasing dependence on the pulp and paper industry that prompted a restructuring of the sawmilling industry. The restructuring was also influenced by the commercial growth and government support of the pulp and paper industry in the region, which invariably meant directing Crown wood away from sawmillers to the pulp and paper mills. The abrogation of the Nova Scotia Small Tree Act in 1965 also contributed to the restructuring process. The only good sources of sawlogs in the 1960s were the areas where the act had been enforced, and they were now made available to the pulp and paper companies.[10] One outspoken individual affected by this process was Murray Prest, a sawmiller from Mooseland in Halifax County.[11]

Many of the large sawmillers with large tracts of land willingly sold out to the pulp and paper companies. Other sawmillers and sawlog contractors were either forced out of business or became pulpwood contractors. Another aspect of the pulp and lumber connection involved log exchange, which assumed crucial importance as the sawlog market tightened. In New Brunswick, Eddy had become dependent on the Bathurst Pulp and Paper Company for part of its sawlog supply. This resulted from Bathurst having managed to tie up the local Crown limits for pine. Eddy, like so many other sawmillers, was in the weaker position when selling woodchips to Bathurst. It was thus doubly dependent and vulnerable, and in the end the sawmilling operations were sold.[12]

Elmer Bragg and Wilson McClelland bought out Eddy's milling operations in 1968, though the hardware and plumbing businesses were retained. Eddy also kept its timber lands until about 1990, when most of them were sold to the government of Nova Scotia. Eddy kept about 6,000 acres around lakes and cottage lands. This sale was done on highly favourable terms.[13] The government had just lost a large deal for about 90,000 acres to a Dutch company brokered by 'Island Bob' Douglas, who was buying land (especially islands) for Europeans.[14] After this, the government went after the Eddy lands aggressively. Since that time, both Ches and Bob Eddy have died.

Eldridge decided to leave Eddy when the lumber operation was being phased out in 1968, despite the fact that the company offered him an open cheque to continue working there. By the time that he left, the company owned 120,000 acres. Some time later, he was asked to come back and survey the company lands. This he was happy to do and only charged one cent per acre, as a payback for the past. Shortly thereafter, Eddy started to sell large lots to the pulp and paper companies, which proceeded to clearcut them for pulpwood.

The sawmill squeeze intensified after the Eddy Company left the business. At the top end of the scale, a handful of firms produced large and expanding volumes of lumber, while at the bottom there were hundreds of smaller firms producing less than 1,000,000 (or even less than 100,000) FBM annually. Shortly before the mill closure, Eddy figured in the 3,000,000 to 6,000,000 FBM category.

As Table 2 reveals, the intermediate sawmill operation such as the Eddy Company still exists, though it is now dwarfed in importance by an upper tier of high-volume producers.

Table 2

Composition of sawlog output by mill category

	1966		1980		1996	
	No. firms	% FBM	No. firms	% FBM	No. firms	% FBM
Over 6,000,000 FBM	4	0.19	5	0.27	17	0.77
3-6,000,000 FBM	8	0.14	10	0.23	11	0.11
2-3,000,000 FBM	16	0.18	12	0.16	6	0.04
1-2,000,000 FBM	34	0.22	20	0.14	11	0.04
Under 1,000,000 FBM	325	0.27	300	0.20	246	0.04
[Under 100,000 FBM]	[190]	[3%]	[207]	[3%]	[182]	[-]
Total annual FBM	222.6 mill		188.3 mill		422.3 mill	

Source: Nova Scotia Forest Production Survey, 1966, 1980, 1996.

Executive Director of the Nova Scotia Forest Products Association, 1968-79

When Eddy decided to end its lumber operations in 1968, the Nova Scotia Forest Products Association (NSFPA) approached Donald Eldridge to become its full-time executive director. This association was more than thirty years old, having been founded as a voice for the modernizing wing of the sawmill sector. John Bigelow was its first part-time secretary, and the NSFPA had been run by its volunteer executive and board of directors ever since. The hiring of full-time staff marked a major step in the evolution of the group. It was partly a response to the increasing complexity of the forest industry during a time of rapid change. There were many factions in the forest sector at the time, and the government of Nova Scotia had decided to support the NSFPA as a potential organizational umbrella. On first being approached by the association, Eldridge declined the offer. Then the directors met with him again, and he relented on the condition that they move their office to Truro, where he lived. At the time, the office was on Quinpool Road in Halifax, three or four doors away from prominent sawmiller Jim Wilber, who kept a close eye on the organization. Eldridge thought that Wilber would object to his condition, but the other members apparently overruled him, and 'they called my bluff.'

Eldridge's biggest challenge with the Forest Products Association was to bring the organization together to speak with one voice. The sawmillers were already well represented, and the pulp mills were joining, with Stora and Scott opening their mills in 1962 and 1967 respectively. 'We wanted to pull the pulp mills in,' Eldridge recalled. This was confirmed symbolically in 1971, when the first pulp and paper man was elected president. The NSFPA also wanted to include the pulpwood contractors, and many of them joined the association. This was a volatile combination. Eldridge relied on an elaborate committee structure to contain the differences between the factions. Then he used the board of directors to address issues of common concern. The association was to be the industry's 'eye' on the forest. This role was dramatized in the mural that hung behind the podium at NSFPA annual meetings in the 1970s. However, it was a constant challenge to find the right balance for a given issue. Eldridge explained it to the membership this way: 'As your Executive Director, I must try to keep as well informed as possible on any matters that affect the forest industry. This, as you know, involves many, many meetings involving a number of varied interests. You get to feel at times that you are a jack of all trades and in between all of this you are called upon, and quite rightly so, to be a sounding board for all the various groups within our Association and I can assure you that there are a number of interests and groups in our Association.'[15]

Eldridge was confident that he had the abilities to direct the group. As a professional forester, he could talk the language of Mersey, Stora, and Scott

Donald Eldridge presenting the executive director's report to the annual meeting of the Nova Scotia Forest Products Association, mid-1970s. Note the 'eye on the forest' in the backdrop.

and their woodlands staff. As a lumberman, he was well known to the sawmill sector, landowners, and contractors. He had also served on committees with the Maritime Lumber Bureau, chaired its grading authority, and worked on the Canadian Standards Association lumber section.

Just as Eldridge was moving into the association, it scored a political coup that would dramatically alter its future course. This involved the Gas Tax-Access Road program. In this scheme, the provincial government provided an annual grant to the NSFPA, which dispensed it to forest operators on a cost-shared basis for the construction of logging roads on private forest lands. The government based the total annual grant on an estimate of the provincial tax paid on fuel consumed by operators in off-highway forestry work. The program rationale was that forest operators received no benefit for the fuel taxes that they were forced to pay. This offered a means of cycling those receipts back into the sector. The scheme was devised in 1966 by a group of NSFPA sawmillers/directors in a meeting with Deputy Premier Ike Smith. Eldridge arrived at the association just in time to reap the benefits. As the delivery agent for the access road program, the NSFPA gained in three respects. First, it was a powerful mark of trust and influence to be chosen by the DLF for this job.[16] Second, it earned an income for the NSFPA through the administrative overhead charges that the province paid. Third, and perhaps most importantly, the access road program helped to broaden the association beyond its corporate base. Since grants were only open to NSFPA members, many woodlot owners and logging operators joined to secure eligibility, thereby swelling the membership rolls. An Access Road Committee was struck to process the applications and award the 'assistance,' and a $100 administrative fee was charged to process each application. Set at $300,000 for the opening year (with a $7,000 ceiling per road grant), the overall budget increased by 50 percent in 1968 and continued to climb over the years.[17]

Within the NSFPA, Eldridge constantly had to walk a tightrope between the sawmillers and the pulpmen. He believed that the pulp mills came together more tightly than the sawmills, but the lumbermen had pride and a history of family enterprise on their side. By the time that Eldridge joined the association, wood allocation was beginning to tilt in the direction of the pulp mills, and over time they grew increasingly assertive. The association's Wood Allocation Committee and Chip Committee put up some resistance, but it was to little avail in the long run. The woodlands managers and foresters of the pulp and paper companies were surprisingly hard-nosed in their positions. In fact, Eldridge thought that many good foresters were brainwashed over time to the corporate line. Bob Murray at Scott expressed this outlook aptly. He used to say that his first job was to have enough wood to keep the mill going. 'Talk to me about silviculture after that.' Eldridge was not without sympathy for this perspective. He had a business background himself and had experienced the pressures of staying competitive. However, it complicated his job in balancing NSFPA member interests.

Ultimately, Eldridge was unable to resolve the underlying differences between the sawmillers and the pulpmen. But he was more successful in rallying the support of the two constituencies for common concerns. The most dramatic of these was in opposing the pulpwood-marketing proposals advanced by the small woodlot owners (explored in detail in Chapter 7). But there were other proposals. In 1970 the NSFPA called for the establishment of a 'Forestry Commission.' It was envisaged as a board of six representing forest sector organizations and working under the Forest Improvement Act. It would advise the minister and develop programs for marketing, silviculture, and fibre supply.[18] In effect, it was the association's alternative to the pulpwood-marketing authority being promoted by the woodlot owners.

Although he claimed to have 'a great love for the small fellow,' Eldridge shared with the industry a genuine distrust for the attempts by the extension people at St. Francis Xavier University to organize rural woods producers. He was especially suspicious of the newly formed Nova Scotia Woodlot Owners Association (NSWOA) and one of its organizers, Rick Lord, a forester from the Eastern Townships in Quebec. In Eldridge's eyes, the woodlot owners were blinkered by a focus on wood price and nothing else.[19] This view is captured in his comments, detailed below, to the association's annual meeting. Significantly, Eldridge was unable to grant the same legitimacy to the woodlot and environmental constituencies that he so readily extended to corporate and professional interests.

In the year before he left the association in 1978, Eldridge spoke pointedly to the association's annual meeting about the state of its political mandate. He began by painting an embattled picture. The NSFPA had to deal 'more and more with governments at three levels – municipal, provincial and federal. There is no question in today's society, the dice are loaded against industry or the free enterprise system as such. Industry is fighting to survive.'[20] Eldridge was clearly an unconditional supporter of capitalism and the competitive business system, and he believed that the association stood for that message. The central agent in the forest industry was thus the entrepreneur who spent the money and worked full time. Eldridge was less sympathetic to small woodlot owners, whom he viewed as peripheral players cutting and selling wood on a sporadic basis. The same applied to the environmentalists, who had no direct economic stake and therefore, in his eyes, no understanding of the forest environment.

At the annual meeting, Eldridge went on to review a series of associational concerns from the previous four years. His comments offer an apt picture of his free enterprise, anti-government ideology at work. His first concern was the government ban on spruce budworm spraying. Having been defeated on this issue, Eldridge explained that 'we haven't done a very good job on this one, not all our fault, but partly.' His second concern was wood allocation, for which he described the future as uncertain because of the unresolved tension between the pulp and paper companies and the sawmillers over fibre exchange. The third concern dealt with the prospects of a new pulpwood-marketing bill, which would have allowed stronger organizing and bargaining rights to small woodlot owners. Here Eldridge claimed that the NSFPA was 'back to the old situation of trying to fight for the protection of the rights of our own members to govern their own destiny. It would appear this association is not going to give up this right for another body [the small woodlot owners] to act on our behalf.'

Eldridge then moved on to the topic of 'government legislation as it affects the forest Industry.' He noted that 'it is still with us and we have to constantly be on guard in respect to the kinds of legislation that are introduced into the Assembly by other departments.' His fifth point dealt with cooperation with the Department of Lands and Forests, and he concluded that 'this is perhaps the only one that we have made any real headway on.' The sixth concern involved the implementation of the newly signed federal-provincial agreement on forestry. This was a government-to-government initiative, and the NSFPA had been kept at arm's length. Here

Eldridge 'hoped that the powers that be who have kept this under so tight and close wraps do not come out with any bombshells to cause our Association and the industry harm.' A final concern was the 'involvement of a certain group of, I think, well-intentioned people with the forestry situation in Nova Scotia, they go under various names and groups, their members are small in numbers as compared to the general population, they own very little land but they are very articulate; they can and do use the media.' He referred, of course, to the environmental groups that had recently emerged on the budworm spray issue. Eldridge challenged both their credibility and their constituent base. 'Just who are these people talking for? This is the question the elected representatives of the people will have to ask themselves when making decisions that are going to affect the whole forestry complex in Nova Scotia for years to come.'[21]

His concerns for the well-being of the NSFPA continued after Eldridge left the association. He was succeeded by Lorne Etter, a member of a well-known lumber family, who served the association with distinction after 1978. When Etter retired, however, Eldridge worried about the succession. In particular, he tried to warn the board members about one prospective new candidate, Claudette Terrio. She had previously sold insurance to sawmills and was known to many directors, but Eldridge was aware of a court case linking her husband with commercial fraud. His concerns fell on deaf ears, however, and the directors gave Terrio the position. This was a decision that they would regret. During the next few years, she took over $30,000 out of NSFPA funds.[22] This was a time of transition for the association in another respect too. Although it had operated, during Eldridge's and Etter's tenures, from a modest second-storey walk-up accommodation in downtown Truro, the board had approved the construction of a spacious log-frame headquarters overlooking the Trans-Canada Highway near Hilden.[23] Not long after the official opening, the financing burden threatened to overwhelm the NSFPA, and the provincial government was forced to step in with a special grant to relieve the mortgage burden.

Eldridge and the Environmental Challenge

Since Eldridge's career coincided with the emergence of environmentalism in Nova Scotia, his relations with this movement are of more than passing interest. Indeed, environmentalists were responsible for shaping some of the most important forest policy questions of the era. In one sense, Eldridge was a typical corporate man, dismissing the critique of industrial forestry as technically ill informed and socially marginal. Yet he was a dedicated sportsman and resource conservationist, with a woods cabin in Guysborough County. These contradictions may be reconciled in a short exploration of the environmental challenge.

Nova Scotia environmental activism appeared in name in the 1970s. As in many parts of Canada, this emerging political force was strongly shaped by the early issues that it embraced.[24] Some of them were local expressions of national and international themes. The breakup of the oil tanker *Arrow* in 1970 polluted the shores of Chedabucto Bay and dramatized the vulnerability of a coastal province. In addition, generations of accumulated coal and steel sludges in Sydney were finally recognized as one of the country's most dangerous toxic dumps in the middle of a major city. And the impact on Nova Scotia lakes and forests of acid rain originating in the Great Lakes industrial belt was a growing concern during this time.[25]

As important as these issues were, forest-based controversies played perhaps the most crucial and continuing role in shaping Nova Scotia's embryonic environmental movement. More than the others, they had the potential to fuse a province-wide network of community and regional activists facing common problems and seeking common solutions. Moreover, the forest sector offered a rich vein of environmental concerns that kept public attention focused for more than two decades. This began with the spruce budworm infestation of the mid-1970s that threatened the spruce-fir softwood forests of northern and eastern Nova Scotia. It should be remembered that the damage wrought by the insecticide DDT was the focus of Rachel Carson's *Silent Spring*, the book that launched North American environmental awareness in 1962. When the government of Nova Scotia approved an extensive aerial spray campaign in 1976, it provoked a remarkable mass political protest, which began in the kitchens of rural Cape Breton and spread across the province to media outlets and the floor of the legislature. Known as Cape Breton Landowners Against the Spray, it provided a crucial mobilizing umbrella as well as strategic leadership.[26] Ultimately, it succeeded in forcing the reversal of Cabinet's spray policy, largely by redefining the issue from one of insect control to one of human health. Given that both New Brunswick and federal forest authorities had been deeply committed to budworm spray programs since 1951, this was a formidable achievement.

In the early 1980s, the battle lines shifted from chemical insecticides to herbicide sprays. They were used to defoliate and suppress hardwood growth on the new spruce plantations being created at public expense to ensure the next generation of forest fibre supply.[27] The Cape Breton Landowners Against the Spray went to court seeking an injunction against mass herbicide applications. Although the case was lost in 1983, the intensity of the continuing controversy was instrumental in the Nova Scotia government's establishing a Royal Commission on Forestry in 1982. It was charged specifically to advise on the future use of chemical spray agents as silvicultural tools. This provided a public forum for extended debate on the forest environment. Ultimately, the commission endorsed the use of chemical agents, 'properly used,' and found no safety or ecological grounds for public concern.[28] However, the political opportunities that the inquiry afforded for research, organization, and publicity were formidable, and the provincial network of environmental organizations was significantly advanced as a result.

By the 1980s, the environmental agenda in Nova Scotia forestry also challenged industrial timber-harvesting practices and clearcut logging in particular. The previous decade had seen a substantial investment in woods equipment ranging from skidders and forwarders to tree harvesters. This equipment transformed logging from a labour-intensive process to a mechanized one and greatly increased logging productivity. The economics of this production required clearcut operations of ever larger dimensions. This, however, was only part of a much wider silvicultural system that involved mechanical site preparation (following logging), replanting with seedlings, and ongoing herbicide treatments.[29]

During the 1980s, the anti-spray campaign evolved into a wider critique of 'industrial forestry.' British Columbia was labelled the 'Brazil of the North' for its extensive clearcutting practices, and the Crown pulp forests of Nova Scotia were not far behind. This critique accelerated with the 1987 release of the Brundtland report on sustainable development, and it was evident that the political legitimacy of industrial management was fast slipping away. The result was the rise of 'sustainable forest management' certification in the 1990s. Based on the concept of third-party audit of forest management practices, it became a prime environmental battleground in the era following the Rio Earth Summit.[30]

Over this twenty-year period, environmental interests moved from the uninvited periphery of the policy community to a recognized and entrenched stakeholder position. At the outset, its efforts were sustained by dedicated volunteers in local-issue groups, often labouring with little statutory or regulatory support against a closed and antagonistic corporate and state system. However, the relentless accumulation of pollution episodes, advocacy campaigns, and political alliances struck with producer and consumer groups has legitimized the environmental presence in contemporary politics.

It is common to distinguish a series of layers in the environmentalist community. One of them is the sport and wildlife segment. It has deep historical roots in a largely rural province such as Nova Scotia, where military officers advocated game laws in the nineteenth century. Numerous local fish and game associations are linked to networks such as the provincial Wildlife Federation. On the non-consumptive side, the provincial Federation of Naturalists and the more recently formed Trails Federation represent outdoors enthusiasts such as birders, hikers, and campers. Both segments hold a general affinity to the natural environment. While the first is often described as conservationist (committed to regulated harvests) and the second as preservationist (committed to wilderness protection), these two categories are seldom exclusive.

A separate and more recent political tendency is commonly described as environmentalism. This ideology springs from a belief in the primacy of nature as defined against civilization and the repudiation of humans as the predominant species. It comprehends nature in ecological terms, as a complex web of interacting organisms capable of sustained reproduction. Environmentalism provides a framework through which pollution can be understood not simply as the isolated and accidental breakdown of the industrial machine but also as a systematic degradation of the ecosystem through the use of oil, arsenic, uranium, or chemicals. More particularly, it includes the end-of-pipe pulp effluent that decimated the fish populations at Boat Harbour, the aerial budworm sprays that killed thousands of birds on Cape Breton Island, and the clearcut-harvesting practices that 'converted' vast tracts of Acadian forest into softwood monocultures.

In Nova Scotia, popular environmentalism began in the 1970s and was expressed through a series of local and regional groups. Cape Breton Landowners Against the Spray, the North Shore Environmental Web, the South Shore Environmental Protection Network, and Citizens Against Boat Harbour are four cases in point. A fifth, the Halifax-based Ecology Action Centre, provided an informational clearing house for local actions across the province. On paramount forestry issues such as the budworm and herbicide decisions or the royal commission inquiry, this segment was joined by two further sets of groups: the outdoors lobbies, and the forestry sector groups with environmental dimensions.

In this light, Eldridge's perspective can be more clearly located. His was the outdoorsman's outlook on nature: respecting the animal world, knowledgeable in its lore, and guided by what John Reiger calls the 'code of the sportsman.'[31] For its adherents, there was only one correct way to take game, its commercial trade needed to be outlawed, and favoured grounds were celebrated for their untouched character. This was organized and perpetuated at the club level across the province and was closely linked to the DLF by district and regional wildlife officers. Provincial forests, particularly Crown forests and sanctuaries, were key fish and wildlife habitats, and insofar as their health was compromised the sportsman's future was in doubt. For much of the twentieth century, this was unthinkable, as even degraded forests offered cover and shelter. However, the industrial clearcut and plantation, which first emerged on the modern scale about 1966, was an entirely different prospect. It was against this that Eldridge reacted in the later phases of his career. At the same time, there was no necessary link to or resonance with modern environmental arguments, which looked well beyond wildlife, or even habitat, to ecosystems. This lay beyond the horizon of either the forest industry or the bureaucracy as then constituted.

Donald Eldridge presenting an award at an NSFPA meeting. Note the presence of Conservative Party leader John Buchanan, third from the left.

Deputy Minister of Lands and Forests, 1979-87

By 1978 relations between Nova Scotia commercial forest interests and the provincial government were at low ebb. A pulpwood-marketing board had been established over strenuous industry opposition. Moreover, Halifax had refused to support extensive aerial spray campaigns against the spruce budworm, despite the prospective loss of millions of cords of softwood. Some political gesture of reconciliation was clearly in order. In 1978 the Liberal premier of Nova Scotia, Gerald Regan, asked Eldridge if he would accept the position of deputy minister in the Department of Lands and Forests. The previous deputy, Bob Burgess, had retired, and his replacement, who had come over from the Department of the Environment, didn't seem to be working out. Before Eldridge had a chance to reply, Regan called an election, and the Liberals were defeated by the Progressive Conservatives. The new premier, John Buchanan, then asked Eldridge again, and he accepted the post. The basis of his decision was the hope of continuing his industrial advocacy role more effectively within government. In his words, 'We are not being given the ear we should be in industry. We should get our views across. There are problems going on here that politicians are not paying attention to, and so I thought, well, when the opportunity came along ... I could be of more value to industry and to my profession by being on the inside ... than I can be ... hollering at these fellows from a distance.'[32]

His time as deputy minister, however, was disruptive and frustrating. On a personal level, Eldridge maintained his home and residence in Truro and commuted every day to Halifax. He left home at 6:45 a.m. and arrived at work about 8:10 a.m. In the evening, he left between 6:00 and 7:00 and was home an hour later. Eldridge believed that the government got its money's worth from him as he completely severed his business and social connections with his community. He had been active in community work as chairman of the Truro Tree Commission and as a member on various committees for the Kiwanis Club. He also curled and played golf a little but lost interest in them because his job did not allow him to spend time on those things. It completely dominated his life: 'So, any time you hear of a deputy minister being short of time ... I have all the sympathy for the deputies or ministers. Their lives become hectic and not their own.'[33]

It was therefore not surprising perhaps that Eldridge left the NSFPA with some hesitation and regret. As he told the association's 1980 meeting,

> I felt very sad in having to leave that environment. It was a job I liked well and it is a job that I had not planned on leaving but I couldn't resist the opportunity afforded me to move into this position ... I have pledged myself to working for the entire forest community – for the small independent contractor ... the smallest landowner, right up to the largest pulpmill. I would also like to ... [point out] that in my new job, being a hunter and fisherman myself, I would take into consideration closely their need and attempt to bring to you some of the problems they have.[34]

The 'entire forest community' clearly corresponded to the constituencies with which Eldridge had worked closely: the small independent contractors whom he had hired while employed by the Eddy Company; the small landowners from whom he had bought stumpage or forest lands; and the pulp mills that had bought much of the Eddy lands in the end. Clearly missing were the environmentalists and the small woodlot owners.

Once in the government, Eldridge was frustrated by the operations of a large bureaucracy. Nothing in his previous experience had prepared him for this challenge. His criticisms were typical of any free-enterprise critique of government and its bureaucracies. One thing that struck him first in government was 'the lack of appreciation of the buck.' Although part of the government bureaucracy himself, he referred to it as a foreign body: 'These people ... throwing 50-60 thousand dollars around as though it was peanuts ... out in industry, that kind of money was big, big bucks, and it didn't matter whether you were a small company or a large company, you didn't throw that kind of money around.'[35]

A second source of his frustration in government was the sheer amount of work that Eldridge faced. His first address to the NSFPA as deputy minister gives an impression of the bulk of work that he confronted while at the same time confirming his commitment to serve his former industry clients:

> I can tell you this, that in Lands and Forests, it is a multi-faceted type of job. There are all kinds of jobs coming at you. You just have to turn your head and there are two more coming at you from other directions. You really don't have too much time to think in that job. So, if I don't get a chance to carry through on some of the jobs you ask, would you please try to understand that it is a very busy job ... [and that] it is not [for] lack of wanting or trying ... [that I do not] carry out your requests.[36]

A third source of frustration was related to bureaucratic red tape. Eldridge soon discovered that his department was enmeshed in a far wider state decision-making system, including Cabinet and its policy board and management board committees. New to the public sector, he sometimes seemed overwhelmed by the scale and complexity of the processes. Somewhat ruefully he observed that 'things are going along really at a snail's pace compared to industry.'[37]

The final source of irritation was the challenge mounted by various factions hostile to the DLF/industry development strategy. Eldridge saw these conflicts primarily as motivated by personality. Vince MacLean, the minister of lands and forests for the Liberal government under Gerald Regan, who imposed the spray ban on spruce budworm, acted so because he 'was a school teacher and didn't have the business background.' Eldridge's own minister under the Progressive Conservatives, George Henley, by contrast, 'was quite business oriented [and] brought in a policy ... for business.'[38] Another challenge to the DLF's forest policy came from the Provincial Forest Practices Improvement Board, which operated independently of the department to develop good forest management strategies in Nova Scotia.[39] The board clearly favoured an approach that better balanced the sawmilling industry, the wildlife sector, and recreational and environmental concerns. Eldridge had earlier been part of the board structure, but he now argued that the positive potential of the Forest Improvement Act had been destroyed by interpersonal complications:

> Certain personalities got involved in it, and it started to swing as an environmentalist or an activist group ... took the pulp and paper out of it. Hugh Fairn, who was chairman of it, went to the forest products meeting, and he took a swing – a broadside swing – at the industry and said you fellows don't pay attention to your cutting and so forth. We'll put you in your place, that type of thing. Well, he became the most unpopular person in the whole forest industry, and in a sense he hasn't recovered as yet from that. He was so out of place with his comments at that particular meeting. I don't know whatever enticed him to do that, but he did more harm to the environmental movement there ... From then on it was downhill for the Forest Improvement Act.[40]

On the other hand, Eldridge considered the Nova Scotia Royal Commission on Forestry to be a positive result of his term of office. For some time, the NSFPA had called for an inquiry, and its executive had met with Cabinet for this purpose. Appointed in 1982, the commission held more than a year of public hearings before reporting in 1984. Eldridge and his staff spent a lot of time preparing the DLF submission. In fact, Eldridge praised his staff for this document, considering it 'probably the best document that Lands and Forests ever put together, and [it] will probably be a bible.'[41]

Under the conceptual rubric of a 'conservancy' approach to forestry, the commission's report endorsed most of the points made in the DLF submission. One of the report's recommendations was the abrogation of the Forest Improvement Act and its replacement with a Forest Enhancement Act. This vested full control of forest management in the hands of the Department of Lands and Forests, which had recently signed the first of a series of federal-provincial forest development agreements with Ottawa. The commission also proved quite unsympathetic to the woodlot owners' movement and the environmentalist critics of aerial spray. Finally, the report led to the creation of a new post of Commissioner of Forest Enhancement, which later fell to Eldridge himself. However, the result was to shuffle Eldridge into a marginal role in the forest sector, while providing him with a place to play out the last three years of a colourful career.

Commissioner of Forest Enhancement, 1986-90

The royal commission report of 1984 provided the basis for Eldridge's appointment as Commissioner of Forest Enhancement in June 1986. Although the report recommended that an Auditor General of Forestry be appointed to oversee and report to the legislature on forest practices on Crown lands, the actual legislation looked quite different. It made provision for the appointment of a Commissioner of Forest Enhancement who was to report to Cabinet through the minister. This allowed Cabinet to receive the commissioner's report in camera and to keep it confidential if so desired. In fact, this proved to be the fate of all three reports that Eldridge wrote, and they consequently had no impact in the public debate. When he passed away on 7 June 1995, his reports had still not seen the public light.[42]

According to the act, the duties of the commissioner were fourfold: to facilitate the implementation of forest management programs; to report annually to the governor-in-council through the minister on the implementation of forest management programs; to coordinate the activities of the Forest Advisory Council with those of the government; and to perform functions assigned by the minister or the governor-in-council. Eldridge regretted his limited mandate from the outset. In 1987 he ruefully conceded that implementation rested not with the commissioner but with the Cabinet. Nevertheless, he was enthusiastic about his new assignment.

When the premier asked him whether he was interested in the job, he laid down two stipulations, both of which he claimed were met. First, he wanted to retain a deputy minister's status and to operate from Truro. Second, he asked for a small office, 'with a couple of field staff, maybe a couple of professional field people, and some technical people underneath them.'[43] He received secretarial help and a travel budget.

As Commissioner of Forest Enhancement, Eldridge had several roles. His main task was to look at the cutting programs of the large companies, especially Stora Forest Industries, given that it operated principally on Crown lands. Overall, during his term as commissioner, he was satisfied with Stora's performance. He thought that Stora and the other large companies were the best providers of employment, revenue, and forest management: 'If you wanted real forestry and massive forestry to take place in the province, you really had to go to the larger people because they had the expertise, they had the machinery, and they had the trained personnel to do the job ... I still say we could speed up forestry quite a bit more ... by going to the larger people.'[44]

Eldridge was particularly impressed by Stora's efforts at forest regeneration, and he believed that one of the best fibre supplies in the province would come out of its Crown leases in eastern Nova Scotia. His favour for the large pulp companies is clearly reflected in his reports quoted at the close of this chapter. By encouraging good Crown land management, he hoped to let private landowners 'see from our example.'[45]

Comparing the pulp and paper companies' forest management efforts with the province's Group Venture program, Eldridge sided with the former (though recognizing that the latter had provided some benefits for the awareness of forestry and silviculture):

> Now the government is going with this group venture effort and has spent a lot of bucks unlocking a few doors, but if they had spent the same amount of money by giving it to the larger companies or making it available to the larger landowners ... they could have gotten more for their bucks – more work done for their bucks than they have, but they have made an awful lot of small landowners aware of forestry by going to group ventures. There is no doubt in my mind today if the government stopped supplying the money to group ventures that, of the eighteen to nineteen group ventures going, probably only three would survive.[46]

Eldridge's harsh assessment of the Group Venture program contrasted sharply with his views on the low stumpage payments paid by the pulp and paper companies for Crown wood. Eldridge readily admitted that these rates were too low, but he argued that 'you can't hit a company right between the eyes for something you did early on and just switch around because you can do real damage if they happened to be in an economic slump, and in newsprint this can almost be the difference of being in the black and red.'[47] He nevertheless thought that no government measures were necessary to ensure that small woodlot owners be assisted to make up for the initial advantages accorded to the pulp and paper companies.

Eldridge's support for the forest management regimes of the pulp and paper companies was not, however, unqualified. In the case of Stora Forest Industries, though the company was working within the agreement with the government, Eldridge thought that there were 'little things coming out that I see that are not just in accordance with the agreement.'[48] One was that Stora's reports on cutting practices and schedules were routinely made after the fact and failed to inform local DLF field staff in the spirit of the act. Often field staff arrived at prospective cutting sites and roads only to discover that harvesting was already under way.

One issue that particularly concerned Eldridge was the perceived abuse of the province's hardwoods, especially since the practice destroyed wildlife or game habitats. He observed some huge clearcuts of hardwoods at the time. He believed that some of these areas should be preserved or they would end up clearcut and converted into softwood plantations, like 'the area that has been cut up in Deny's Mountain, Cape Breton, ... where there [are] ... 10,000 acres ... that have been clearcut. Some of this wood was taken out as sawlogs and pulpwood, but most was chipped up for fuel. Forested areas with hard maple or rock or sugar maple, and yellow birch, should be used for sawlogs or maple syrup production.'[49] Once again, Eldridge thought that the pulp companies were insensitive to the impact of their cutting on fish and game.

It also riled his instincts that hardwood trees were being chipped for pulp, as many not only provided habitats for wildlife but also served as potential fire breaks. Yet Eldridge could also see the hardwood conversions from the perspective of the pulp companies. Hardwood stands often had the best soils and growing conditions and were more accessible to the mills. When flying over Matties Settlement, one could even see the Stora mill in the distance. Eldridge even thought that many company people knew the damage they inflicted on the forest environment: 'I think some of them know this, but they like the big massive areas they can look after cheaply, and I guess maybe if I were a forester for a company I would probably have a blinder look until I was able to produce wood fibre for the mill.'[50]

On some occasions, Eldridge confronted disturbing incidents of blatant neglect of wildlife in forest management. In one case, he recalled that biologists

> had laid out a nice hardwood corridor of trees to be left for the wildlife. I went with them – went up to look at this area – and behold, the corridor had been cut – completely gone, and they weren't happy at all, and neither

> was I. I brought it to the company's attention, and ... the story they got back was the contractor hadn't enough wood to fulfill his quota, and so they had to give him extra work that fulfilled the quota. Besides that, they felt the corridor was too wide. Well, you know, that kind of attitude, that kind of talk, doesn't endear me to their practices.[51]

The last set of problems that Eldridge confronted was related to the services and subsidies provided to contractors and small woodlot owners by the federal/provincial forestry agreements. In some respects, this seemed like selective criticism. Infractions alleged against small operators were not unique, but his faith in the larger operators was much more firm. Another problem that Eldridge found was what he called 'a little hanky-panky going on between the contractors.'[52] They inflated the acreage of the areas planted and thereby charged the government a higher fee.

Eldridge had several solutions for the problems that he confronted as commissioner. Recognizing his weak mandate, he was not averse to proposing legislative and regulatory measures, though he confined this attention to Crown lands. With respect to the controversial hardwood conversions, he suggested that Cabinet could 'throw in an amendment to the Forest Enhancement Act and maybe limit the size of clearcuts, and maybe a better evaluation [could] be taken of these hardwood stands that are being cut.'[53] In some specific cases, he intervened personally. Discovering irregularities in seed provision for site renewal, he went directly to the management committee of the federal-provincial forestry agreement. In another case, where silvicultural contractors were inflating the reports of areas being treated, he 'blew the whistle, ... and they corrected that one pretty fast.'[54]

Eldridge took a different approach to the pulp companies that owned freehold lands. Here he exploited their sensitivity to publicity. He was inclined to both praise and condemn in this respect. Bowater Mersey's wildlife policy, for example, developed by Acadia University biologist Don Dodds, Eldridge praised publicly. Conversely, if a company was doing 'a bad job, I would tell them. So, I guess I know about the public relations part of it and the amount of money they spend. I can hit them where it hurts them rather than go to what I can do through the act.'[55]

Eldridge also hoped to use his personal experience and clout to influence his colleagues in the DLF. His annual reports would not be the only means of communication. He would also speak regularly to his successor in the DLF.[56] He also used his personal touch and acquaintance with the forester community to push his points. He knew all of the woodland managers of the major pulp companies on a first-name basis, and 'it doesn't hurt me to call them SOBs or whatever – if they are not doing the right thing, and I have never been known to back away from making comments to these fellows.'[57]

As indicated earlier, the major aspect of the poor cutting practices and hardwood conversions to softwoods that touched Eldridge was their impact on wildlife and wildlife habitat. As a hunter himself, he repeatedly stressed the importance of hunting and fishing to the provincial economy. Compared with forest revenue, Eldridge stated, 'The government takes in a million and a half dollars in licences ... [from] hunting and fishing, and we take in maybe less than a million ... from stumpage on Crown lands ... So, in other words, recreation and so forth could be developed much larger and in different areas, and they would bring in more revenue if ... the recreational areas are not all obliterated and clearcut.'[58]

This is the point where Eldridge found a way to reconcile his preference for the forest management pursued by the large pulp and paper companies and their negative impact on wildlife and wildlife habitats. The solution was for the provincial government to expand the Crown forest estate and to make it available for hunters and recreationalists. The point is made consistently in Eldridge's reports as Commissioner of Forest Enhancement, reproduced below, even to the extent of urging the Nova Scotia government to buy a massive block of forest lands from the Eddy Company.

As commissioner, Eldridge continued to comment on some of his old concerns even though they were not strictly part of his mandate. One prominent issue concerned the organizational efforts of small woodlot owners to achieve a strong position in the pulpwood market. In his final report, Eldridge commented,

> My recommendations to government are to walk very carefully, to consider the so-called work of the Primary Forest Products Marketing Board in developing what the board believes woodlot owners really want, and to identify the difference between the needs of the woodlot owners and those of a few opportunists. Don't bring in regulations or legislation which will infringe on the traditional rights of the individual woodlot owners. In my opinion, the government of Nova Scotia would be doing the province and the woodlot owners a big service if the money being spent currently to divide the forestry community was directed to finding alternative markets for the products of the forests.[59]

At the outset of his appointment, Eldridge was optimistic about his prospects of changing forest management in Nova Scotia. He believed that the Forest Enhancement Act offered some meaningful provisions for review and reporting. He clearly trusted that his long career in private and public life would serve him well in his new capacity. In the end, however, he was sadly disappointed and more modest about his own achievements. In his last report, penned in June 1990, he concluded:

> Turning next to my successor as Commissioner of Forest Enhancement, it is imperative this position be filled as soon as possible. Further, it is also imperative that if this position is to mean anything and carry on with what it was developed to do under provincial legislation, it must now get its fair share of funding. In 1986 enough funding was budgeted to get the office of the Commissioner started but it was soon discovered that part of the funding was diverted to expand other parts of the Department of Lands and Forests. Realizing that dollars were scarce, I fully expected that this funding would be returned in the following year but this again failed to materialize. My parting recommendation to you as Cabinet is to appoint a fully qualified person with a forestry background with the ability and leadership to do the job and the funds to carry through.[60]

The government responded in the opposite direction. Upon Eldridge's retirement, the Office of the Nova Scotia Commissioner of Forest Enhancement was quietly closed.[61]

Conclusion

Donald Eldridge's conception of forestry, though changing with his career, was consistently coloured by his position as an industrial forest advocate. When working for a large sawmilling company in the 1950s and 1960s, Eldridge bought mature timber lands and cut them. The objective was to make a profit, and this sometimes involved cutting to a diameter limit (also prescribed by legislation) to maintain a continual supply of sawlogs. More frequently, it meant high-grading (taking the best and leaving the rest). As executive director for a forest industry lobby group in the 1970s, he showed concern for the sawmilling sector, but he was primarily a defender of the industrial forestry agenda (with its clearcut methods) of the pulp and paper industry. This continued in his position as public servant in the late 1970s and 1980s, though with an interesting twist. As a government official, Eldridge was critical of some of the negative aspects of industrial forestry, the most important being the destruction of hardwood forests and wildlife habitats, thereby also damaging the provincial hunting economy (in which game licences brought in more revenue than tree stumpage). He addressed this quandary by personal appeals and consultations with industrial foresters, but most importantly he urged the provincial government to protect forests and wildlife by expanding the Crown's forest estate.

Eldridge's hard-nosed support of private enterprise and industrial forestry, and his distaste for government interventions and spending (though Eldridge himself was a public servant), were remarkably consistent throughout his career. As deputy minister of lands and forests, he worked against the grain in an era of increasing state intervention in all aspects of social and economic life. He was less inclined to challenge the dominant position of the pulp and paper companies in the woods market and the negative environmental aspects of industrial forest practices than to criticize the organizational efforts of small woodlot owners and their 'wasteful' and 'inappropriate use' of taxpayers' money. The environmentalist sentiments in the province he disposed of as irrelevant because they were expressed by people who did not have a personal economic interest in the province. Such feelings clashed with the wider objectives of public forestry (however confined it was in Nova Scotia) and its administrative and technical complexities. His personal reprimands and recommendations to woodland managers of pulp companies were confined to the ways in which harvesting practices impacted on wildlife or game habitats. These were important not only to Eldridge personally but also as a source of revenue from game licences.

At the end of his career, Eldridge was marginalized. His appointment as Commissioner of Forest Enhancement carried little weight, and his annual reports were filed with Cabinet without any further exposure. Excerpts from these reports are included here for the first time for public scrutiny. They show little change in his perspective. In them, Eldridge continued his defence of the pulp and paper companies, his distaste for government programs in support of forest management, and his attack on other forest interest constituencies.

Appendix

Excerpts from the *Annual Reports*, 1988-90, Commissioner of Forest Enhancement (Donald Eldridge)

I am very concerned that the forest community has not done enough to make the public aware of what could really happen to our forests if more monies are not put back into the forest resource. In administering the growing of trees, we must be held accountable for our actions, but we must not allow ourselves to be looking back over our shoulders at what we have done. We should be looking ahead to see what the trees on the ground are doing. Hopefully, that will be the small part that I will play in the scheme of things. I will keep my eyes on the trees to see that they are growing and being replaced by a better quality and more abundant forest that will enhance future benefits for Nova Scotia. Now and [in] the future, we must meet the needs of our forest industry, our hunters and fishermen, and all those others who use the forests. I will attempt to be a watchdog for the public.[62]

Clearcuts continue to be a problem with the public, and the Forestry/Wildlife Guidelines and Standards for Nova Scotia as published have done little to alleviate the public's perception of what is taking place in the forests.

Polls show the public are still disturbed by large clearcuts they observe. Very definitely an intensive communications program between the professional community and the public is necessary. The polls show the public don't believe politicians or pulp company executives, so it has to [be] the professional foresters who carry the ball to the public. This group should be front and centre.[63]

I am still not satisfied with some of the tactics being used by some pulp and paper companies and large producers when they buy large chunks of stumpage from private landholders. These lots are stripped clean. In a good number of cases the land is not replaced but is left to naturally regenerate. Natural regeneration is difficult on lots which have been completely stripped. It has been said before and I repeat it here that guidelines are not being adhered to in this regard, and there should be regulations under the Forest Enhancement Act to see that such lots are replanted by the 'land-stripper' harvesting company. A few of these lots are being replanted under the new federal/provincial agreement, but a good many are not.[64]

There is one major problem that has not been addressed. It is the follow-up on areas planted two to three years ago to see what has happened to planted stock. The number of acres planted has been put into the computers. Computer projections indicate that stock is growing at a good rate and tell us that in a specific year we will have a designated volume of wood. In actuality this is not true. Some areas have fifty per cent mortality rate, and one special case had an eighty per cent mortality rate. This is not acceptable. If special follow-up field surveys are not made and the required number of trees replanted to maintain a fully-stocked stand, then our computers will be providing incorrect information.[65]

There is still good sawlog wood fibre going into pulp. The methods of extraction and machines for extraction are probably the main causes of this problem because they are not geared for picking and choosing between sawlogs and pulpwood. They are geared for mass production of so many units per hour or day. This area has to be properly addressed before the proper allocation of wood fibre can take place.[66]

Large blocks of land are continuing to be cut (approximately 100-acre blocks) on which stumpage is purchased from farmers and private landowners in a block sum and then clearcut. There is no law to make anyone replant such an area. This is wrong. These companies and contractors should not be allowed to do this to our lands and our province. Once again in dealing with lands, it is not who owns the lands, but how they are used. It would appear that the Forestry/Wildlife Guidelines are not working in this particular section, and a regulation must be brought in before too much of our province is clearcut. Remember that trees produce sixty per cent of the world's oxygen. Each person needs 400 kilograms of oxygen per year to breathe. We are told it takes three hectares of land bearing trees to produce this much oxygen for each person. It becomes quite apparent we need trees to survive.[67]

There are proponents for and against the Crown's purchasing more forest land. There has been a substantial increase over the last few years in the posting of 'no hunting' signs on private lands in Nova Scotia. If the province is going to sell hunting licences, then it must be prepared to provide a place to hunt. This type of area is shrinking very quickly. Since the private sector owns seventy-five per cent of the timberlands in the province and a great deal of this is annually being withdrawn from public hunting, a definite conflict could develop within the next few years.

The possibility exists that one or two of the largest private landholders in this province may be putting their timberlands on the open market during the next year. Having personal knowledge of these forest lands, I would strongly recommend that the government of Nova Scotia make a concerted effort to purchase this top-grade forest land. It would provide a base for timber growth and for unrestricted, recreational areas open to the general public.[68]

As Commissioner, it is appreciated that the provincial government listened to the public comment and the 1988-89 report from this office with regard to the recommendation to purchase the forest lands of the George Eddy Company Limited. The Company had close to 122,000 acres of land for sale. Only a few people realized how close this land was to being sold to out-of-province interests. The province purchased this land with the support of the private sector and the elected opposition, adding a very large block of private land to Crown in one of the largest purchases made for a good number of years. These lands are located in twelve of Nova Scotia's eighteen counties. Generations of Nova Scotians to come will benefit from this purchase. The Nova Scotia government is to be commended for this action.

I am still recommending that the Department of Lands and Forests increase its budget for forest land purchase. As the hue and cry to protect our environment continues to grow it will be necessary to rebuy lands as they come up for sale. In my opinion, this is a positive method of obtaining lands rather than the more negative method of expropriation.[69]

It cannot be said too many times that since seventy-four per cent of Nova Scotia is owned by the private sector, one can expect a certain amount of these lands will be on the open market each year. If it is the government's policy to own one-third of the forest land in this province, then it must provide the money each year to successfully purchase some of the private lands that come up for sale. A budget of $1,000,000 should be the minimum amount set aside for such purchases.[70]

The eighteen Group Ventures have now come of age, and it is time to wean them off assistance programs. The landowners need to pay their own way. This should be done smoothly over a period of time during which they receive an increased price for their product. Then the Government should step back one pace at a time and give them autonomy. This would also result in approximately a half million dollars being freed for other forestry uses.[71]

I am concerned with the unrest in the forest community due to the Primary Forest Products Marketing Act. For approximately the next two years problems can be expected with the proposed organization of woodlot owners which may result in conflict, confrontational meetings, unrest between the Group Ventures, truckers, silviculture contractors, large sawmill owners, and the pulp and paper companies. The bottom line here is that the province could be the loser in this struggle, which could result in a lack of trust and confidence that would take years to heal. This is a sad note for me, as a person who has seen the coming together of the Forestry Community to practise better forestry in our beloved province, and the interaction of private industry and government through the consultative process.[72]

Reproduced courtesy of the Nova Scotia Executive Council.

6
David Dwyer: Social Forester in Nova Scotia

The noted forester Jack Westoby coined a memorable phrase when he titled his 1975 talk 'Making Trees Serve People.' From his vantage point at the United Nations Food and Agriculture Organization, Westoby articulated a growing concern that progress would be achieved when 'forestry has become everybody's business, not just the business of the forester alone.'[1] This has always presented a challenge to the profession. If today foresters are still uncertain how to respond to the social dimensions of their work, such issues were seldom if ever raised in the immediate postwar generation. For this reason, David Dwyer stands as a notable and revealing exception. His career in Nova Scotia crossed four decades, during which time he worked vigorously and persistently toward a wider and socially informed forestry practice. This was accomplished almost entirely in the service of the provincial Department of Lands and Forests, a setting that both supported and constrained his efforts.

Dwyer was led in several fascinating directions in his career: exploring the economics of private woodlot forestry; developing new tools for small private forest management; supporting efforts toward more accountable public forest practice regimes; and bringing new community forestry initiatives to life. Since these efforts frequently ran against the grain of conventional practice, Dwyer's experiences further highlight the particularities of Nova Scotia forestry. They also demonstrate that alternative perspectives were both available and competing for attention within government forestry circles. Here another key feature of Dwyer's career in government forestry becomes evident. As a staff forester in a bureaucratic structure, Dwyer was free to exploit the creative potential of his mandates in various jobs. But he was also constrained by the overarching political and administrative framework of forest policy in Nova Scotia. While it was often possible to nudge this framework in desired directions, the ultimate powers of agenda setting, budget allocation, and dispute settlement lay beyond his direct control. In such cases, the failures are often as revealing as the successes.

This chapter shows that Dwyer's early years in Nova Scotia forestry were closely associated with Lloyd Hawboldt. As a young graduate of the Maritime Ranger School, Dwyer was first employed with the Division of Forest Biology, where he was influenced by some of the most innovative field foresters of the era. In addition to Hawboldt himself, there was Simon Kostjukovits, an émigré forester whose contributions were noted in a previous chapter. It was here that Dwyer first witnessed the techniques and results of intensive field research in unravelling the birch dieback problem and in devising new and improved cruising techniques. He then joined Bélanger and Bourget, the Quebec-based consultants who directed the historic Nova Scotia forest inventory of the 1950s that was so important to modernizing provincial policy. Consequently, Dwyer had already enjoyed a diverse and rich apprenticeship under prominent mentors by the time he returned to school for his forestry degree. Not surprisingly, his experience there was different from that of the teenagers who arrived directly from high school. His classroom instruction could be weighed against years of field experience. Summer seasonal jobs with Mersey Paper and the DLF were added to an already substantial résumé.

By the time that Dwyer returned to the Department of Lands and Forests as a graduate forester, all of his formative influences had germinated for more than a decade. The early years of field research opened an interest in science, history, and forest ecology that returned in his choice of senior paper topic. Unusual for the day, Dwyer tackled the topic of hurricane blowdown rather than the conventional logging report that most students undertook.

While Dwyer absorbed much from his early experiences with Hawboldt, Kostjukovits, and Belanger, he also began to push his own forest interests in new directions. He developed an early appreciation of the 'social context' missing from much of postwar forestry. It was the difference between approaching the forest principally in terms of site, species, and merchantable volume and approaching it in terms of social relationships of ownership, access, use, and return. All Nova Scotia foresters knew that private woodlots accounted for the greatest proportion of forest cover. Not all, however, accepted this as a defining reality and a central variable to any effective forest program. Fewer yet saw it as a positive feature that could enhance Nova Scotia's forest practice (as opposed to a barrier and a burden). Over the course of some forty years, Dwyer was drawn increasingly into these issues. While doing so sometimes put him out of step with the professional (not to say bureaucratic) mainstream, it put him in the vanguard of much new forestry thinking that has since made rapid strides elsewhere.

Perhaps it was this independent bent that led Dwyer to a series of absorbing positions over the years. He practised field forestry as a District Forester, worked with private landowners as an Extension Forester, gravitated into policy work as an executive officer with the Nova Scotia Forest Practices Improvement Board, and nurtured the provincial forest group ventures program as an innovative experiment in cooperative management. Following retirement from the DLF, Dwyer sampled international forestry for the first time as a consultant for projects in Latin America, southern Africa, and Indonesia.

Beginnings

It was during his early high school years that David Dwyer first considered a career in forestry. As a city boy, he first encountered the woods through the Halifax Boy Scouts Camps and Lorne MacPherson, the warden of the Waverley Game Sanctuary. These encounters were reinforced by time spent on a family friend's farm and forest land at Blomidon. As it turned out, one of the scout leaders was David Dyer, a professional forester with the Department of Lands and Forests (DLF). Several summer jobs with the department followed for Dwyer, and by the time he completed grade eleven he had decided on a woods career. This was a time when the profession was relatively young in Nova Scotia. Although the forestry school at the University of New Brunswick (UNB) had turned out graduates since 1910, the number of practising foresters increased slowly in the province. In 1946 there were only five foresters in Nova Scotia's DLF and perhaps twelve in the entire province. Tracts of Crown forest were tended by forest rangers, often sawmillers or outdoorsmen whose appointments were determined on partisan grounds by the government of the day. This situation began to change in 1946 when the Maritime Forest Ranger School was established in Fredericton; the first trained crop of rangers, or forest technicians as they came to be called, graduated a year later.

Dwyer began to attend the ranger school as a nineteen year old in 1948. He had set aside the possibility of UNB since returning war veterans enjoyed an admissions preference and there was a high demand for places. The ranger school also had its standards. To qualify for the program, applicants had to be male, twenty-one years old, and have a grade ten level of high school along with one year of woods work experience. Most candidates had worked earlier in government or industry and spent two months in the program and two months at work on a rotating basis. These forest technicians got a good practical education. They learned how to build roads, fire towers, and telephone lines and how to take forest inventories (cruising timber), perform surveys, and even operate sawmills. The emphasis was on operational matters.

At the same time as the ranger course was becoming more professional, a growing number of university graduate foresters were being hired in the postwar years, by both industry and government. These Dwyer saw as the planners and work organizers (though still practical to some degree) in the forestry field. They were often new arrivals who took up positions senior to those of the rangers. Perhaps inevitably, this bred certain resentments. At the same time, both streams shared something in common. Many of the foresters and forest technicians were Second World War veterans. They were tightly knit, well known to each other, and often friends socially. Most importantly, they filled the expanding ranks of both the forest industry and the civil service. This generation became a dominant force in the maritime profession up to the mid-1970s.

Throughout his time at the ranger school, Dwyer continued to work with the Nova Scotia DLF, and he returned on a full-time basis in 1951. Lloyd Hawboldt, the Provincial Entomologist, was building the Division of Forest Biology within the department, and it was to this group that Dwyer was attached. The team included Ken Greenidge (physiologist), Doug Redmond (pathologist), George MacGillivray (entomologist), Dwyer, and Hawboldt. Dwyer was the all-purpose assistant to the team. As described in an earlier chapter, one of its first assignments involved the birch dieback problem. It was jeopardizing one of the most commercially valuable hardwood species in eastern Canada, as yellow birch was the prime source for veneer plywood. During the war, it had been used to build thousands of light-weight Mosquito bombers, and K.C. Irving had built a thriving business in his Canada Veneers. Consequently, the postwar birch dieback was a significant concern. Despite the insights and the scientific reports that followed the root-washing work at Lake O'Laws, 1950 was the last year for the research division.

Dwyer spent the next period assisting Hawboldt and Kostjukovits in developing new measurement techniques for forest inventories. Once again the applied research yielded important results. By the end of the field plot studies, the group had compiled a set of normal yield tables used in the upcoming Nova Scotia forest inventory.[2] It was also during this time that Dwyer first encountered the Antrim woodlot, a 173-acre property in western Halifax County that had reverted to the Crown. Kostjukovits and Dwyer did the first cruise of Antrim in 1951. It generated the baseline data that (together with subsequent cruises) allowed Dwyer to produce his influential study of woodlot economics more than twenty years later.[3] As for the site-quality, normal yield tables, it was only later in Fredericton that Dwyer fully appreciated the path-breaking character of this inventory work. Their techniques were not yet being taught at the UNB Faculty of Forestry during his degree years (1954-9). For Dwyer, such insights were a measure of the talents of the Petrograd-trained Kostjukovits, to whom Dwyer dedicated his study on the Antrim woodlot.

Early in 1954, the DLF sent Dwyer to Quebec City. There he spent four months working for Bélanger and Bourget, the forestry consultant firm hired to conduct the Nova Scotia forest inventory. The photo mapping of

Cape Breton had been completed during the summer of 1953. Dwyer's job was to assist in interpreting the photos and transferring the forest types to large base maps that displayed major geographical features. This later led to an article in the *Forestry Chronicle* on the use of the pocket stereoscope in photographic interpretation.[4] That summer, Dwyer served as the DLF liaison to the Bélanger field crew, which conducted a straight-line cruise in Cape Breton, the first of three regions to be inventoried. Diameter-height-age data were taken, and the normal yield tables were applied.

During the next two years, the balance of the province was covered similarly. The inventory process marked a milestone in the evolution of the DLF, providing as it did the first authoritative modern survey (and the first comprehensive field survey since B.E. Fernow's effort of 1908). Although it took years to digest fully the results, the way was now open to design a forest policy and a management regime in accord with professional standards. This included the question of matching harvest and renewal levels, for which the inventory yielded some difficult conclusions. It was clear that the harvest rates ran well ahead of sustainable levels, according to standard allowable cut calculations. The official survey report put the problem bluntly: 'One immediate corrective measure for the present overcutting is a reduction in production. It is clear that any such reduction must be applied primarily to the sawmill industry,' while the softwood pulp sector was held constant. Furthermore, 'if recuperative measures are to be considered, restrictive cutting must be instituted.'[5] These grave findings would shape Dwyer's working environment for decades to come, but only after Dwyer secured his professional standing as a forester.

Professional Forester

It was while working in Quebec that Dwyer decided to enrol for the degree in forestry, entering the UNB program in the fall of 1954. He earned one year's (five courses) credit from his previous training as a forest technician. His entry year coincided with the massive damage of Hurricane Edna, which hit the Atlantic provinces on 11 September. In Nova Scotia, it left the equivalent of two years' volume of sawlog harvest lying on the ground. It also left a major impression on Dwyer, who later combined his interests in history and science by writing his senior research paper on hurricane damage to Nova Scotia forests.[6] Here again Lloyd Hawboldt played a role. Dwyer had been interested in forest blowdown as a historical phenomenon since his earlier work on growth and yield. One of the quarter-acre plots that they studied was in the Tobeatic country in western Nova Scotia. The team noticed several large mounds in the woods, some with trees growing out of them, and matching pits beside the mounds. They reasoned that these were blowdown mounds created when the trees and their root structures were upended by windstorms. In some cases, the living trees growing out of the mounds could be dated, indicating the approximate time of the blowdown.

It was this problem that Dwyer sought to revive for his fifth-year research project. Normally, the project entailed a logging report on a wood-harvesting operation. Dwyer put the case for a silvicultural topic to the dean of the faculty, Miles Gibson. With the approval of the logging and silvicultural professors, the historical study of hurricane blowdown was under way. Just as most logging reports were based on summer work experience, Dwyer put together his data while cruising for the Mersey Paper Company in the summers of 1957 and 1958. Mersey forester Ralph Johnson provided some blowdown data from the company records, and Dwyer spent many a long evening working over these records in the bush. At one stage, Nova Scotia writer Thomas Raddall allowed Dwyer to examine the Simeon Perkins diary, enabling him to pinpoint the timing of one late-eighteenth-century storm. A summary of Dwyer's findings appeared in the quarterly newspaper *Forest Times* in November 1979.[7]

During the intervening summers, Dwyer worked first for the DLF on the forest inventory (1955 and 1956) and later for the Mersey Paper Company, which had just been purchased by the English Bowater Corporation (1957 and 1958). The latter summers in Liverpool were to change his life in another way. It was there that he met his future wife, Bea MacDonald, the daughter of Mersey sales manager J.H.S. MacDonald. There he earned the considerable salary of $300 a month, though the standard wage for a student forester was $150 a month. Dwyer speculates that his new family relations, and the prospects of his joining the company permanently, might have been special considerations. His income may also have been based on his ranger diploma. Although the prospects for a permanent post seemed to be strong, neither of the Dwyers wished to return to Liverpool and 'sell their souls' to the Mersey, as it was locally known.

The forestry degree program was quite a contrast to that of ranger school. At UNB, the students were exposed to much basic scientific theory, in a variety of specialties, but there was very little practical application. This may help to explain why summer jobs were so influential in shaping job-placement prospects after graduation. The first year focused on courses in sciences and mathematics, with forestry courses beginning in year two and dominating those that followed. Dwyer was given forestry credit for several of his ranger courses, so he was allowed to sample some non-forestry options not normally available within the five-year BScF degree. In some respects, his past employment must have given him a level of experience beyond his years. The work with the Hawboldt research group provided field experience and mentoring of a high calibre. Involvement with the Kostjukovits site-quality, normal yield research put him well ahead of the traditional cruise and inventory practices being taught in Fredericton.

Moreover, his summer experience on the Nova Scotia inventory, with Belanger and Bourget, placed him on the cutting edge of field survey work. In academic terms, he may have been to graduate school before he settled into his first degree.

Looking back on the curriculum, Dwyer was struck by the lack of socioeconomic training for future foresters. He recalled a basic course in economics but no specialized option for forest economics. Then (as now) sociology and political science had no formal place in the program. Instead, the curriculum was still heavily oriented toward 'forest engineering' subjects such as mensuration, logging plans, road building, and wood products. While the silviculture professor, Doug Long, was highly regarded, the single course in silviculture could never match in influence the many courses related to logging work. This reflected the expectation that graduates would move into timber-harvesting positions, either with the woods departments of forest companies or with the field services of government departments. Dwyer's interests already extended beyond the technical core of forestry knowledge, essential as it was. The university environment, in which he could indulge a wider constellation of interests, he found constantly stimulating.

Following graduation in 1959, Dwyer rejoined the Department of Lands and Forests. The deputy minister, Wilfrid Creighton, indicated that two positions were available: one as an Extension Forester with Lloyd Hawboldt, and one as District Forester based in Musquodoboit Harbour. Dwyer chose the latter, and his responsibilities covered the eastern half of Halifax County, beginning at Dartmouth and ending at Sheet Harbour. The eastern boundary was extended to the Guysborough County line in 1961 (see Map 2). Excluded from this was the Liscomb Game Sanctuary, which was handled from New Glasgow. At the time, he was one of a dozen District Foresters located around the province. Ultimately, the Department of Lands and Forests planned to assign one forester to each of the eighteen counties.

At that time, much of the District Forester's work involved the administration of Crown lands, with which eastern Halifax County was well endowed. There were essentially two ways to administer the Crown forest resource, and both were controlled by the District Forester. The first was through the issue of timber licences. This began with a cruise of selected areas and advertisement of the standing stumpage. Tenders were received and awarded, the timber was cut, and it was scaled if time permitted. Large jobs were a greater priority, while smaller jobs (ranging from 10,000 to 20,000 board feet) could be accepted on the word of the operator provided that field checks were made. Here the stumpage price was often kept quite low, with ten dollars per 1,000 FBM not uncommon, as the local operators worked on shoestring margins. The second method involved contract logging, in which a crew was hired to cut designated Crown limits. The logs were then tendered at roadside. Here it was possible to make a bit more money given greater certainty about the product. There was also a third, lesser licensing procedure that affected hardwood fuelwood. Here, however, the volumes were extremely modest, with individual licences limited to eight, ten, or twenty cords per year at most. Any amounts above that required a timber licence.

However, this only begins to capture the duties of the District Forester. In addition to being responsible for Crown timber sales, he was responsible for the traditional DLF duties of fire detection and suppression and game patrolling. In certain traditional circles, firefighting was still considered the acid test of forestry skill, and the worst that could be said of a man was that he had let a fire 'get away' from him. The department was also assuming new functions, as reflected by two new pieces of legislation. The Provincial Parks Act authorized the minister to acquire land and develop recreational amenities.[8] For District Foresters such as Dwyer, this involved road development work as the parks network expanded rapidly in the 1960s.

In a related development, local pressure led to the establishment of the Martinique Bird Sanctuary near Musquodoboit Harbour as a protected space for ducks and geese to winter. To this must be added enforcement of the Beaches Protection Act.[9] The Crown jurisdiction ran seaward from the mean high-water mark and included adjacent lands such as sand dunes. The act was a response to the mounting damage from contractors removing vast volumes of sand. Many prime beaches were degraded by this practice, and in extreme cases, such as Cow Bay, the Silver Sands beach disappeared altogether. There was also considerable land titles work connected to ungranted Crown land. For many reasons, the boundaries of rural Nova Scotia properties are notoriously inaccurate. In some cases, residents found that their houses were not located as intended on deeded (i.e., private) land but on Crown land. However, on a report from the District Forester, the Crown lands division could grant a lot to a homeowner.

Dwyer's district consisted of two distinct segments, spatially and physically. One ran along the Musquodoboit Valley, a heavily forested interior belt supporting numerous farms and sawmills. The other consisted of the Atlantic coastal strip, with more sparse forest cover, running east from Dartmouth. This took in the small fishing villages as well as the town of Sheet Harbour, the site of Canada's first sulphite pulp mill in 1885 and several subsequent mills. Overall, Halifax District represented the provincial forest industry in microcosm. At the hinge of these two distinct regions lay Musquodoboit Harbour, where the district office was located in an old garage facility. To handle this considerable workload, Dwyer could draw upon the assistance of half a dozen forest technicians. Three of them worked in the valley, with two handling Crown timber matters and one responsible for fire. Over on the Eastern Shore, two additional technicians handled similar work, while a sixth arrived with the Sheet Harbour extension in 1961. With radio connections and relatively short distances separating the various sites, staff could be deployed flexibly in handling various tasks. Dwyer remembers these years as a time when District Foresters enjoyed considerable autonomy. Policy advice could be obtained from the deputy minister's office in Halifax when needed, but local decisions were largely made locally.

No DLF district was entirely self-contained, because concerns inevitably slipped across the boundaries. One of these concerns was pulpwood pricing, which led to an inquiry headed by R.J. MacSween in 1963-4. A group of forty or fifty woodlot owners in the Musquodoboit Valley called a meeting to discuss a presentation and invited Dwyer to assist them. They wanted to demonstrate that under prevailing market conditions private pulpwood sellers faced a commercially impossible choice: to accept no return in stumpage for their trees, or no return for their labour in cutting the trees. Together with neighbouring District Foresters Ralph Hale and Ron Day, Dwyer prepared a questionnaire for the woodlot owners to circulate among their membership. Among other items, it sought data on the age of trees being cut. Dwyer recalls a comment by the district agricultural representative to the effect that 'everyone in the valley is out counting rings on stumps.' In the end, the woodlot owners' brief (submitted under the name of the Musquodoboit Valley local of the NS Federation of Agriculture) figured in the findings by Commissioner MacSween, who confirmed the contention involving stumpage/labour shortfalls.[10] While these District Foresters were eager to assist their woodlot activists, they were equally happy to avoid identification on public policy controversies such as this.

In Halifax County, the District Forester's role was transformed dramatically at mid-decade. Negotiations took place between the Scott Paper Company and the province, which would commit virtually all available Crown timber lands to a massive pulpwood licence for fifty years.[11] Since Halifax County held the second largest volume of Crown forest in Nova Scotia (trailing only Guysborough County), this held major ramifications. As Dwyer heard it from Deputy Minister Wilfrid Creighton, he would lose virtually all of the Crown land in Halifax County to the Scott agreement. The entire face of Crown forestry would change. This was also a pivotal moment for the provincial forest industry as a whole, as it confirmed the ascendance of the pulp and paper segment at the expense of the hitherto dominant sawmillers.[12]

Faced with this prospect, Dwyer began actively considering alternative positions. There seemed to be a possibility of joining the DLF Division of Extension, an appealing option since he was also keenly interested in woodlot forestry. Indeed, Dwyer recalls Lloyd Hawboldt observing many times that he was already doing too much extension work! Actually, he did work closely with the district ag rep, Peter Stewart, since for many purposes they shared the same clientele. This led Dwyer into another career opportunity, after he started dabbling in radio broadcasting. Stewart was responsible for a daily five-minute radio spot on station CHNS in Dartmouth. This he used for agricultural news capsules, and he suggested that Dwyer air forest topics on alternate weeks. As it happened, some CBC Halifax producers were aware of the show, and, when the much larger farm show at CBC Radio needed a new host, Dwyer was approached late in 1964. In the end, he took a one-year leave of absence. He joined CBC Radio to work as a commentator on the daily farm broadcast, which aired from 12:30 to 1:00 p.m.

This was both challenging and exhilarating, and it took Dwyer into new territory in covering farm issues. He was largely responsible for filling the thirty-minute program with news, weather, business announcements, and interviews. The tasks were many and varied. Each Thursday he prepared a Maritime regional 'clip' for the CBC cross-Canada survey broadcast at week's end. Friday was thankfully an easier day, since the survey feature could be relied upon to take up a share of the show. In those days, the farm show had its own dramatic serial, revolving around a fictional farm family known as the Gillins. Each day a five-to-seven minute segment chronicled the life and times of this Maritime family on their land. The writer, based in Yarmouth, would interview the three correspondents for material before drafting a script. Then each month, the cast of half a dozen actors assembled in Halifax to tape a series of new instalments. In 1965 the Gillins discovered that their woodlot offered any number of fascinating features, and Dwyer was able to inject a considerable amount of forestry content into this popular segment of the show. He also learned to be flexible when inserting the Gillins into the lineup, as doing so afforded a five-minute cushion when needed to grapple with late-breaking stories.

As the year drew to a close, Dwyer had to weigh his options. Ultimately, he chose forestry over broadcasting and returned to the Department of Lands and Forests at the close of his leave. His radio successor was Sandy Cameron, a former ag rep and CBC reporter who later went on to serve as lands and forests minister and later still Nova Scotia Liberal Party and opposition leader. These 'radio days' had brought a number of insights, however. One involved the challenge of conveying forest issues to the public. Dwyer was often struck by the difficulty of successfully interviewing foresters. Many of them didn't seem to know what to say. This relates in turn to a wider observation: despite an often deep professional camaraderie, 'foresters as a rule don't like to deal with the public.' (Extension foresters, and extension forestry, are obviously exceptions here.) For many foresters, Dwyer suggests, the lure of the profession may be the relative isolation of work in the woods.

On the other hand, the medium of radio and television seemed to offer tremendous potential to reach a public audience and communicate important messages about resource management. Looking back on this experience,

Dwyer wondered why the Extension Foresters did not take greater advantage of radio in extending their reach. His personal commitment to radio and television remained strong, however, carrying over to the 'Land and Sea' television broadcast reproduced in transcript form below.

Extension Forester

When Dwyer returned to Musquodoboit Harbour the following year, it was as an Extension Forester in Hawboldt's division of the same name. There was a long tradition of extension activities in the postwar Department of Lands and Forests, with which Dwyer fit comfortably. During the early 1950s, the foresters created a travelling display known as the 'circus.' This exhibit, which could tour the province during the summer fairs, was built at the Shubenacadie Forest Depot by Charlie Mackenzie. Under a plywood and canvas roof, they showed movies and presented displays featuring live animals, fire pumps, and an old-fashioned portable sawmill, among others. A variation of this is still done at the Provincial Forestry Exhibition. In this way, the department could present its concerns about firefighting, logging, and wildlife to a wide cross-section of Nova Scotians. In addition, the department published a series of printed 'bulletins' on professional, technical, and commercial topics, many of which were aimed at woodlot producers. These covered forest crops ranging from Christmas trees to blueberries. They tended to take a self-help approach to raising the value of rural lands by making information available to those able to take the initiative.

The decade of the 1960s was perhaps the heyday of extension work. By this time, there was also strong support at the ministerial level. With a background in agriculture, Conservative Minister of Lands and Forests Ed Haliburton (1959-67) appreciated the value of extension forestry.[13] Specialist staff were recruited, and by the late 1960s six provincial Extension Foresters (with five assistants) were at work throughout the province. David Dwyer (Halifax-Hants), Ralph Hale (Pictou-Antigonish-Guysborough), Ron Day (Cumberland-Colchester), Tom Ernst (South Shore), and Roy Wright (Annapolis Valley) all worked in the field, while Gary Saunders developed educational material in Truro. Curiously, there was no Extension Forester based in Cape Breton. (The shared joke was that Hale could handle that area in his spare time.) Dwyer considered Haliburton to be a strong booster of their work. On one occasion in 1960, he sent the whole division to Quebec for a CIF forestry conference. In addition, the extension specialists from all over eastern North America began to convene annual woodlot extension seminars in 1966, and Dwyer represented the DLF every year until 1975.

He recalls with satisfaction that the Extension Foresters were given a free hand to develop new ideas. One of the most innovative concepts became known as the Whiz Wheel. It was a simple tool designed to make timber cruising accessible to ordinary woodlot owners. The prototype, designed by

Ron Day (with Kostjukovits and Hawboldt), was dated in 1967. It was utilized, among other settings, in the 1971 (third) cruise of the Antrim woodlot. The theoretical grounding of the Whiz Wheel went back to the site-quality, normal yield compilations of the early 1950s. Drawing upon just three essential variables (height, age, and stand density), the Whiz Wheel produced volume estimates of sawlogs (FBM) and pulpwood (cords per acre). From the outset, the Extension Foresters recognized its immense significance as a practical tool for woodlot owners. The Whiz Wheel, pictured in Figure 2, worked in the following way. The user took three 'readings' from the forest stand in question: one for height (in feet), one for age (in years), and one for density (percentage of canopy cover). Taken together, height and age could be correlated to establish a fourth variable, site class rating, on a scale ranging from II to V. (There are no site I locations in Nova Scotia.)

Once established, the site class dictated the particular yield table on the wheel, from which volumes were derived. The user then worked with the height and site-quality numbers, rotating the wheel to bring these two figures into focus in its perimeter windows. The result highlighted a series of volumetric measures, now correlated with the density coefficient. It was possible to read off sawlog volumes in thousands of FBM per acre and roundwood in cords per acre. In this manner, the Whiz Wheel provided basic volume data. The next step was annual yield calculations, integral to sustained management.

The fate of the Whiz Wheel is highly significant. Dwyer recalls that the District Foresters were never very impressed, though the wider extension forestry community had the opposite reaction. When Dwyer took it to the annual meeting of North American woodlot specialists, it proved to be very popular. In fact, some were interested in promoting it, but 'we [at DLF] had the copyright.' In the end, this was perhaps unfortunate given the ultimate fate of the wheel. Today Dwyer may possess the only surviving prototype from 1967, and the instrument goes entirely unmentioned in official government reports. The wheel was never promoted further by the Division of Extension in the years before its disbandment in 1971.

Always a supporter of the wheel, Dwyer attempted to revive it more than a decade later under the auspices of the Provincial Forest Practices Improvement Board. Following its mandate to promote better management, particularly in the private forest sector, the board sponsored the production of a manual to outline technically sound forestry practices in plain language. The result was *The Trees around Us*, a 200-page illustrated guide published in 1980.[14] Dwyer recalls that by design '*The Trees* aimed to tell people how to do for themselves.' As an executive officer to the board, Dwyer hoped to revive the Whiz Wheel and have it included in this book. In fact, Chapter 4 describes the very operations that the wheel was designed to support. While

Figure 2

Whiz Wheel

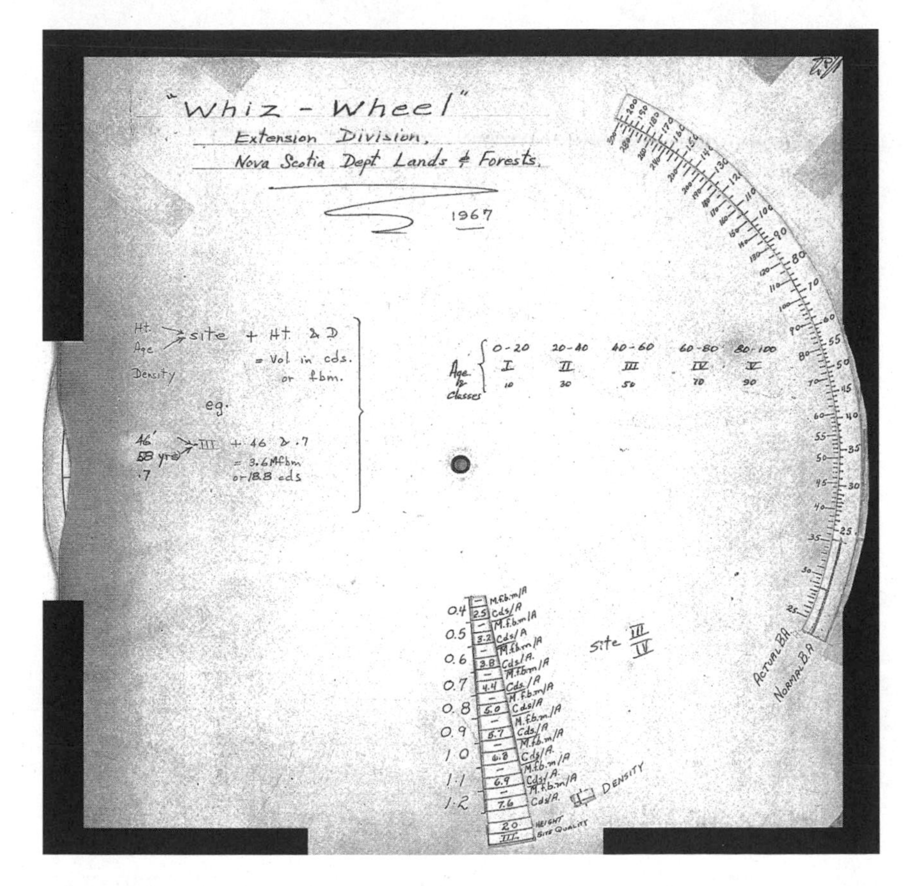

its absence may have been related to the broader editorial problems with which this volume was beset (discussed further below), it is also clear that the political 'moment' for promoting woodlot self-management in the DLF had faded badly by 1980, and with it had faded the prospects for self-help tools.

That the wheel never found its way into wide use is a telling commentary on the limits of extension forestry of this era. It seems aptly to reflect the context in which George Bernard Shaw described professions as 'conspiracies against the laity.' It is striking that, within less than a decade of the wheel's appearance, the promotion of practical self-management skills had been swept aside. In their place, successive federal-provincial forestry programs conferred a virtual monopoly for 'management planning' on accredited forester professionals. Along the way, the DLF Division of Extension,

David Dwyer at work with the Whiz Wheel.

which constituted a vanguard of outreach activities, lost its autonomy and vigour. This is aptly expressed in the fate of the Whiz Wheel. While its technical foundation lay in the site-quality, normal yield research conducted and reported by Hawboldt and Kostjukovits in the 1950s, the popular application of the wheel was never promoted in the 1960s. Yet by the 1970s the basic research had been assimilated into the professional toolkit of technical forestry in Nova Scotia. Here it was further revised and extended by DLF personnel to include formulae for basal area and crown closure.[15] Ultimately, the department published 'revised' normal yield tables for softwoods,[16] work that helped to earn the principal author an honorary degree from the UNB Faculty of Forestry in the 1980s.

In its prime, the Division of Extension program's strength lay in its diversity. Certainly, the Extension Foresters didn't always think alike. Tom Ernst's specialty was Christmas tree growing, which had been a DLF concern since 1953, when Hawboldt wrote *Christmas Trees Are a Crop*.[17] The original hope was that the commercial returns from successful tree growing would encourage a broader interest in woodlot management. The challenge was to persuade growers to cultivate high-quality trees that could be graded and command higher prices in the US market. Ernst argued for tougher standards and more intensive cultivation to meet the increasing competition in the American market.

Dwyer was more interested in small private woodlot forestry. In addition to dealing with private landowners, he was the driving force behind the development of the Antrim woodlot as a DLF demonstration project. This property consisted of woodlots that had reverted to the Crown before the Second World War. As a result of the Kostjukovits normal yield studies begun in 1951, an initial inventory of the woodlot was on record. When Dwyer moved to Musquodoboit Harbour in 1959, he rediscovered Antrim, which lay within his district. The aim was to produce red spruce sawlogs, possibly cutting some each year, to demonstrate a high-value return from a small woodlot. In 1960 a road was begun, and logs were cut under contract. In this way, production costs and sales data were also recorded over the ensuing years. Working with Kostjukovits and Day, Dwyer repeated the inventory after the 1960 growing season. A decade later, in the autumn of 1970, this was done again by Extension Foresters Dwyer, Day, and Hale, accompanied by their technicians. The result was a very impressive database, a rare experiment in small woodlot management and economics in Nova Scotia.

It fell to Dwyer to tease the analysis out of the figures, resulting in his *Twenty Years of Forestry on the Antrim Woodlot, 1951-1970.*[18] Dated 1974, it was in fact released in August of the following year. This was the largest publication sponsored by the department to date. It was dedicated to Kostjukovits in recognition of both his theoretical and his field contributions to understanding Antrim, and Hawboldt contributed the Foreword, in which he reaffirmed the value of demonstration woodlot projects. Early in 1973, Dwyer was invited to the University of New Brunswick to present his findings to the forestry professors and forest ranger instructors. He expressed some surprise, after his slide presentation was followed by a screening of the 1971 CBC Television production on Antrim, that there was so little discussion from his old professors.

The release of the Antrim study was not without controversy, and some DLF officials harboured reservations. For one thing, the conclusions challenged a long-standing assumption of official Nova Scotia policy: that owning and managing a private woodlot was a sound commercial investment. This was an article of faith in many quarters of the forest sector. Dwyer recalls Deputy Minister of Lands and Forests Wilfrid Creighton advising him on joining the department in 1959 to look for a woodlot as a source of supplementary income. Yet ten years later Dwyer knew, from the study of existing records, that Antrim was not making any money. He also wondered about the most appropriate forms of records to indicate full costs of woodlot economics from a private investment perspective. At this point, he encountered W.E. Hiley's study of the Dartington Estate, a small English woodland managed for profit since 1926.[19] It was Hiley who offered valuable pointers on record keeping and profit-loss, which found their way into the many appendices to the Antrim book.

As far as woodlot economics were concerned, Dwyer's conclusions were sobering. Put simply, the data showed that there was no money in woodlot management at prevailing prices. With an Antrim return of thirty-seven cents per acre per year, it was evident that woodlot forestry did not pay sufficiently to be viable under the conditions of the day. The implication was either that small woodlot supply would decline as lots were either mined or left untouched, or that the cost-price relation had to change in favour of a higher stumpage return. This analysis was also broadcast in 1971 in a half-hour CBC Television interview that Dwyer did for the Maritime series 'Land and Sea.' A concise and informative presentation, it is reproduced below in transcript form.[20]

Although the print run was substantial and distribution was extensive, the Antrim study was never intended to be a household reference in the private woodlot sector. Its dense technical calculations prevented that. Nevertheless, it had greater significance in forest policy circles, where the implications of the findings were well recognized. The marginal economics of the Antrim operation offered implicit support to the Woodlot Owners Association, which spent the 1970s battling industrial buyers and government agencies for improved pulpwood returns. Neither was the point lost on the federal officials negotiating the terms of a new intensive forest management regime with their Nova Scotia counterparts. One senior Ottawa figure, Bob Boutilier, was the lead federal negotiator on the Forestry Sub-Agreement signed with Nova Scotia in 1977, opening a new era in intensive silvicultural management. Boutilier disclosed to Dwyer that Ottawa had several reasons for pressing for a substantial private lands component to the agreement: small woodlots were highly accessible sites and often superior sites for growth; they held available wood; and the Antrim study had demonstrated the lack of return under prevailing conditions. Such evidence from Antrim helped to convince Ottawa that new fiscal incentives for silviculture were imperative.

Already, however, Dwyer's career was taking another turn. The early 1970s saw the abrupt decline of the Division of Extension program. This began with the DLF reorganization announced in 1971, which saw Extension disappear as an independent unit. The architect of the new design, Lloyd Hawboldt, was already wearing a second hat as policy thinker for the new deputy minister, Bob Burgess (appointed in 1969). In 1970 Hawboldt was appointed assistant to the deputy minister. He convened a meeting of his five Extension Foresters at the White Spot Motel in New Minas. There he set out the charts illustrating the draft reorganization, the first major departmental review since 1958. The DLF was regrouped into three branches: one for administration, one for planning and services, and one for operations (including the newly regionalized field staff). It was obvious that extension was finished as a discipline. The foresters argued strongly that this was a bad move, but the director replied that, if the department did not restructure itself, the job would be done by someone else.

There was a sense that Cabinet was unhappy with the department. The deputy minister was considered a 'Crown land man,' and Crown forests were now largely committed to pulp and paper lessees. In this context, the Extension Foresters were certainly off in left field. Furthermore, it could be argued that the woodlot initiative had largely passed from departmental hands to the new Woodlot Owners Association and the St. Francis Xavier extension program under Rick Lord (profiled elsewhere in this book). By the terms of the reorganization, the extension specialists were folded under the District Managers of Forest Resources. Divisional status was gone, and extension forestry slipped into a relative obscurity from which it never recovered. The staff also dispersed gradually in many directions. Hale became a District Manager, while Ernst became a Christmas tree specialist. Dwyer stayed on in extension work until 1975, when he joined the Provincial Forest Practices Improvement Board. Day moved to the Division of Parks, and Wright remained in extension in the valley.

Dwyer views this reorganization and the ensuing events as an important signal of shifting priorities. On the one hand, it was a function of the increased power of the pulp and paper companies. Extension forestry no longer commanded priority. Contact with the small private woodlot owners was de-emphasized at the very time when woodlot owners were organizing their own association and mobilizing politically. Not only did it seem that 'education for the masses' was passé, but also the policy priority had shifted from private woodlots to Crown-owned forests under lease to companies. As mentioned earlier, Dwyer believes that as a group foresters are not particularly suited to dealing with people. Part of the lure of the profession is the isolation offered by time in the woods. Extension foresters were an exception to this rule, and their decline served further to isolate the department from the public.

It also signalled a new pattern of administrative control. Under Deputy Minister Wilfrid Creighton in the 1950s and 1960s, the district people had been given responsibility to make decisions, provided that they kept Halifax informed. In the 1970s, the DLF 'began to shift from being a department to being a bureaucracy,' with more centralized decision making and top-down control. Field staff had less responsibility, and the Halifax headquarters, installed in more elaborate downtown Halifax office blocks, assumed a far more direct role in operational activities. Finally, and despite this stiffening internal hierarchy, it was becoming increasingly evident that the DLF stood low in the provincial bureaucratic pecking order. Rather than a professional management service for one of Nova Scotia's most valuable resources, Lands and Forests figured as a tool for the Stanfield strategy to attract large foreign forest businesses and investment to Nova Scotia. The political energies behind the Nova Scotia Pulp Limited lease (signed in 1959) and the Scott Maritimes lease (signed in 1965) came not from DLF but from the Department of Industry and not from officials so much as Cabinet ministers.

It was also at this stage that the Nova Scotia Forest Products Association (NSFPA) was transformed into a lobby vehicle for the pulp and paper industry. For thirty years after its founding in 1934, the NSFPA was a group of small- and medium-sized sawmillers. As an association, it was notable for promoting a grading system for lumber products and higher returns in offshore markets. It was an informal group whose energies ebbed and flowed over the years, with various provincial lumbermen taking turns as president. The executive volunteered its time over and above work commitments to provide a voice for the lumber industry in dealing with the provincial government. In the 1950s, its members were nicknamed the 'basement bums,' in reference to the annual dinners convened in the basement of the DLF's Shubenacadie Depot. However, many of the smaller sawmills were disappearing as prime logging stock became more scarce and Crown lands were increasingly tied up by pulp companies. Although the NSFPA was still led by sawmillers in the late 1960s, it also included the major pulp and paper companies, and its Pulp and Paper Committee became an increasingly active centre. The NSFPA also administered an influential forest road-building grant program funded by the provincial government. Since membership in the association was a prerequisite for applying for grants, the membership rolls expanded rapidly to include many woodlot owners. For the first time, it was able to open a full-time office and appoint a full-time executive director, Don Eldridge (profiled elsewhere in this book). While significant tensions split the sawmill and pulp segments in the late 1960s, the DLF was fully committed to the pulp and paper economy built by the Stanfield government.

This opened a period that Dwyer describes as 'the wood wars.' It revolved around a battle for ownership and control of the wood fibre supply. The government of Nova Scotia had already fired the first salvo. In the eastern and central regions, virtually all available productive Crown land had been committed to the Stora and Scott Maritimes mills, and a major lease had been agreed on with Bowater Mersey before the company changed its mind and declined.[21] On private lands, the Nova Scotia Woodlot Owners Association was established in 1969, and it was already campaigning for organized pulpwood marketing along agricultural lines. (This is explored extensively in the next chapter.) Faced with intractable opposition from the pulp industry and the DLF, the battle took more than a decade to resolve.

Moreover, during this time, the share of total provincial wood supply from small producers dropped by half, to only 35 percent. This in turn aggravated the rivalry between pulp and lumber producers. Sawmillers were growing increasingly desperate for logs, and many small- and medium-sized mills were shutting down. Then in 1974 concern began to grow about the impact of spruce budworm infestations on the softwood stands of Cape Breton. In fact, there were few 'uncommitted' tracts of significant Crown land in central or eastern Nova Scotia. (Dwyer profiled one of these, the Stanley lands, in a 1975 article.)[22] Finally, the DLF entered the wood wars with its proposal to purchase 1,000,000 acres of private forest lands. This was justified as a solution to the impending wood shortage. It would also fulfil a long-standing ambition by Crown authorities to control 50 percent of Nova Scotia's forest. By 1975 the minister, Maurice DeLory, was publicly advancing this scheme. Such factors dominated the politics and policies of the DLF for years ahead. They also affected the next stage of Dwyer's career.

Provincial Forest Practices Improvement Board

In 1975 David Dwyer became involved with the Provincial Forest Practices Improvement Board (PFPIB) as its executive officer. The board was rooted in the Forest Improvement Act (FIA), passed ten years earlier as part of the first major policy reform since the war. As seen earlier, the 1958 inventory revealed not only that the forest was being overcut relative to sustained yield but also that there was a serious imbalance between sawlog and pulpwood harvests. The Small Tree Conservation Act of 1942 was clearly insufficient, and its replacement was inspired by Swedish mandatory forest legislation. The FIA paved the way for new harvesting regulations, a manual of best forest practices, and formalized policy advisory channels to the DLF.[23] At the core of the new system were the district-level Forest Improvement Boards. By 1970, nine had been established (some with Extension Foresters as their secretaries). Each board embraced a broad range of forest interest groups, with authority to formulate forest management practices to be followed within the district. However, their work was severely handicapped by the government's unwillingness to proclaim the act (in response to a strong lobby from industrial interests). The act was amended in 1968 to establish the province-wide PFPIB, but proclamation was again delayed until 1976, when sections 9-12 took effect. Not surprisingly, the FIA remained irrelevant during the decade-long hiatus. Lacking statutory power, the district boards lapsed, and even the provincial board found it difficult to sustain interest.

In Halifax one day in 1975, Dwyer learned that the reactivated board was advertising a full-time executive officer position. Encouraged by several board members, he applied for the position. Familiar with the original act, but still unclear about the extent of the renewed commitment, he recalls questioning the members closely about their plans. Offered the job, he accepted it and began work in the spring of 1975.

By the time that Dwyer joined the provincial board, it was at the centre of a growing political dispute between its supporters and its opponents. Back in the 1950s, the Nova Scotia Section of the Canadian Institute of Forestry (CIF-NSS) had strongly supported a form of FIA-type legislation. At that time, the older generation of foresters provided a high standard of

David Dwyer explains the use of an increment borer to Hugh Fairn, chairman of the Nova Scotia Forest Practices Improvement Board.

input and exhibited a high morale within the institute. CIF meetings could generate valued policy input, especially 'when the foresters were not being watch-dogged by their employers.' In Dwyer's experience, this disinterested professional commitment did not survive the era of the wood wars. The newly employed generation of foresters seemed to him to be much more politically sensitive, with a corresponding decline in CIF-NSS input. As he once put it, 'You might say that our age class distribution [in the NSS] was rather poor after 1970.'

At the time, however, Dwyer was very active in the section, and he speaks highly of its work. Foresters agreed that drastic action was needed to introduce enhanced forest management in the province, and they continued to press for the implementation of the act in a major forest policy brief in 1971.[24] However, the policy climate was by then evolving rapidly, with major new developments connected to the DLF reorganization, pulpwood-marketing arrangements, the spruce budworm outbreak, and new federal-provincial forest management programs. Under these conditions, the viability of the act was increasingly challenged, and professional opinion also began to diverge. Foresters in industry often articulated their employers' positions on policy measures, and the CIF-NSS consensus dissolved. For example, Dwyer observed that the pulp and paper representative on the PFPIB, a corporate forester, often changed his positions and abandoned agreements after further consultation with his industry colleagues.

Despite this more contested climate, several key board members remained committed to the full implementation of the act and sought to revive the provincial board. In 1975 the board convened field meetings to further acquaint members with forestry practices. Later in the year, a delegation travelled to Sweden to examine the work of the Swedish counterpart boards. Subsequently, a delegation travelled to the southern United States, where it encountered a booming forest industry based largely on privately grown pine plantations. Back at home, pressure was put on the minister to proclaim the outstanding provisions of the act. Before this could happen, at least one key technical issue required clarification. Critics of the act pointed to section 9, which prohibited the felling of healthy immature spruce, pine, hemlock, and yellow birch. In 1976 the board convened a meeting to clarify this issue. Debate turned on whether chronological age was an adequate criterion of maturity or whether stand conditions played an equal role. In the former case, spruce could be protected to a designated age, say forty years. In the latter case, a heavily stocked sixty-year-old stand could still be considered immature, since its growth potential had yet to be released by thinning. The board settled the question with alternative criteria based either on age (less than sixty years) or average diameter (under seven inches at breast height.)[25]

Even then, the proclamation of the act remained in doubt. It was poised between one set of firm supporters (including woodlot owners and environmental groups) and an equally determined set of opponents (including the pulp industry and DLF officials). After such a long stalemate, it was pressure by the federal government that prevented its complete abandonment. Ottawa insisted that Nova Scotia have new forest management legislation in place before the federal-provincial forestry agreement (under negotiation between 1974 and 1977) could be signed. While the department had long since cooled on the act, the agreement was a prize of great value since it opened the way to intensive silvicultural management on a wide scale. Thus, it was a hesitant Nova Scotia government that proclaimed the Forest Improvement Act in 1976.[26]

Full implementation of the Forest Improvement Act, however, was another matter. Despite its linkage to the federal-provincial agreement, the act was subject to increased industry criticism. Bob Murray, with the Woodlands Department of Scott Maritimes, worked against the act and opposed the modest beginnings of the Group Venture program. The Nova Scotia CIF reversed its earlier endorsement of the act as well. It was around this time that Dwyer resigned his lifelong membership in the CIF, declaring that the foresters had been programmed to follow an industrial agenda rather than a path beneficial to the provincial good. Ironically, Dwyer's former colleague Doug Redmond was the CIF's acting executive director at the time of his resignation.

Meanwhile, the Provincial Forest Practices Improvement Board was facing its own problems at this time. The board had received two bids for the preparation of its forest manual, with considerable distance between them ($40,000 versus $70,000). The contract went to the lower bidder, but the ensuing draft was judged unsatisfactory. Feelings on the board were mixed. The chairman, Hugh Fairn, worried that the manual was being sabotaged. Others, including Dwyer, thought that the weaknesses of the manual necessitated a wholesale revision. Its errors and inadequacies were rectified by an additional expenditure of $20,000 to $30,000 for a rewrite. Members of the editorial committee, including Lloyd Hawboldt (then in retirement), Ian Miller (of the Canadian Forestry Service), lumberman Murray Prest, and Hugh Fairn worked on the manual over an entire winter. (Dwyer was also involved to some extent.) The result was *The Trees around Us*, a major 206-page publication with a print run exceeding 20,000 copies. Published in hardcover, it was widely distributed around the province. In the end, Dwyer believes, the manual turned out to be an excellent publication, technically sound and written in simple language for ordinary people. However, the energies of board members were being worn down. A combination of the production problems with the manual, the continuing opposition of the NSFPA and other industrial interests, and the evident lack of support from the provincial government almost triggered the board's resignation in 1977-8. However, Fairn persisted, and he won the trust and support of the members.

Dwyer, meanwhile, was having second thoughts about the political viability of the Forest Improvement Act. It was clear that, for woodlot forestry to become viable, one of two things had to happen. Either stumpage rates would have to rise, or the government would have to provide direct financial assistance for the small woodlot owners. Given the strong opposition of both the industry and the DLF to the former, Dwyer saw the latter as the more likely alternative. It was also clear that the DLF was not willing to tolerate the board as a more powerful authority over forestry on the ground. He finally concluded that enforcement of the FIA would be socially unacceptable, in light of the independent nature of Nova Scotia landowners and their resentment toward government interference. Although its provisions were still sound, the FIA seemed like an idea whose time had passed, and it was hard not to see the publication of *The Trees around Us* as a last hurrah. Dwyer also began to realize, from friends in government, that it was time to rejoin the DLF for career considerations.

The Group Venture Program

The federal-provincial forestry agreements provided David Dwyer with the opportunity to shift positions and to assist small woodlot owners in a different way. In 1973 Ottawa had announced a new strategy for economic growth through joint general development agreements (GDAs) with the provinces. Under the authority of an umbrella GDA, subsidiary agreements could be negotiated on a sectoral basis, with forestry as a leading contender. Delegations of officials from Ottawa and Halifax spent several years developing the framework for the 1977 deal.

There was a tremendous amount at stake in these negotiations, since the future shape of provincial forest management was being cast. Contrary to the DLF's massive land purchase plan, the federal government was committed to a private lands program in Nova Scotia. Chief federal negotiator Bob Boutilier argued the case. The small woodlot owners possessed the best lands for forest growth. Their woodlots were the most accessible, and they had the best growing sites. Dwyer likes to claim at least some credit for the monies made available to small private woodlot owners through the agreement. The Ottawa people quoted his Antrim study to point out the futility of woodlot forestry in current markets. This analysis helped to convince Boutilier that monies had to be made available for the small woodlot owners in the province, as well as to Crown lands and large corporate landowners. The question was how.

One option involved a private land management program for individual landowners. It was consistent with the cultural assumption of fiercely independent rural producers, and it continued the approach of the Agricultural and Rural Development Act (ARDA) years.[27] Another option, designed by DLF policy chief John Smith with input from Dwyer and others, envisaged small owners banding together in groups to achieve greater efficiencies in both silvicultural treatment and harvesting operations. The concept had several precedents. Both John Smith and David Dwyer had accompanied the PFPIB to Sweden in 1975 and had viewed firsthand the vigorous woodlot organizations that produced and marketed forest products while practising advanced silviculture. Dwyer was also familiar with the cooperative forest ventures in Quebec. While each option had its supporters within the DLF, the private lands program was structured to include both individual management and group venture options. However, the rivalry persisted in budgeting, ideology, and departmental politics.

Overall, these agreements have pumped hundreds of millions of dollars into forest management in Nova Scotia. For Dwyer, they represent the second major watershed (after the pulp licences) in the modern history of the DLF. To forestry officials, the sudden but massive infusion of federal money after 1977 was like 'manna from heaven.' With Ottawa paying sixty cents of every dollar, the department increased its staffing twofold, bringing a new generation of foresters and technicians into the service. Yet the sudden expansion also caught the department by surprise. It was a time of intense flux, when basic choices were being made as priorities were set, and programs were fleshed out and delivered on the run. Virtually all forest interests fought, with varying degrees of success, for a share of the new riches. Pulp companies looked for public financing for intensive silviculture, woodlot owners looked for management programs, the business of silviculture consulting and contracting boomed, and the group ventures came into being.

In the initial years, it sometimes proved difficult to spend the money. The proper construction of forest roads, for example, could take three years: one to grub out the road, one to shape it, and one to gravel it. Only 30 percent of the first year's budget was spent on schedule. Dwyer thought that the threat of future wood scarcities was fabricated, or at least exaggerated, at the time to justify the money coming in and to ensure that the money would continue coming in.

In 1978 the board was winding down, and Dwyer was assigned to the Group Venture program. In some respects, it was a natural continuation of his interests in extension work and woodlot management. Particularly in light of the continued opposition of both pulp and paper industry and government to higher stumpage prices, the group ventures offered a means to help organize small producers for effective forest management.

The concept called for the delivery of a publicly funded woodlot forestry program based on associated groups of small landowners as members and shareholders. It began as a pilot program in Pictou and Cumberland Counties, where the West Pictou and North Nova group ventures were established by 1976. They provided the prototype for group organization and the core staff model of forester and secretary. Dwyer's job was to provide organizing and operating advice to the new groups as they sprang up around the province. It was soon evident that the potential was great. The 1977-82 agreement called for eight groups to be established by its close. However, the first eight were operational by March 1979, and more were ready to go with the provision of additional funds.

Within the DLF, the support of John Smith and chief administrator Gary Rix was critical to the survival of the venture program, whose early years were always tenuous. It turned out that other senior people in the DLF, such as Deputy Minister Donald Eldridge (1978-86) and Private Lands Director Slim Johnson (1978-85), opposed the venture program in principle. Such opposition was not unexpected. The two men shared an industrial forest background. Eldridge had served as woodlands manager of the E.B. Eddy Company for more than twenty years, prior to his decade with the Forest Products Association. Johnson had spent most of his career working for a pulp and paper company in northern New Brunswick, before moving to Newfoundland as an implementation officer for the first forestry subsidiary agreement there.

Given this background, Johnson was a curious choice to direct a private lands program. Two DLF officials more experienced in this work were assigned beneath him. Arden Whidden took over the individual management side, while Dwyer took on group management. Many DLF staff were puzzled and even resentful that an industrial forester was put in charge of a private woodlot program. Neither, as it happened, did the director win the confidence of the small woodlot owners. His hostility to the groups was well recognized in the field, and Dwyer often felt embarrassed when Johnson addressed woodlot groups.

The pulp and paper companies were also against the group ventures. By this time, they were ideologically opposed to anything that furthered the organization of small woodlot owners, so again such opposition was not unexpected. Scott Maritimes was particularly hostile. One sawmiller in the Musquodoboit district decided to publicly oppose the Conform group venture in that locality. Like many central Nova Scotia lumbermen, he depended on Scott as a key supplier of sawlogs, taken as part of Scott's clearcuts and then sold to area mills as part of an integrated logging program. Dwyer has no doubt that the man went public in the face of threats to his future log supply. But as time passed, some of the companies came to accept the idea, particularly if wood was in short supply. In Mersey country, Arcade Comeau (who managed La Forêt Acadienne group venture) and Victor LeBlanc (Mersey's assistant woodlands manager for the Mersey-Sissiboo District) were partners in a Christmas tree venture. They convinced Mersey that the group ventures could be advantageous. On this matter, Jack Dunlop of Mersey was one of the first to come on stream. However, Dwyer recalls Dunlop declaring that group ventures were only acceptable under three conditions: a new deputy minister was needed at DLF, Stora and Scott should not be allowed to cut pulpwood in the western part of the province, and there should be no interference with or talk about the price of pulpwood. Subsequently, some group ventures (although not strictly allowed under the terms of the program) negotiated delivery quotas with pulp and paper companies.

Opposition also came from the silviculture contractors, who believed that the group ventures (with their own staff) would squeeze them out of woodlot management contracts. The same fears were expressed by consulting foresters, who were in the business of drawing up individual forest management plans funded by the federal-provincial forestry agreements. Some sawmillers were opposed; they may have been pushed by the pulp companies on which they depended for sawlogs. The protesters voiced their concerns at public meetings and by lobbying Halifax on an individual basis. In Dwyer's view, most of the fears proved illusory. There was enough work and funding to go around for everybody.

The most difficult battles were those within the department. Dwyer recalls one meeting particularly vividly, since he feared at the time that it might cost him his job. Some time in 1982, with eight groups already functioning and another eight applications well advanced, a meeting was convened in the office of the Conservative Minister of Lands and Forests George Henley. A number of woodlot owners backing a new group proposal were in attendance, as were the minister, the director of forest operations, the private lands director, and Dwyer. Drawing on official briefing materials, Henley explained to the woodlot men that the only obstacle to approving more applications was the exhaustion of funds under the agreement. Convinced otherwise, and frustrated by what he perceived as a bureaucratic prejudice against the group program, Dwyer differed strongly with his seniors. He interjected to explain that he had confirmed the availability of sufficient uncommitted funds to launch two more groups and possibly three. Although Henley took the matter under advisement and promised further investigation, Dwyer wondered whether he had publicly spoken too freely. He was reassured several days later when Henley stated that in this case he considered Dwyer to have met his full responsibilities as a public servant. However, the funding battles went on.

Stability for the Group Venture program was eventually confirmed through the political support of a new minister, Ken Streatch. After the 1984 election, money was located for six more groups, bringing the system to fourteen in total (with four more added by 1986). Here Dwyer entered his 'political period,' drawing on local connections to lobby on his program's behalf. Many years earlier, he had inspected the Streatch family woodlands in the Musquodoboit Valley and given advice to the minister's father. A day before Streatch was due to rule on the ventures program, Dwyer sat down with him at home in his kitchen. Dwyer asked, 'How many times have you heard a Nova Scotia woodlot owner say, "Damn it, I would rather have my wood rot than turn it over to the pulp and paper companies." Quite a few times! Well, how do you get at the wood?' Dwyer suggested that the ventures program would be one way. Give a group of woodlot owners the opportunity to manage their woodlots effectively, with government support, and they would do so. Tom McInnis was another Cabinet minister whom Dwyer enlisted to support the ventures program. McInnis had been involved as a lawyer for the Conform group, and he was at the time working for a group venture to be set up in his own constituency on the Eastern Shore.

The groups proved to be very popular, and they grew in number far faster than anticipated. There were, however, growing pains. One of the early group ventures in Pictou County, for example, spent over half of its budget on operating costs ($70,000) and only $60,000 to $70,000 on the ground. Some ventures had to cope with other problems in the management of operations, since administrative skills were often at a premium and the small staffs functioned on shoestring budgets. Some struggled because they faced a poor forest base or had difficulties finding markets. Others, however, were extremely successful. Conform, which had no difficulty in finding members, offered one viable model. It concentrated on building up a capital and machinery stock, and its shares have been split three times. The machinery is financed through bank loans, which are paid off by the federal monies that Conform charges for services rendered to its members. An alternative model can be found in La Forêt Acadienne, which preferred to see its assets on the stump. It employed forest workers or contractors with their own machinery and avoided heavy financial overhead.

These were issues that Dwyer had to deal with on a day-to-day basis in his work with the group ventures. Some would argue that the ventures were a cause and calling to him rather than a job responsibility. He does not deny this. Without a determined campaign, the program may well have foundered in the face of its powerful opponents. Dwyer believed that he operated in a hostile environment within the department, with two powerful senior figures unsympathetic at best and actively obstructionist at worst. Administrative and financial roadblocks seemed to be needlessly thrown in the way of a program that was proving to be extremely popular.

For Dwyer, these years combined feelings of exhilaration and frustration. Grassroots involvement with the groups represented the cutting edge of extension forestry in the 1980s, and the level of community response was most rewarding as thousands of shareholders/members joined the program. Yet the bureaucratic battle seemed to be relentless and dispiriting. Not only did the latter run against the formal policy espoused by government ministers, but it also had the effect of undermining and neutralizing the potential for program success. During late 1984 and 1985, Dwyer exchanged a series of blunt correspondences with the private lands director and the deputy minister, detailing his growing frustrations.[28] By the close of 1985, he was seriously considering the early retirement option being offered by the provincial government. Early in February 1986, Dwyer wrote to the Minister to announce and explain his intention to resign. He reviewed the problems of 'contradictory direction' and administrative interference. Over the past year, he wrote, 'I have been confronted time and again, when meeting with Group Boards of Directors and Staff, with letters and reports of phone calls which they had received that were either intimidating, insulting, or downright abusive. In all cases, these communications were allegedly intended to support the Department's policy with respect to the Groups and in all cases they were interpreted as efforts to undermine the Groups' activities.'[29]

In effect, Dwyer's resignation pointed to the failure to resolve longstanding political differences within the DLF pertaining to the merits and priorities attached to the fast-growing Group Venture program. Nevertheless, Dwyer had managed to nurture an impressive system of woodlot management embracing thousands of members. When he left the department in May 1986, the groups were well established, with their own province-wide association soon to come. It was a fulfilling conclusion to a provincial career dedicated to extension and woodlot forestry, during which the DLF had been his steady home for almost forty years. Perhaps it was time to branch out.

Overseas Consulting

As his departure date approached, Dwyer attended a two-day workshop in Halifax that dealt with 'how to retire.' One of the speakers represented the Canadian Executive Service Overseas (CESO), which specialized in matching retired professionals as volunteer consultants with project assignments in the developing world.[30] In 1988, Dwyer applied to CESO. The following year, he was contacted in relation to a private land forestry program in Swaziland, and this led to a four-month (May-August) field assignment in 1989.

Dwyer's project involved the Emafini Estate Forest, a private 100-hectare eucalyptus and pine holding of the Swazi Kingdom. Since 1977, it had been under the supervision of an appointed manager. A timber sale and road-building contract signed in 1984 had collapsed, and Dwyer was to serve as a consultant to the manager in implementing a successful management plan, tender process, and contract agreement for logging and silvicultural treatments. After aerial photography and a ground cruise, an inventory was prepared and a work plan outlined. Next the tender was advertised, and by July all bids were received. The successful bid was formalized in an agreement, by which the contractor would establish a portable sawmill on the estate. In the process, the estate manager and two of his staff were trained in cutting, scaling, and milling procedures. By late August, the first load of lumber had departed the estate.

From a consulting project point of view, all essential objectives were achieved; as Dwyer observed in his CESO report, though, 'This project can be considered only a start to developing a private land forestry program that could be undertaken in Swaziland.'[31] This project was his first exposure to fast-growing hardwood species in a dry climate. In one respect, his experience suggested that the fundamentals of woodlot forestry are applicable to a variety of forest ecologies.

Just over a year later, Dwyer was approached again, this time with a project assignment in the province of East Kalimantan, Indonesia. Here his client was the International Timber Corporation Indonesia (ITCI), which sought expert advice on the management of a planting and nursery operation in the Mahakan River district. Dwyer spent March and April 1991 at ITCI's field camp at Kenangan and at ITCI's headquarters, where his final report was presented to the company's board of directors. During the field study, he examined the forest management program then in place, as well as a series of selection logging and plantation (clearcut and planted) sites. It was, as he reported, far from the stereotypical rainforest destruction often portrayed in the West. ITCI worked consistently within the national forest management policy, which Dwyer considered 'appropriate and advanced.' ITCI distinguished two types of forest. The first was the 'production forest,' in which volumes exceeded twenty cubic metres per hectare. The second was the 'conversion forest,' in which volumes fell below that level. In the former, selection cutting of large-diameter trees was limited to six per hectare, while in the latter clearcut and plantation methods were used. One-quarter of the total ITCI holding was in the 'production' category, while the balance was in 'conversion.' Native tree nurseries provided seedlings for enrichment planting when logging reduced the number of commercial species below twenty per hectare. Meanwhile, non-native tree nurseries provided eucalyptus and acacia for replanting the conversion stands.

Dwyer was also impressed by the overall approach of local management, particularly in the ecological sensitivity of local officials. ITCI forester Suwardi Suwasa adhered to a philosophy of 'the three rights': the right species, on the right site, using the right technique.[32] At the same time, it was evident from the inventory and harvesting data that ITCI was in a serious overcut situation relative to the annual allowable cut. Dwyer pointed out, however, that a similar situation of overcut applied in Nova Scotia in 1975, and this problem was met by intensified silvicultural treatments similar to those proposed by the Indonesians.

Conclusion

Like any field forester, David Dwyer faced the challenge of making creative use of opportunities with which he was presented. Unlike many field foresters, he proved to be highly adept at reflective learning: absorbing knowledge, synthesizing distinct streams, and applying them to original problems. He had the good fortune to associate with a series of innovative free thinkers whose contributions he was able to disseminate. As a forester, he also knew his own mind. He did not hesitate to cut across the grain when circumstances allowed. Over the course of his career, Dwyer sampled a variety of jobs and handled a plethora of projects. While he did not always initiate these projects, or control their outcomes, his contributions made a difference and left their mark.

Dwyer's career was spent entirely in the public service, and it touched five decades of the Department of Lands and Forests. Dwyer arrived when the DLF was in the early stages of professionalization, hiring its first major cohort of field foresters (as part of the postwar bulge from UNB) and grafting them onto an evolving field structure. Dwyer became one of them, spending the second half of the 1950s at forestry school achieving the crucial credential. He also experienced the results of the first wave of postwar forestry policy – including Ottawa's underwriting of inventory and research in a province ill equipped to finance and deliver programs to desired standards.

His first field assignments as a forester coincided with the transformation of the Nova Scotia forest economy from lumbering to pulp and paper. Some of his most important duties as District Forester involved the management of Crown timber permits and log sales to the lumbermen of the Musquodoboit Valley and Eastern Shore. Yet by the end of 1965, virtually all of 'his' Crown limits had been committed by the Nova Scotia government in the Scott Maritimes licence. His colleagues in the central and eastern parts of the province experienced a similar shift. It involved more than a simple change of clientele from sawmills to pulp mills. District staff shifted from the role of Crown forest lessors to monitors of company cutting plans under long-term leases. Much of the discretion of 'Crown management' was eliminated in the process.

It was at this point that Dwyer started searching out alternative routes: exploring his growing interest in mass media and later returning to the DLF in an explicit Division of Extension role. Here the clientele were the Nova Scotia private woodlot owners, those tens of thousands of small holders who collectively held title to half of the provincial forest. This offered far greater challenge and fascination than the command and control levers of Crown forest administration. It was here that Dwyer found his ultimate vocation: working first to understand the dynamics of ownership and management and then to encourage it in positive directions within the framework of provincial policy.

In forestry or elsewhere, policy regimes seldom change sharply or cleanly, and Dwyer's experience with the Provincial Forest Practices Improvement Board offers an interesting glimpse of a 1965 initiative whose political trials and tribulations postponed the possibilities of serious implementation for so long that the entire enterprise came under challenge and ultimate decline. First, the political balance in the industry had changed decisively following the industrial pulp leases of the Stanfield period. Second, the 'new' harvesting and silvicultural regime was overtaken by the cathartic struggles over spruce budworm spraying. Third, a new era of forest management was about to dawn with the 1977 federal-provincial agreement. Sensing this, Dwyer accepted the opportunity to move into the Group Venture program, the innovative new program for small woodlot cooperation that was made possible by the new agreement.

Indeed, its impact on the department, and on the shape of the forest industry, cannot be overestimated. The agreement defined new horizons for program spending, hiring, and policy making. In effect, a new policy regime was embedded in the FRDA agreement and the detailed program manuals that governed its administrative delivery.

Dwyer spent the last eight years of his DLF career promoting the group ventures within the new management setting. Here he encountered the ongoing political tensions among bureaucrats and between organized interests. Initially, the group program had few friends within either the department or the industry. To many DLF foresters, the groups might have been seen as the creation of a rival delivery mechanism external to the department. This was hardly unique, of course, since large sums of silvicultural improvement money were distributed indirectly by the terms of the Stora and Scott Crown leases. However, the novelty of the Group Venture program (and the familiarity with the pulp licences) may have obscured this fact. Elsewhere the group program faced intense political opposition from the traditional enemies of woodlot forestry, which extended into senior levels of the Private Lands Directorate and the department as a whole. Overall, the advancement of social forestry faced complex challenges in an era when the Forest Products Association leadership occupied the DLF executive suite.

On some occasions, Dwyer's contribution was decisive, leading to decisions, discoveries, or new initiatives that would not have happened otherwise. Dwyer revived the Antrim property, overseeing its treatment and writing up the results in a way that placed woodlot forestry in a new light, triggering important policy debates in the process. He helped to restore momentum to the PFPIB during a troubled period. His untiring advocacy of the group program advanced it more quickly than would otherwise have been the case in such an adversarial environment and extended the system to the point that it could survive two decades of agreements and the termination of state support in 1995.

Given the diversity of his engagements, the parameters of Dwyer's career were predominantly 'operational.' His was the dilemma of the intermediate official, more able to channel influence 'downward' to the grassroots than upward to elite circles. Like most professional foresters, Dwyer never made the shift to a senior managerial position. Thus, he was not involved in the controversial decisions over forest spraying, pulp licensing, or federal-provincial funding.

Appendix

'On the Antrim Woodlot' (Transcript, CBC Television, *Land and Sea*, October 1971, Peter Brock [Producer], with David Dwyer)

DD: The Antrim Woodlot is designed for quite a number of reasons. Our main objective is to produce red spruce sawlogs. Now our objective was to find out [whether we] could operate this lot over "x" number of years, over some period of years, at a profit. And could we also demonstrate to woodlot owners what we think is a good way of doing this, of cutting methods, in other words, silviculture, of building roads, in other words, the whole package deal as far as a small woodlot forestry program is concerned?

PB: And you've been keeping careful control of costing every operation?

DD: Since 1961. Although fortunately we have all the records back to 1951. We want to produce a forest in here that we can come in every year, or every ten years, and take off so many board feet, or so many cords or so many cubic feet of wood. And if we just keep going and going and going, and it never runs out. (See Table 3.)

PB: Dave, how do you determine what your annual cut is going to be, how much do you take out each year?

DD: Well actually there are two methods of determining this. The rule of thumb method would be we determine the overall volume here on the lot and we determine the average age of all the stands. And let's assume that we have about a million board feet in here and that the average age is around seventy years. We just divide seventy into the million board feet, and we come up with an average figure of so many board feet per acre per year.

PB: And you get the age [of a tree] by taking a boring like this?

DD: We take the age by getting a boring.

PB: So how often would you come through here and take a boring sample like this?

DD: We like to take a cruise every ten years, for instance in 1951, 1961, and 1971 we've taken an inventory of the lot. Back in 1951 we determined that the average age was so many years and there was so many board feet, so we tried to cut during that period of 1951-60 at the allowable cut figure that we calculated.

DD: We're coming up now to an area that we clearcut in 1963. We were low on areas of young stands. We had no young stands in this lot. So we wanted to do a bit of clearcutting so that we would get some young stands which would produce a crop of wood in 50, 60 or 70 years or so. Well we cut this area out in 1963, and by 1965 or 1966 or so we had good regeneration coming, but we also had a very serious problem because the hardwood stumps were starting to sprout. And these stumps are now producing ten and fifteen young red maple sprouts which you can see here are ten, fifteen, twenty feet tall now, this hardwood, young ... maple and so on. This is hindering the softwood regeneration, and we would like to get rid of it, but if we cut it down with power saws it will just sprout up again. So what do we do with it now?

DD: In this area here, this hasn't been thinned. There's been a bit of cutting in here ten, fifteen, probably forty years ago, little bits of stuff taken

Table 3

Inventories: Antrim woodlot (173.4 acres)

Year	Average age	Volumes		Average annual increment cu. ft.	Average annual increment* cu. ft./ac/yr	Allowable annual cut FBM
		cu. ft.	FBM			
1951	70	416,620	1,169,900	5,952	34.4	207,000
1961	77	586,200	1,780,000	7,613	44.8	280,000
1971	68	400,658	1,049,100	5,892	35.7	170,000

* *Forest acreage lost to road after 1960: 1951 (173 acres); 1961 (170 acres); 1971 (165 acres).*

out here and there, but generally speaking we refer to this particular stand as an uncut stand. Now we've got to come in here and start doing some cutting in here if we want to increase the growth rate on these trees.

PB: This stand over here, where we're coming to, you thinned this out in preparation for further cutting, future cutting.

DD: Exactly, and wherever we've done stand improvement today, it has cost us money today. And whether we'll make any profits on it in the future or not, well one wonders, in order to get an increase in volume. Now the stand was very thick, crowded, perhaps 2,000-3,000 stems per acre.

We can only get an increase in growth if we bring this down to possibly 800 or 1,000 trees per acre. Now there's another factor involved here too. We're also establishing a crop on the ground, getting regeneration in. Before this stand was thinned, there was no regeneration whatsoever. The little tiny seedlings that were coming along would die very quickly. No regeneration would come, there's not enough sunlight in the stand.

DD: This particular stand here was cut in 1966. We came in and we did a partial cut, and took out about twenty percent of the volume. On this particular six acres we cut about 15,000 feet off it plus about twenty-two cords fuelwood. I don't think we'll cut it for another four to five years yet, but we've got good regeneration started. It's getting really going now. We'll get another cut in here in four or five years.

We'll be in here again in about another fifteen years. So we've got three or four more cuts in here before the big trees that you see here are gone. By that time some of the regeneration will be bigger. So it's quite possible that we could keep cutting in here for a long time yet.

DD: This is what I like to see ... and this is what we aim for. I like to see stumps, trees still on the lot, and regeneration coming. The regeneration is replacing ... the stumps, and there are still some good-sized trees in here, to come in and cut in five, six, ten years or so.

DD: Peter, this particular chart that I've drawn up very roughly indicates where we're going, where we've been trying to go, as far as getting these stands on the Antrim Woodlot adjusted, so that we can continue cutting ad infinitum. The vertical axis is acreage, and the horizontal is the age classes, in twenty-year periods. Age Class I is zero to twenty years, Age Class II is twenty to forty, forty to sixty and so on to Age Class VI, which is over one hundred years. (See Figure 3.)

Our objective is to have thirty-two acres in each age class. Age Class I, thirty-two acres, Age Class II, III, IV, and V. Now we've got some stuff that's in Age Class VI which is over one hundred years old, and we'll be cutting that right away. Now in many cases, of course, we are

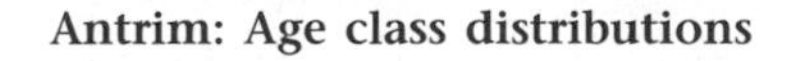

Figure 3

Antrim: Age class distributions

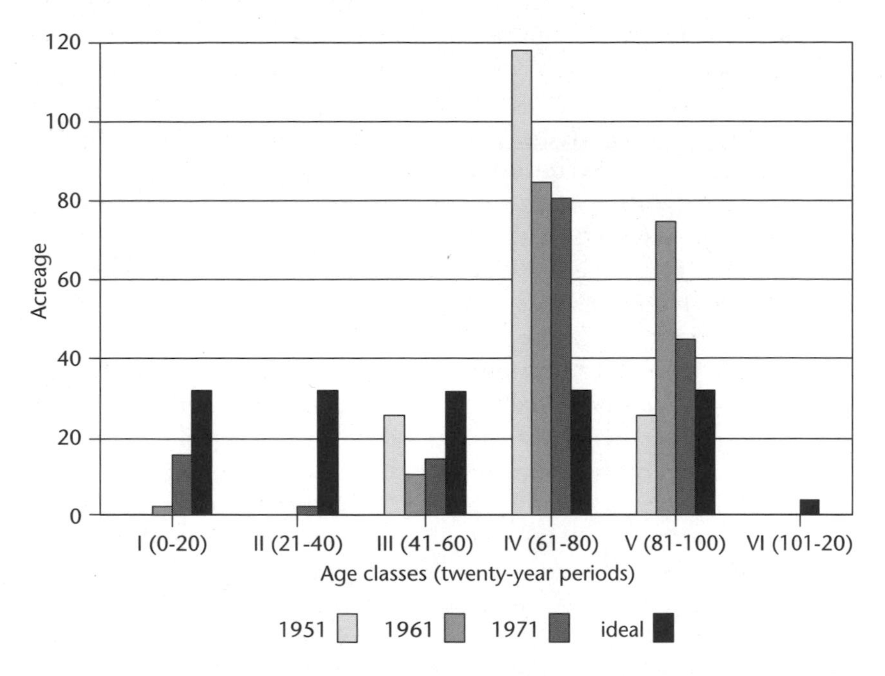

cutting in Age Class V, and we're just knocking the age class back to III or IV, because we're going in and ... just doing partial cuts. Where we've clearcut, that particular area, that number of acres, reverts to Age Class I.

As a matter of fact, it probably takes a whole rotation to do two things: to not liquidate the lot (in other words, to cut our allowable cut) and at the same time to get this age class distribution. So this is a very slow thing to do.

PB: Dave, this road has been one of your big expenses, hasn't it?

DD: Yes, we've probably put $2,000 or $3,000 in this road.

PB: And the road that we're coming to is the latest one then?

DD: This one is the latest one. We want to open this lot up to make it a little more accessible.

PB: You keep exact costs on this particular wood?

DD: We are keeping exact costs on the wood. We try to keep exact costs on every operation we've done, and separate them. We have costs on what this particular stand operation cost us, the logging on it as well as the silvicultural costs, what the road cost us, the culvert costs, and so on. We've got these for the past twenty years.

PB: Twenty years is a long time, Dave. The volume that's standing on the woodlot remains the same. What volume have you taken out?

DD: We've taken out almost 400,000 board feet of lumber, we've taken out nearly 200 cords of pulpwood, and we've taken out over 300 cords of fuelwood from this lot. Now that removal and sale of material have brought us in almost $20,000 in revenues. But then of course you have to take the costs from that.

PB: And you've included the capital costs of roads and ...

DD: The capital costs of roads, the costs of culverts, all these various costs. Everything has been included.

PB: This must be unique in Canada, or maybe even in North America.

DD: Well, this is possible. I don't think that anybody has records this long, over a twenty-year period, of all of these costs. Twenty years ago, that is in 1951, we had a little bit better than a million board feet in here plus some pulpwood. And here it is 1971, we've still got better than a million feet in here. We've maintained this growing stock and only sold, you might say, the interest. It's like a banker. He maintains (if he's smart) his capital, which in this case is the forest, and he's only using the interest, which in this case is the extra growth that we've put on. And we've cut this extra growth over the twenty-year period.

PB: The real crunch comes when you try to [determine] what the profit has been.

DD: That's the crunch. And the rough figures I have here, and I'll round them off. As I've said, we have sold, and our gross returns are, around $20,000. Our costs for logging come to around $15,000. Therefore we've made a bit of money on the actual logging. But then if you consider taxes, the capital costs of your roads and so on, actually we've only made, over a twenty-year period, just a little bit better than $1,000. And when you work that in terms of the number of acres, this comes to thirty-seven cents per acre per year that we've made on this woodlot.

And I think that in some respects this is a rather serious situation, because we've got six million acres in Nova Scotia that are in this small woodlot category. And these acreages, these small woodlots, are producing well over 50 percent of the products for our industries.

If they're not able to practise good forestry practices, then they are up against it. And they have no choice but to go in and take all the material they can ... (See Table 4.)

PB: capital and interest ...

DD: capital and interest together. In other words, they'll have to liquidate, they've got to sell their land and make a profit by holding it and selling it, or go in and liquidate the actual capital growing stock.

PB: Well working with that sort of profit, which is really, I think, staggeringly low, it's no wonder that things are operating this way. In the

Table 4

Antrim woodlot: Actual costs

Years	Revenue $	Costs $	Net Income $	Taxes $	Capital cost $	Balance $
1951-60	4,308.11	2,181.19	2,126.92	201.29	446.05	1,479.58
1961-70	14,941.14	12,376.56	2,564.58	525.78	2,233.41	(194.61)
1951-70	*19,249.25*	*14,557.75*	*4,691.50*	*727.07*	*2,679.46*	*1,284.97*

Annual profit: $1,284.97 / 20 yrs. = $64.25 / year

Annual profit per acre: (from 173 acres) = .37 / acre / year broader picture, Dave, do you not think that in the past primary products, whether from agriculture or from our forests, the real costs have never been involved?

DD: I think this is true. I don't know about agriculture too much, but in forestry I don't think that we've considered all our costs. We've looked at it from this point of view: I go in, and I do a logging operation. I buy gas, I have to hire a little bit of labour and so on, but I don't charge my tractor, my horse, or my saw up against the operation. It just disappears. In other words, I'm not charging depreciation against the operation.

PB: Or renewal or maintaining the capital, the trees themselves. I think in Canada we've generally been profitable because the trees have existed. Now we're getting to that point where –

DD: The trees have not cost us any money to grow. And now we've got to get down to the business of actually putting money in in order to grow them. If a farmer wants to put in a crop, he's got to fertilize and lime and so on. We've got to do basically the same thing. Only we can do this, with timber being such a long time to grow, we've got to maintain that capital stock because if we don't maintain it and it's gone, then we've got troubles. Because we'll have to sit there and wait for perhaps 50 to 100 years before that capital stock gets back. So this means that we have to put monies into silviculture, maintain that capital growing stock, and use the interest.

PB: Well this would apply in a general way not just to woodlots but to the whole forest.

DD: But I think there are two things necessary to make this situation a little better than what we've got today, and it's not a very good situation, as far as I'm concerned. One, I think we have to raise the value of the product. In other words, raise the value of it standing on the stump.

PB: We have to raise it closer to its real value.

DD: Exactly, exactly. And the other thing is, we have to put money inputs. It's the responsibility, I think, of everybody. If everybody wants trees, then it's everybody's responsibility to help pay the tab to maintain this growing stock and to better this growing stock. Meaning, of course, that monies have to be put especially into stand improvement, silviculture, roads, proper inventories, all these things that are involved with managing a woodlot.

PB: Well there's an old saying that once you start spending your capital, you're on the way down.

DD: Well, this is what I think we're doing to some extent in Nova Scotia. We're spending our capital, especially in these small woodlots.

7

Richard Lord: Forestry and Landowner Organizations

Many of the foresters in this book took on formidable challenges during their careers. Schierbeck struggled to pioneer a professional forest service in the interwar years. Hawboldt nurtured a research capability in a department oriented to field administration. Yet none faced a greater task than Richard (Rick) Lord in organizing and defending the interests of private woodlot owners in the age of pulp forestry. In certain respects, Lord was an heir to the tradition of Bigelow, who saw profitable timber harvesting as the pathway to enlightened management. Lord was also a contemporary of Dwyer, whose woodlot studies and group venture promotions affected woodlot owners in a different way. There are also notable differences, however. Where Bigelow was a practical forest economist and Dwyer was a Division of Extension specialist, Lord emerged as a gifted social organizer and strategist of what we now call the 'non-governmental sector.' One of his chief protagonists in this was Donald Eldridge, who spent the decade of the 1970s directing a rival forest sector interest group.

Lord occupies a nearly unique position as a graduate forester in postwar Nova Scotia who was never employed by either government or industry. Professionally, this imposed a sometimes solitary path. While he was active in the workings of the Canadian Institute of Forestry in Nova Scotia, his job obliged him to sit on opposite sides of the table from both civil service and corporate foresters. Particularly during his time with the Nova Scotia Woodlot Owners and Operators Association (NSWOOA), these conflicts of political interest were sharp. Yet it is a measure of Lord's talent for organizing a mass constituency and advancing its claims that the woodlot owners were able to survive so long in a manifestly hostile environment.

Several themes emerge from his career experiences. One is a commitment to autonomous structures and collective action in pursuit of progress for subordinated groups. There were many interests in the forest sector that had plans 'for' woodlot owners but far fewer committed to action 'by' those owners. Through the efforts of Lord and others, the NSWOOA became the first province-wide 'popular' movement to appear in Nova Scotia's forest sector. In a parallel way, Lord was a formative influence for the Nova Scotia Christmas Tree Growers Association, which succeeded in establishing a stable business foundation for independent producers, in contrast to that in pulpwood. A second theme involves the effective synthesis of ideas and practices from technical forestry with those of social organizing. This has defeated many academic and government programs over the years, and the contrasting success of both the woodlot owner and the Christmas tree movements says much about the discipline of mass politics. The notion of public 'education' is much abused in professional circles, which are accustomed to treating lay elements as passive and even dull receptors. Lord's experiences argue otherwise. The fact that hundreds and later thousands of ordinary rural folk responded to invitations to band together for self-improvement within their own organization says much about the limitations of technical communication and the institutional bias that it carries.

Formative Years

Born and raised in Ontario, Rick Lord's early ambition was for a career in engineering. However, an initial year at the University of Ottawa persuaded Lord that civil engineering was not in his future. He then spent time working in the Far North in several settlements on Hudson Bay. Perhaps it was the absence of trees that turned his interest toward forestry. There were four university faculties of forestry in 1960. As Lord saw it, British Columbia was too distant, Toronto was too familiar, and Laval was ruled out on linguistic grounds. Consequently, he arrived in Fredericton in September 1960. On the strength of his grade thirteen standing from Ontario, he entered the second year of the Faculty of Forestry at the University of New Brunswick. The program was geared toward training professionals for industry or government work. There wasn't a single course addressed to small private woodlot forestry, the area in which Lord was destined to spend most of his career.

Lord remembers his summer jobs more than the curriculum as having been particularly influential. The first season was spent working at the Acadia Forestry Station, a government research and experimental farm near Fredericton. In retrospect, the experience may have put him off civil service forestry altogether. The following summer was spent on a timber-cruising crew for the Mersey Paper Company. Looking back, Lord describes this as the closest thing to traditional field forestry that he ever did. Despite a career filled with forest-related jobs, in both the voluntary sector and in private business, he never considered himself a professional forester in the strict sense.

During his final school year in Fredericton, Lord worked casually for Bruce Kelly, then the director of the New Brunswick Forest Extension Service. This involved shearing Christmas trees on Kelly's private tree lot. It was Lord's introduction to the Christmas tree industry, to which he would devote much of his working career. Lord spent his final summer working for the Forest Extension Service in the countryside surrounding Bathurst. While the local woods economy was dominated by the pulp mill in town, there was also a large but economically marginal rural woodlot population. His job involved working with the Acadian small landowners, demonstrating Christmas tree production techniques and encouraging tree cultivation as an additional source of woodlot income. The job stretched his rudimentary French to the limit. Returning to the region for a Forest Extension Service conference nearly two decades later, Lord was surprised to find that several of the longtime residents recalled his student days in their area and had continued the Christmas tree cultivation that he had initiated.

The combination of extension work and Christmas tree growing had clearly made an impression by the fall of 1963. During his senior year, Lord chose as his thesis topic the social and economic issues affecting the mainly subsistence woodlot owners in Gloucester County, with special attention to the role of Christmas tree production as an income stream. The focus was the settlement of St. Sauveur, not more than twenty miles from Bathurst. It was a colonization community, established twenty-five years earlier when the Roman Catholic priest from St. Isidore led a group of young parishioners some ten miles into the bush, seeking new prospects against the teeth of the Depression. The goal was agricultural clearance, and the measure of its success was the severely depleted forest base that Lord documented in his report. In fact, much of the field acreage was returning to scrub bush, leaving little of merchantable value. The problem, he argued, was not strictly one of forestry: 'If these woodlots are to be made productive, long-term forestry management is required. It will be necessary to build up a sound growing stock in a manner which will sacrifice immediate yields in favour of long-term increased production. For just this reason, such a program appears impracticable under existing conditions and with the present form of ownership. The potential advantages of long-term management are difficult to explain to people whose standard of living is largely at the subsistence level, and who are concerned primarily with finding some immediate means to supplement their present income.'[1]

Given the obstacles to full-scale tree farm management, Lord canvassed several alternatives. A Christmas tree program offered one possibility, albeit partial, to generate an income stream while promoting cultivation and management skills. Also anticipating a growing industrial demand for pulpwood, Lord pointed to the Quebec model of setting aside accessible Crown lands as forest reserves for the benefit of forest syndicates composed of small private producers. The normal Crown stumpage rates would apply, with the Department of Lands and Forests overseeing the logging operations. Moreover, 'proper management of the forests is assured while the settler is provided with a substantial continuous source of wood.'[2] Finally, he noted the potential for commercial blueberry farming, again stressing the need for technical advice and for some form of managing-marketing syndicate to coordinate the commercial end of an ongoing business. In the conclusion, Lord noted that the predicament of St. Sauveur was not uncommon and that the severity of these problems deserved much more attention. As it turned out, he himself would be a major contributor. Already there were foreshadowings of his future work: in a willingness to take on tough field situations (complicated in this case by language), an approach to woodlot problems as mixed forestry and agricultural concerns, a recognition of the need for organizations to achieve results and overcome deficiencies of information and scale, and a tendency to look outside the familiar for possible models of change.

There seemed to be an almost boundless opportunity in full-time job offers for graduates at that time. Most of Lord's fellow students had ambitions for careers in industry or government. But by now Lord had a taste for working with small private woodlot owners and producers. Consequently, he chose a job at MacDonald College, the agricultural faculty for McGill University. The Department of Woodlot Management, to which he became attached, was primarily a research unit. It managed the Morgan Arboretum, on the tip of the west island in Montreal, and did educational work with private landowners. Significantly, this was being supported by an agricultural college, not by a forestry school. The chairman was A.R.C. (Arch) Jones, and Dan MacArthur was another senior person in woodlot affairs.

Lord joined a team conducting a research study of owner attitudes and practices in the west end of Quebec's Eastern Townships. Funded by the federal Department of Forestry, it covered Huntingdon and Soulanges Counties, which offered an interesting basis for comparison. Huntingdon was a heavy maple products area with a majority of Anglophone residents, while Soulanges was predominantly Francophone. Huntingdon encompassed the westernmost wedge of Quebec, south of the St. Lawrence River and bounded by the forty-ninth parallel, and Soulanges was immediately 'above' it on the north side of the river, extending from Montreal to the Ontario border. The study covered the full range of woodlot products: Christmas trees, maple syrup, pulpwood, and others. It also explored the reasons for acquiring and owning woodlots, the incomes derived, and the treatments applied to the forests. The data were based on personal interviews, which Lord obtained by calling on residents in their homes. The results were coded and processed, with the help of a bit of Fortran, on an early computer in a basement lab at McGill. In the end, several reports were produced. Lord found relatively low levels of management on these small woodlots. He concluded that the better-stocked Huntingdon area offered greater potential for intensive forest management practices, as yields could be increased substantially with corresponding income flows. For Soulanges, he recommended reforestation of heavily cut land and management assistance where owners requested it, while warning that a comprehensive program could be counterproductive where sparse forest conditions or extremely low incomes threatened the prospects for success. One characteristic of the study area was particularly striking: 'In Quebec, farm woodland area approximates 32 percent of total farm land, yet sales of cash products presently provide only 5 percent of total cash receipts. The majority of woodlot harvests are unplanned, with no consideration of sustained yield or future production. Detrimental practices such as clearcutting, highgrading and pasturing are common.'[3]

Lord would confront similar circumstances later in the Maritimes, where it would become evident that a complex set of factors perpetuated such marginality.

Organizing the Woodlot Owners in Nova Scotia

After a couple of years at MacDonald College, Rick Lord responded to an advertisement that seemed to offer further involvement in woodlot forestry. The Department of Extension of St. Francis Xavier University was seeking field-workers as part of its campaign to organize and educate private woodlot owners in Nova Scotia. The university was well known as home to the 'Antigonish Movement,' a cooperative program of local economic development that dated from the 1920s. In later years, the Department of Extension became heavily involved in woodlot owners' issues in the eastern part of the province. Beginning in the early 1950s, parish priests on Cape Breton Island encouraged local cooperative groups for the marketing of forest products. At first, the results were mixed. Then came the 1959 deal to build a pulp mill at Port Hawkesbury. The government of Nova Scotia leased 1,000,000 acres of provincial Crown forest in the seven eastern counties to Nova Scotia Pulp Limited (NSPL), a subsidiary of the Swedish forest firm Stora Kopparberg.[4] This was based on the very low stumpage rate of one dollar per cord, a level first applied in Nova Scotia back in 1929 when the Mersey Paper Company was established. For years, local leaders had pushed for a forest manufacturing outlet in the eastern region to create jobs, utilize surplus wood, and spur forest improvements. But once the terms of the Crown lease were finalized, the private woodlot sector became alarmed at its pricing and supply implications. Since the Crown stumpage rate was low by Canadian standards, it was feared that this would push down private woodlot stumpage as well.

This spurred new efforts, beginning in 1959, to organize a woodlot owners' association. Over the next several years, hundreds of study sessions and organizing meetings took place. An executive was elected in 1960, and membership continued to grow.[5] This woodlot owners' movement aimed at far more than pulpwood marketing alone. Equally important were education and assistance to members for sound management planning to generate primary wood products for the long run on a sustained yield basis. The 1958 forest inventory results pointed to the need for enhanced forest yields to meet future needs. Consequently, this promotional campaign enjoyed wide institutional support and cooperation, with District Foresters and agricultural representatives joining the local activists, the Federation of Agriculture, and the St. Francis Xavier Department of Extension in spreading the word.[6]

In the opening years, NSPL planned to rely on private pulpwood supplies for its mainstay. As the mill site was being completed in 1961, NSPL began contracting for regular private deliveries. The woodlot movement leadership approached the company to negotiate a group supply contract for the membership. It would set the volumes and prices for a significant share of private wood, which the mill required, particularly in its early years. When the mill management declined to negotiate, the association sought extra leverage through provincial farm-marketing legislation. Under the leadership of Wendell Coldwell, a longtime activist, the (first) Nova Scotia Woodlot Owners Association (NSWOA) approached the provincial government, seeking to establish a pulpwood-marketing board under the authority of the Natural Products Marketing Act of 1933.[7] This provided the statutory authority for the organized marketing of a number of farm commodities. The immediate problem for the inclusion of pulpwood was that it was not listed in the act as one of the eligible commodities, so the NSWOA persuaded the minister of agriculture to sponsor the necessary amendment. Yet when this was accomplished, and the woodlot owners renewed their campaign, the government did not grant the expected approval to organize. Instead, it struck a royal commission to further investigate pulpwood prices and marketing arrangements throughout the Atlantic region. In effect, industrial interests were resisting this 'farm marketing' strategy, and the government Departments of Lands and Forests, and Industry, were raising doubts of their own.

Reporting in 1964, the MacSween Commission found that pulpwood prices had been stuck for ten years at the level of fifteen dollars per cord and that the stumpage return to woodlot owners was next to nothing.[8] It acknowledged the marketing boards already functioning in northern New Brunswick and Quebec, and it showed particular respect for the woodlot associational structure in Sweden, which encompassed some 125,000 members. Crucially, MacSween advised that a well-established producers' association be established *in advance* of any decision on marketing structures. Once the provincial government accepted the report, the need for a new organizing drive was clear, and it was this project that brought Lord to Nova Scotia to build the second NSWOA.

There were actually several initiating sponsors for the new drive, including Coldwell's NSWOA, the Nova Scotia Federation of Agriculture, and the St. Francis Xavier Department of Extension. An application was submitted to the federal Agricultural and Rural Development Act (ARDA) program in 1966, and a one-year pilot project was approved early the following year.[9] It was widely thought that the earlier Coldwell campaign had been slowed and weakened by the absence of full-time field staff. Consequently, the ARDA project called for three field-workers, under a project director, and assigned them to the eastern, central, and western parts of the province. Lord was hired as one of the three field-workers who would animate the project.

The team brought together a diverse set of skills. The director, Reverend A.A. MacDonald, was a sociologist with deep roots in both the region and the university. John McLellan was a farm boy from Cape Breton who had also been studying, by coincidence, at MacDonald College. Tom MacPhee was an experienced Nova Scotia lumberman, and Rick Lord was the young graduate forester. Each field-worker was based in a different region. McLellan went home to Cape Breton to concentrate on Inverness County, MacPhee was assigned the central region and took on Pictou County and the Musquodoboit Valley, while the newcomer Lord was dispatched to the distant frontier of western Nova Scotia to tackle Lunenburg and Queens Counties. Thus, Lord found himself moving in 1967 to Bridgewater, on Nova Scotia's south shore. It has remained home to the Lord family for the past thirty years.

However, Lord's western region was a long way from Antigonish, where the cooperative model and the community ethic were well established. In addition, St. Francis Xavier University was virtually unknown in the predominantly Protestant counties of the South Shore and the Annapolis Valley. Undeterred, Lord launched his one-man operation in 1967. He began in the traditional fashion of community development, getting to know the Extension Foresters for his districts and the Extension Agriculturalists. They, in turn, helped to introduce him to interested people and potential activists.

In each county, the process was similar. By word of mouth and by letters of introduction, the contacts were established. First kitchen meetings were arranged in local neighbourhoods to discuss[10] woodlot affairs and gauge opinion. The next step was the local community hall, which brought together a wider group. At each step of the process, the agenda remained open-ended, aimed at establishing people's concerns and determining the level of interest in moving further toward formal organization. Since the

Nova Scotia Federation of Agriculture supported the woodlot movement, its local branches were also useful networks. Gradually, the initiative passed to the ad hoc groups, which set agendas and established study committees. Only later did they move to the question of forming an association.

As one of the three pilot areas, Lord's home county of Lunenburg was one of the first to get organized. It was quickly apparent that pulpwood marketing was neither as urgent nor as popular a priority on the South Shore as it was farther east. The culture of German individualism was suspicious of group businesses, and the Bowater Mersey Company cast a long shadow as the main pulpwood buyer in its own backyard. Instead, Lord discovered that the issue attracting the greatest interest was Christmas tree production. It offered a substantial cash return, with stumpage worth two to three times that of pulpwood. It could also be produced over a short rotation of eight to ten years, as opposed to thirty-five to fifty years for softwood pulp. The woodlot owners wanted technical information about growing and commercial information about buyers and outlets, but there was absolutely no interest in organized marketing. The first rule of community organization was to ask the people and address their concerns. As a result, Christmas trees headed the agenda. This also happened to fit snugly with Lord's previous research work. Besides, the local Extension Forester, Tom Ernst, was strongly committed, and the issue seemed to promise the greatest degree of cooperation. Consequently, the Lunenburg group of woodlot owners was first and foremost a Christmas tree group, and it adopted as its name the Lunenburg County Christmas Tree Growers Association (LCCTGA). Over the next quarter-century, the local industry grew dramatically, and it accounts today for some 20 percent of Canada's average export of 5,000,000 trees per year.

Initially, the pilot program was funded for a single year, though its obvious success led ARDA to approve a two-year extension to March 1970. Under this plan, the field-workers tackled adjacent counties, with the eventual aim of covering the entire province. Already in the spring of 1968, a 'province-wide' meeting brought delegates from six newly formed associations to discuss a variety of policy questions.[11] By the time of the second general meeting in January 1969, this number had doubled to twelve. Here delegates chose to establish the Nova Scotia Woodlot Owners Association (NSWOA), and they elected an executive and a board of directors. The founding goals of the association testify to its breadth of interests:

(1) To enable Nova Scotia Woodlot Owners to speak and act as an organized group on matters affecting the forest industry.
(2) To formulate and promote woodlot development policies which will increase woodlot productivity and income.

(3) To promote the organized marketing of any forest product when its producers consider it necessary.[12]

Lord's role was also transformed significantly in 1969 when he shifted from being a field-worker to being a project director. Throughout this period, the membership continued to mount, and awareness spread. A newsletter known as the *Woodland Ledger* began in the spring of 1969. A periodic column dealing with woodlot issues was produced and carried by many weekly newspapers, and an agreement was struck with CBC Television to produce a half-hour film about the association, to be shown in the autumn on the national broadcast *Country Canada*.

As the project moved into new media, officials in the provincial Department of Lands and Forests raised certain questions of procedure. This pointed to the potential problems for an activist movement that depended on government grants. The ARDA contract called for a Consultative Committee (including federal, provincial, and project officials) to meet quarterly to review progress. At its first session, Lord was informed that all initiatives not specifically itemized in the budget or annual project plan should be approved by the Department of Lands and Forests as the administrative agency of record. In correspondence with Deputy Minister Bob Burgess, Lord challenged this proposal, arguing that it would 'severely restrict the effectiveness of the day to day operations of the program and would leave the Director in an impossible position.'[13] After further discussions, Burgess asserted that the department claimed the right to 'censor' the CBC film before its release. He also requested that 'controversial matters not be presented' in the newsletter.[14] Not surprisingly, this drew a firm response from St. Francis Xavier Director of Extension Father George Topshee. It demonstrated, however, the growing sensitivity by senior Department of Lands and Forests officials to the impact of a mass organization that challenged, by definition, some of DLF's basic policy coordinates. This incident foreshadowed the adversarial overtones that beset their relationship for more than a decade.

As the project moved gradually from organizing campaign to functioning association, the staff functions were also transformed. Where the Woodlot Project Director had been accountable to the ARDA sponsors, the NSWOA secretary-manager was accountable to an elected executive and board. This was a subtle but important distinction. The first phase had been as much a 'bottom-up' as a 'top-down' process. Clearly, the staff organizers would never have succeeded in attaining the founding threshold without a strong response by activists and leaders at the community level. In addition, the important membership issues had been defined and ranked during the phase of discussion and decision, so that the new association's work program was broadly determined even before the founding meeting.

Furthermore, the organizing function could never completely recede, since a healthy mass-membership organization required constant renewal, never more so than in the formative stages. There remained several county groups to launch, contacts to build, and key people to recruit. At the same time, the pioneer groups had to be nurtured, since several experienced a decline in membership as the price of pulpwood remained low.

Lord summed up the situation in a 1970 report: 'in most cases local associations and their members have realized that little could be accomplished in any of the major areas of concern until a provincial body had been established and had had time to work out its general objectives and specific policies. The role of the local associations in developing policies on their own for local implementation and in seeking broader application of such policies by placing them before the provincial association has only rarely been realized ... This remains as a major objective of the overall program which is yet to be achieved.'[15]

At the same time, however, the new organization had a mandate to deliver and a series of initiatives to pursue, in a context of scarce resources both human and material. The secretary-manager's role was critical to this, and not surprisingly the association chose Lord to continue in the senior staff role. Several key initiatives arose from resolutions approved at the first province-wide meeting in 1968. These included calls for a tripling of the public funds made available for private forest stand improvements, an independent grading system for Christmas trees, a purchasing system for Christmas tree producers to relate to buyers, an official log rule for Nova Scotia, and more vigorous dissemination of policy information to woodlot owners. From the outset, the woodlot movement articulated a rounded set of issues and concerns, which expanded during the early years.

Further evidence is found in the standing committee structure established at the NSWOA founding meeting in 1969. To handle association business between annual meetings, committees were struck for the following purposes: constitution and finance; Forest Improvement Act; forest taxation; pulpwood marketing; woodland tenure and consolidation; sawlogs; and Christmas trees. A series of NSWOA positions resulted, and they were put to the provincial government over the years to come. For example, bylaws had to be prepared and incorporation arranged. Secure ongoing finance was a priority, and a series of proposals was put to the DLF and the architects of the federal-provincial agreement for the association to deliver programs to woodlot owners. The NSWOA endorsed the FIA as a conservation measure and pressed for its early proclamation. It also recognized the centrality of the tax regime to harvest and management choices, and it launched a review of existing measures.

More than anything else, pulpwood-marketing reform was the signature issue of the new organization and its most critical early initiative. But small owners were also alarmed by suggestions that the DLF aimed to expand Crown holdings by buying private woodlots. In response, they called for a government-backed loan program to finance private woodlot expansion, modelled on the Timber Loan Board already available to sawmillers. If this was an ambitious agenda for such a young organization, it also indicated the impressive breadth and depth of woodlot owners' interests. This needs to be underlined in light of the criticisms of opportunism and short-termism that the NSWOA and Lord soon encountered.

The Woodlot and Professional Forestry

The reactions of professional foresters to the appearance of the NSWOA are highly revealing. In Nova Scotia, the woodlot sector has long been treated as the unwanted guest at the stakeholder table. It is undeniably a central link in the forest chain, collectively accounting for more than half of all forested land. As recently as 1967, woodlots provided 60 percent of the total commercial wood harvest in the province.[16] Yet official accounts tend to acknowledge this begrudgingly as part of a 'unique tenure pattern' arising from the historical accident of two centuries of settler land grants and sales prior to the Great War. In reality, it was far from unique. Woodlot forests exceed 90 percent of total forest land in Prince Edward Island, 47 percent in New Brunswick, 14 percent in Ontario, and 8 percent in Quebec.[17] Yet given their relative importance to Nova Scotia, the persistent absence of positive woodlot forest policies is striking.

In professional forestry circles, debates about woodlot or 'farm forestry' were far from new in the 1960s and 1970s. Particularly in times of scarce supply, small suppliers have always attracted attention. In the immediate postwar years, the Canadian Society of Forest Engineers (CSFE) revived its Farm Woodlot Research Committee and canvassed a range of policy and management issues. In 1950 the *Forestry Chronicle* devoted a series of articles to the topic. An editorial declared that woodlots 'offer the best opportunities for the immediate application of sustained yield forest management' and suggested that 'central marketing through specialized cooperatives is one key to this situation. This marketing personnel would have to be qualified to lay down, in cooperation with provincial government authorities, a long term management plan for each woodlot within its area of control.'[18] This culminated in a special woodlot forestry issue in the autumn of 1951.[19] It is notable that, of the leading private lands provinces, only Nova Scotia was not represented in this survey. The debate broadened to explore diverse approaches: technical extension work by state forestry services, demonstration woodlots, public education, woodlot cooperatives and landowner committees, and the tree farm movement already active in the United States. In this last case, companies offered extension programs to franchise this service, providing management planning in return for supply contracts.[20]

Despite the CIF's clarion call for woodlot action, there was little immediate response in Nova Scotia. There a rural fibre market was already firmly entrenched, and it centred on brokers and agents who purchased private wood on behalf of sawmillers, pulp processors, and pulp exporters. Often merchants, sawmillers, or village notables, these middlemen filled contract targets set by their principals, and they often enjoyed monopoly rights to local purchase.[21] While this resulted in the lowest possible prices, it also left owners with no incentive to invest in stand improvement, thereby furthering the degradation of the forest resource. It was this structure that the St. Francis Xavier campaigns sought to challenge, and, while Hawboldt's Division of Extension provided technical support on request, there was no specific program to impart management skills to woodlot owners. In the few cases where it was mentioned, the department simply noted that District Foresters were available to advise woodlot owners on disposing of their crops.

Even this passive DLF cooperation began to change with the rise of pulp processing in the 1960s. Since it would depend heavily on woodlot supply in the initial years, Nova Scotia Pulp Limited promoted a tree farm program through its woodlands division.[22] Over the years that Halifax refused to act on Coldwell's marketing proposal, NSPL gained a lengthy window of opportunity to entrench its buying and harvesting network across eastern Nova Scotia. At the same time, it refused to discuss a group supply contract with the NSWOA.

Neither was Hawboldt's research enthusiasm ever drawn in a woodlot direction. This followed from a combination of social and technical judgments on his part. On the one hand, it was true that uneven-aged (i.e., selection) management in Europe 'is commonly reserved for comparatively small management units, parks and farm woodlots.'[23] Yet when it came to Nova Scotia, selection silviculture was deemed 'too complex to develop and maintain.'[24] Hawboldt shared the emerging professional consensus that the rigours of intensive silviculture were beyond the capability of the owner: 'When I consider the woodlot owner in this context I see that he has problems. He cannot compete costwise with the productivity of mechanized operations. Individually he does not have sufficient acreage to undertake viable management programs involving reforestation, silviculture, access roads. Nor is his tenureship of sufficient duration to warrant such inputs for someone else's benefits. But despite these problems industry is dependent for 40 percent or more of its primary product on the woodlot owner.'[25]

Consequently, Hawboldt concluded that 'the industry has a distinct obligation to ensure the viability of these [woodlot] ownerships for self preservation. This obligation must encompass cooperative development of planned access roads, mechanized harvesting and management advice with groups of woodlot owners in a mill's procurement area.'

This was highly ironic given the virtual ignorance in government circles of woodlot enterprise conditions. The first modern survey of woodlot owners' holdings and activities occurred in 1970, and the economics of the Antrim property were only reported in 1974.[26] In the meantime, the absence of positive data left the way open for prejudicial assumptions about inherent potential. In 1970 the deputy minister of the DLF pointed out 'that our average small woodlot owner is around fifty-five years old, that his average length of tenure is nineteen years, and that only thirty percent of the owners work on or near their holdings – and you have a disturbing glimpse into the future.'[27] Removed completely from social context, these 'facts' were assumed to speak for themselves. Then, following from the implied qualities of passivity and disinterest, he proposed solutions that further marginalized small owners by promoting woodlot acquisitions by sawmill operators and the provincial Crown!

This demeaning view of the grassroots landowner, which is completely unsupportable given the history of the NSWOA, continues into a discourse on public education: 'We can only persuade. To say to John Farmer that he should practise silviculture to optimize fibre production for the good of his province is to ask for trouble. Even to offer silviculture incentives is not going to be enough: John Farmer may not know what silviculture means, let alone believe it's any good ... So education is needed.' Whether this view was grounded in ignorance, wishful thinking, or professional conceit, such statements illustrate the barriers and biases that Lord and the NSWOA faced in establishing legitimacy with provincial forestry authorities who were firmly in the grip of an alternative worldview.

The DLF perspective on woodlot dynamics contrasts sharply with the one held in the NSWOA. Lord's perspective was nuanced but realistic. He acknowledged that the constituency was diverse, with a variety of outlooks. Elderly owners of long standing had little interest in gearing up for management. Recent woodlot buyers were often recreationally oriented, with little interest in selling wood. Much of the current woodlot production came from one-time sales of standing timber or stumpage to contracting harvesters. Lord conceded that only a small minority of owners were committed to regular cutting and sustained management. For him, this was the challenge for forest policy: 'This waste of our current and future resource can no longer be tolerated. Current owners must be provided with the mechanism and incentives to make continued ownership and active management attractive, and thus reduce the pressure on people to neglect their holdings or sell.'[28]

The Battle for Organized Marketing of Pulpwood

Following the inaugural annual meeting of 1970, the association turned its attention to the marketing question. The ensuing controversies absorbed the lion's share of Rick Lord's attention for much of the decade. Eight years had elapsed since Coldwell's initial approach to the marketing issue, and the private corporate supply system was now firmly entrenched. Having met MacSween's key condition for a province-wide organizational infrastructure, the NSWOA turned expectantly back to the marketing question. It was, as Lord remembers, 'our naive period' when the way seemed open to a straightforward registration for pulpwood bargaining under the Natural Products Marketing Act. In talks regarding the St. Francis Xavier contract extension, held with the DLF minister and deputy minister, Topshee and Lord looked optimistically toward a pulpwood-marketing arrangement within a year, which they hoped would leave the association financially independent.[29]

An NSWOA delegation led by President Tom Smith, A.A. MacDonald, and Rick Lord met Premier Ike Smith and senior DLF officials in March 1970 to formally request the establishment of a pulpwood board under the act. Almost immediately, the political battle was launched. Minister of Lands and Forests George Snow met with the big four processors[30] and the executive director of the Nova Scotia Forest Products Association (Donald Eldridge) to solicit industry views. Then, in a revealing instance of partiality, Deputy Minister of Lands and Forests Bob Burgess intervened to orchestrate industry opposition to the woodlot group and to offer advice on tactics. This is described by NSFPA President C.H. Sproule, in a memorandum of 26 May 1970:

> Mr. Burgess contacted the Executive Director [Donald Eldridge] on May 1 to inform him of the urgency of the NSFPA doing something very soon as he had spent the afternoon with Father MacDonald, Dick [sic] Lord and Tom Smith, and Mr. Scammell. At this meeting they were pointing out to him the reasons why the WOA should be allowed to form a marketing board. Mr. Scammell is a secretary of the existing marketing board as passed by legislation in 1962. This information was passed on to the President (Mr. Sproule) by your Executive Director, and as we were having a Directors' meeting on May 14, a tentative meeting of the president, Mr. Burgess, and the Executive Director was arranged for the night of May 13, to allow Mr. Burgess to fill in the president on what was going on up to date.
>
> Mr. Burgess stressed the need to get to Mr. Snow [DLF minister] as soon as possible and let him know the FPA is investigating this matter.[31]

Soon the NSFPA was openly opposing the marketing proposal and working to build an industry-wide coalition to this effect.[32]

In a counterstroke, the NSWOA approached the Nova Scotia Voluntary Planning (VP) organization, a policy advisory body to the government of Nova Scotia. The NSWOA approached the organization's Forestry Committee (of which it was a member) with resolutions endorsing a private silviculture program, a woodlot acquisition loan program, and the principle of a pulpwood-marketing mechanism.[33] The third resolution read as follows: 'Be it resolved that the right of pulpwood producers to join a marketing mechanism which will protect their interests be recognized, and furthermore, that the Government of Nova Scotia facilitate through newly enacted or amended legislation, the establishment of a pulpwood marketing mechanism which will protect the interests of Nova Scotia Pulpwood Producers and thus provide for bilateral and co-equal negotiation between buyers and sellers.'[34]

The Forestry Committee refused to vote on the principle of organized marketing, arguing that it should only consider a specific proposal.

Consequently, the next six months were devoted to the drafting of a detailed plan, through a working group that included Lord and officials of the provincial marketing board and the DLF. This was complete by November and was accepted by the Nova Scotia Marketing Board (the regulator of commodity plans) over the winter. The law required the consent of a majority of eligible producers (by supervised vote), and this vote was scheduled for May 1971. Open to all woodlot owners, some 8,500 had submitted applications to vote.

However, two weeks before the scheduled referendum, industry interests again took their opposition to the Nova Scotia Voluntary Planning organization's Forestry Committee. On 2 April, the committee defeated the NSWOA's (deferred) resolution in principle. Then it approved such radical modifications to the plan as to nullify, in Lord's view, its stated intent. The Forest Products Association then took the initiative, convening a meeting on 29 April and inviting VP and DLF representatives. Lord planned to join this gathering as well, since his group was an associate member of the NSFPA, but he was refused permission to attend. In Nova Scotia forestry, it seemed, one of the hallmarks of shaping decisions was control over the invitation list. Several resolutions were passed that effectively rejected the marketing plan. The deputy minister of lands and forests told the meeting that his department would sponsor legislation to have 'forest products' removed from the Natural Products Marketing Act (thus eliminating the enabling provision for a pulpwood board) and to replace it with a 'Forestry Commission' (the NSFPA's preferred alternative).

After a convoluted set of meetings in which the Forestry Committee first approved and then rejected the NSWOA draft marketing plan,[35] it ultimately advised that the vote be cancelled and that the alternative model of a 'Forestry Commission' be accepted instead.[36] Senior Voluntary Economic Planning (VEP) officials took this position to the Nova Scotia premier. Gerald Regan refused to interfere with the scheduled vote, though he also refused to be bound by its outcome. Under advice from the provincial marketing board, the NSWOA altered its plan to limit eligibility to owners of fifty acres or more. This necessitated postponement of the vote, and a re-registration, after which only 4,857 owners qualified. In the final event, the proposal was approved by 86 percent of the 3,483 who voted in July 1971. Clearly, a stalemate had been reached. Woodlot owner support was too strong for Cabinet to reject the marketing plan, but industry opposition was too strong for it to be approved.

Two months later Regan declared that his government would not authorize the plan as stated. Instead, he sent woodlot owners and industry interests into negotiations to design a commission-like body. Halifax lawyer Ian MacKeigan was appointed as a mediator, but he was unable to fashion agreement. He recommended as an alternative a new pulpwood-marketing statute and a separate board. This was duly enacted in 1972 as a compromise between two strongly entrenched interests. Two years had passed since the woodlot owners had initiated their proposal, and extraordinary amounts of energy, time, and patience had been required of the fledgling organization. Most of this fell to Lord as director. His role was to maintain political contacts, update information, arrange meetings, and write briefs (for the membership, the VEP, and the government) at each new twist in the process. Lord reported in 1972 that 'the Association's internal financial resources have been largely exhausted by the prolonged and expensive campaign to have the marketing plan implemented.'[37]

With the agricultural marketing track definitively closed, the NSWOA prepared to seek recognition under the new Pulpwood Marketing Act. While its enactment constituted a victory for the woodlot owners (vastly preferable to an outright defeat or to the commission concept), its imperfections pointed toward renewed political difficulties. Compared to analogous bodies in agricultural marketing or labour relations, the Pulpwood Marketing Board was neither fish nor fowl. Both its legislative foundation and its administrative structure began from scratch, unable to borrow or benefit from either stream of practical experience in procedural design, economic analysis, or judicial interpretation. The rigours of organizing the board imposed a further brake on a political process that had been surging ahead since 1969. Five months elapsed before the appointment of board members in October 1972. Then, somewhat surprisingly, Cabinet set the term of the initial appointments at just one year. In addition, adversarial dynamics were inserted directly into the board. Three members (including the chairman and vice-chairman) were appointed 'at large' by Cabinet, while the NSWOA and Forest Products Association proposed one member each.[38]

In effect, the NSWOA settled into a political struggle that would last more than eight years before a pulpwood-marketing body was finally recognized. This struggle became a preoccupation for the association, at considerable cost to its wider organizational mandate.

The Woodlot Owners Association and the Department of Lands and Forests

Although the pulpwood struggle dominated all others and overwhelmed the woodlot owners' agenda in the 1970s, the association's broader program offered the potential for cooperative and collaborative relations with the forest authorities. The hope for a constructive relationship figures prominently in NSWOA records. This covered policy advocacy, program delivery, and communication. Yet progress was frustratingly slow.

At the heart of this was the continuing political tension between the association and the Department of Lands and Forests. It had been obvious to Rick Lord that senior DLF officials had been, at best, ambivalent about the group during the period of organization. While formal contacts remained open, on both the policy and the operational levels, the relationship lacked the cordiality and goodwill that would signal full organizational acceptance into the forest policy community. Several factors complicated its realization. One was the dependent financial relationship between the NSWOA and federal and provincial authorities. Each year during the increasingly intense pulpwood struggle, the association faced the loss of its modest core funding. This began with the termination of the ARDA contract. Throughout the winter of 1970-1, St. Francis Xavier Extension had pressed the minister of lands and forests for a contract extension. Director of Extension Father George Topshee appealed to Benoît Comeau for support in this 'most crucial' period. However, it was not until ten days before the contract expired – with the staff layoff notices already written – that ARDA renewed support for 1971-2 (and then at half the previous level).

The field-workers had to be terminated to enable Lord to stay on at the helm for another nine months. Each year thereafter, the association approached the minister for a regular operating grant based on a formal budget proposal.[39] Instead, it received a variable grant out of contingent DLF funds and always on a one-off basis. Lord urged the members that 'we must convince government that the investment of public money in our association is a legitimate and fruitful expenditure.'[40] His appeal was to little avail.

Another factor was the sharp intensity of the pulpwood issue, which pitted industry segments against one another, forcing some awkward choices on the DLF. Obviously, the association had challenged the prevailing fibre supply system. The reaction of the pulp and paper firms was not surprising, though the wider industry reaction, including sawmills, was less predictable. Commercial sawmillers grew increasingly resentful and suspicious of the pulp companies as access to Crown stumpage was preempted by the vast NSPL and Scott pulp leases. However, the pulpwood-marketing issue sealed over those fissures, as the threat of higher fibre costs served to unite lumbermen, exporters, and pulp people.

When the Forest Products Association declared war on the pulpwood-marketing proposal in May 1970, President C.H. Sproule (himself a lumberman) called together the big three pulp mills and the members of the association's Chip Committee, whose mandate covered the sale of sawmill slabs and edges (after chipping) as a pulp input. These were the larger sawmills and those seeking a valuable new commercial relationship with the pulp segment. Sproule stressed that, 'whatever we do, we must act as a unified body or front to the government ... There are forces at work that hope to divide us as an Association.'[41] His successor, Scott Maritimes Woodlands Manager R.G. Murray, described the pulpwood-marketing issue as a ploy 'for the immediate gain of a few who wish to control our industry, in which they have a limited investment and no concern for the future.'[42]

Less predictable, however, was the reaction of the Department of Lands and Forests. The birth of a province-wide woodlot association offered unprecedented opportunities for the DLF to reach the owners of half of the province's forests. This was particularly strategic given the wood shortage demonstrated by the 1958 inventory report and the need for enhanced management of private lands for increased yield. Furthermore, in the wake of the elimination of the DLF Division of Extension (in the 1970 administrative reorganization), a vibrant NSWOA might have offered an invaluable instrument for both policy consultation and program delivery. These views were certainly held among some DLF staff at the field level. However, it was the association's involvement in high policy matters that made relations with it a 'political' matter directed at senior bureaucratic levels. Here woodlot owners were still regarded as alien elements. Their funding came from the federal ARDA program, their provincial government allies were located in the Department of Agriculture and in the Nova Scotia Marketing Board, and they sought a farm-style commodity-marketing plan. The deputy minister of lands and forests observed pointedly, in March 1970, 'Now this organization is active and they call themselves the Nova Scotia Woodlot Owners and have left off the word farm.'[43] Moreover, the association was challenging long-standing core policies of the DLF, such as cheap Crown stumpage and the aggressive Crown purchase of forest land.

It was evident to the woodlot association leaders that the delivery of concrete material benefits to their members was a key means of maintaining and expanding the popular base. Moreover, given the association's continuing interest in improved silvicultural practices on small private woodlots, this was readily seen as a potential area for expansion. Particularly after the Department of Lands and Forests abolished its Division of Extension in 1970, the NSWOA recognized an important gap that needed to be filled. While the ARDA III program had a forest stand improvement element, the take-up from private owners had been disappointing. The woodlot association offered to deliver these services in an expanded program of training, technical advising, and financial assistance. This stressed the importance of organizational infrastructures in reaching the small private owners, while also offering the NSWOA a source of earnings.

In contrast to its perceived treatment of the NSWOA, the DLF executive suite remained open, cooperative, and even collaborative in its relations with the Forest Products Association. The contrast was striking, and it remained a source of continual frustration to Lord and other woodlot activists. No issue drove home the point so vividly as the 'roads program.' By one key decision in 1967, the department had transformed the financial and membership potential of the Forest Products Association from a part-time volunteer interest group to a service delivery agent with full-time staff. According to the terms of the forest roads access program, the NSFPA administered a substantial grant for the construction of private woods roads, collecting a 15 percent administration fee in the process. Since eligibility for road assistance was limited to NSFPA members, prospective applicants had to join in order to qualify, and the membership rolls swelled with woodlot owners who would otherwise have little reason to be involved.

Although this form of delegated delivery was unusual for the time, particularly as it confined eligibility for public funds to the membership of a private organization, the woodlot owners viewed it as a precedent to be emulated, particularly given their own tenuous finances but expanding membership base. The opportunity arose as early as 1970, when the Forest Products Association suggested that it deliver a new program (on the roads access model) for silvicultural assistance to its members. In a submission to the premier and the minister of lands and forests, the Woodlot Owners Association endorsed the concept of delegated delivery of silviculture, while challenging the suitability of the Forest Products Association (and the closed membership provision) as sole intermediary. Then, as later, the association made clear its willingness to share in the delivery of silvicultural assistance under the ARDA program to woodlot owners, whether members or not.[44] One week later, Deputy Minister Burgess wrote to his minister to appraise the NSWOA brief. Ignoring all other items discussed in the ten-page presentation, Burgess confined his critique to a proposal for enhanced credit for woodlot expansion. Both simplistic and sarcastic, the memorandum was a clear attempt to present the woodlot owners as short-sighted opportunists seeking narrow special advantages to expand their holdings.[45]

Several years later, in 1974, Lord again submitted a proposal, this time for the association to train silviculture crews that could offer services to woodlot owners. Once again a substantial brief addressed a major question of forest policy, whereby the association would deliver publicly funded silvicultural programs to woodlot owners and in the process achieve a degree of financial independence. The proposal combined provisions for training, silvicultural treatments, and extension services, arguing that 'individual owners will be more receptive to crews, whether for harvesting or silviculture, employed by the Association, than they would be to crews working for either the Department of Lands and Forests or individual companies.'[46] Hearing no response in over nine months, Lord wrote again to Burgess. The lack of concrete acknowledgment, much less encouragement of a constructive proposal, continued the pattern of indifferent or hostile responses and aptly prefigured the outcome. It was not as if the DLF was opposed in principle to delegated program delivery, as the Forest Products Association access road scheme proved. To the NSWOA, and to Lord in particular, this was evidence of consistent partiality.

In the silviculture case, the DLF chose ultimately to support the concept, but not through the Woodlot Owners Association. Ironically, Lord believes that the association was instrumental in the adoption of an alternative delivery mechanism through the forestry group ventures. At one point in the early 1970s, both Rick Lord and Dave Dwyer of the DLF attended an interprovincial meeting of forestry extension specialists in Quebec. At that meeting, they learned about the *groupements forestiers*, a system of woodlot-owned enterprises already functioning in that province. Both men were impressed by this experiment and brought home favourable reports. Lord later endorsed the approach for the private lands program in Nova Scotia. The West Pictou Pilot Project began early in 1975.[47] While he was disappointed that the county-level woodlot owner groups were not included in the program as proposed, he supported the Group Venture program as it has evolved since 1977.

It seems clear that the pattern of organizational alliances around the DLF was already firmly cemented by the time the Woodlot Owners Association celebrated its first full year in 1970. Given the critical importance of the pulpwood-marketing issue in aligning the forest policy 'community,' it is questionable whether the NSWOA could attain a place in the inner circle as long as the issue remained outstanding. Since the marketing issue went unresolved for more than a decade, the Woodlot Owners Association was almost fatally impaired. This fault line proved damaging not only to the association but also to the department, which enjoyed less opportunity for creative policy making as a result. In dismissing the NSWOA as an interlocutor, the DLF lost the opportunity to build networks into the private forest estate and to win credit from some worthy initiatives. It also lost the benefit of a political counterweight to the industrial forest block, which instead was encouraged to exercise regular vetos over unpopular issues. If good professional forestry practice hinges on the presence of countervailing forces, then the drastic tilt of DLF planning toward servicing a single interest was both short-sighted and flawed.

At times, Lord wondered whether his association was battling with the pulp industry or with the provincial government. In some cases, the answer was likely 'both.' This was particularly the case when the provincial government adopted the cheap fibre strategy to promote forest manufacturing. However, it should not obscure the fact that the NSWOA had its share of victories. It was instrumental in blocking the massive Crown forest purchase plan of the 1970s. It helped to keep the Forest Improvement Act alive, and it opened the way for the group venture scheme. Most important of all, the association was the catalyst and champion of the pulpwood-marketing system despite its imperfections.

From Marketing Policy to Pulpwood Supply Groups

This ongoing political conflict passed through several lengthy subsequent stages. In fact, the first woodlot owners' supply group would not sign a contract until 1980. For much of this time, the NSWOA sat on the sidelines, unable to proceed while NSPL mounted a series of prolonged legal challenges to the legislative and procedural foundations of the Pulpwood Marketing Board. Each time the association drew up an application to be recognized as a bargaining agent, NSPL launched litigation that effectively froze further action pending judicial resolution (normally involving both trial and appeal court proceedings). For example, the NSWOA launched an organizational drive in 1973 seeking to bargain for all private woodlot owners across the province. When the board ruled favourably on the application, it was challenged in court by NSPL, which contested the board's legal mandate to recognize the association. Following more than a year of litigation, the company's case was upheld (on grounds of flawed administrative procedure). The board was obliged to begin anew, spelling out a more detailed set of regulations to cover certification and bargaining. It was clear from the Nova Scotia Supreme Court ruling that regional rather than province-wide supply groups would be required, so the second NSWOA campaign focused on the eastern region, its strongest. Here again NSPL chose to litigate to prevent the board from releasing lists of company suppliers and woodlot owners to the organizing committee. After another round of legally driven delays, the Stora Suppliers Division was finally certified in 1979, with the first contract signed in 1980. A second supplier group tied to the Scott Maritimes pulp mill was certified in 1981, and its first contract was signed in 1982.

Inasmuch as the ultimate goal of organized bargaining among small private woodlots was achieved, this decade-long process might be accorded a success. However, the end result was quite different from what the organizers had envisaged some fifteen years earlier. Instead of a single province-wide body negotiating for up to 10,000 producers, there were two geographically limited groups negotiating for a far smaller constituency. Instead of being able to harness the experience of farm commodity marketing, the NSWOA was forced to grope its way through a new and untested marketing mechanism that proved highly vulnerable to legal challenge. Instead of enjoying the political support of a provincial government that conceded the legitimacy of small commodity producers seeking market power, the government seemed to be frozen by the offsetting claims of industry and woodlot owners and allowed the standoff to be perpetuated.

However, this was not the only issue requiring the association's attention. As executive director, Lord was called upon to represent or coordinate representation for the group on a number of advisory bodies, both technical and policy-oriented. One of them was the Provincial Forest Practices Improvement Board, which operated from 1967 to 1978. Another body was the Forestry Sector of the Voluntary Economic Planning apparatus. Despite its negative experience with VEP on the marketing issue, the NSWOA found sector meetings to be a valuable listening post for gathering information.

As the only full-time employee, Lord found that his time was in continual demand in two directions. One involved the external representation and advocacy activities with government as sketched above. There was a continuing need to articulate a woodlot owners' position on issues and to ensure that formal advisory and political channels were made to appreciate this. On the other hand, the voluntary structures at the county and local levels required continuing attention, as befit a mass-membership organization that aspired to popular legitimacy. This was particularly important during the prolonged uncertainties over the marketing question, when many members questioned the value and future prospects of their association.

After fifteen years of continuing service, Lord reconsidered his future with the Woodlot Owners Association in 1982. A number of converging forces suggested that his time was drawing to a close. Clearly, the woodlot sector initiative was in the eastern region as opposed to the western, where Bowater had declared its opposition and a backlash was building against the pulpwood levy in the absence of marketing groups. In addition, Lord found himself increasingly disillusioned with woodlot developments in Cape Breton. As a recipient of funds from the levy, the Stora Supply Division in Port Hawkesbury enjoyed resources far in excess of the association. But rather than making common cause, its leadership seemed to be bent on increased autonomy (which culminated later in the formation of the Landowners and Forest Fibre Producers Association) in competition with the NSWOA. The latter was no more financially sound, for a decade's work. Lord had experienced several spells of unemployment over this time, and the board of directors was again raising the issue of moving his office to Truro. He pointed out to them that, with his position never guaranteed for longer than six months, he could hardly abandon his home and business in Bridgewater, where his wife also worked as a teacher. Finally, his Christmas tree and silvicultural contracting business was expanding and therefore demanding more time rather than less. When the issue came to a head with

Richard Lord receiving an award from Dr. Alex MacDonald, director of the Department of Extension at St. Francis Xavier University, during the woodlot owners organizing program.

the board, Lord departed, with mixed feelings about his own exit and the prevailing direction of the movement.

To this day, he is not sure whether he quit or was fired. However, he did dispatch 'a polemic' to the directors, putting his views on record. As it turned out, the woodlot association had not forgotten about Lord. At the 1986 annual meeting, he was awarded the Glen Higgins Memorial Award (named after the group's second president). By this time, the province had declared a new forest policy, which included a commitment to strengthen and renew the pulpwood-marketing board, while designating private lands the primary source of timber. (Despite intense industry lobbying, the next several years under Minister of Lands and Forests Ken Streatch marked a high-water point for progressive private land initiatives in the forestry field.) Lord told the woodlot association meeting that, 'on every page of the policy, you read things for which the association fought long and hard,' while at the same time warning that 'the challenge to you people is to make sure these ideas get off the policy paper and into practice.'[48]

When the opportunity arose in 1983 to appear before the Royal Commission on Forestry, Lord did so as a professional forester in private employment. His thirty-page brief offered a comprehensive review of private land forestry over the past fifteen years, together with twenty specific recommendations.[49]

By this point, Lord had moved into a new associational role as the first president of the Nova Scotia Silvicultural Contractors Association. This was a relatively new industry segment that was vastly enhanced after the appearance of the 1977 agreement. With over 100 approved contractors recognized under the agreement, it constituted a sizeable interest group. For Lord, it seemed to be a natural extension from tree farming to silvicultural contracting. Although his scale of operations was never huge, it did enable him to keep his Christmas tree crews employed for a longer period each year.

Growth of the Christmas Tree Industry

From the outset, the NSWOA encouraged the development of a diversified mandate that did not depend entirely on success in pulpwood marketing. One eligible direction was to expand into alternative forest products. Here the Christmas tree industry represented a most promising line. Wild Nova Scotia balsam fir had been sold in the American northeast market since at least the 1930s, shipped primarily by rail but also by sea. (Since the late 1960s, truck transport has largely taken over.) One major American firm, J. Hofert, became a sizeable landowner in Lunenburg County, cutting trees as well as purchasing them for export. The government of Nova Scotia passed legislation in 1944 to require commercial buyers to purchase a licence for dealing in balsam trees. Records from the first year reveal that thirteen buyers purchased almost 1.3 million trees, with the largest agent accounting for almost half that number. The historic peak for Christmas tree volumes occurred in 1958, when 3.8 million were sold. The trade was also quite localized within the province, centred in Lunenburg and neighbouring counties but with substantial sales in Antigonish and Guysborough as well.

The trees were primarily wild balsam fir, which grew naturally and were cut and shipped in a short but intense season from late October to early December. Despite the surging numbers of trees sold, there was a discernible decline in quality by the mid-1950s, as the prime natural areas were exhausted and poorer stock was offered. Working with progressive producers who had pioneered more intensive techniques of cultivation, the DLF Division of Extension attempted to promote a more deliberate approach to both the forest and the market. On growing, the message was simple, though the treatments required knowledge and commitment. As one publication title put it, *Christmas Trees Are a Crop,* and this implied continual husbandry. The department's circular covered the arts of growing, shearing, and preparing trees for transport. Another bulletin tackled the subject of the trade itself, tracing the chain of commerce from woodlot to living rooms in New York and Boston. This included the price gradations and the vast quality variations that prevailed in the balsam export market.[50] The authors argued in favour of several measures to improve the position of growers: a standardized two-tier grading system to protect quality stock and eliminate misunderstandings, and an association for Nova Scotia Christmas tree growers to promote the cultivated as opposed to the wild product.[51] Although these were sound proposals, they waited several decades for implementation, and when it happened Lord and the Woodlot Owners Association each played a role. The American firms whose Nova Scotia landholdings grew in the postwar period were also a progressive influence when it came to encouraging cultivation.

The lure of Christmas trees was not hard to see. Unlike sawlog or pulplog material, the Christmas tree could move from young plant to marketable stem in eight to ten years. This was an investment horizon that was meaningful to small producers, because a cash crop could be taken off in less than a decade of investment. Furthermore, the Christmas tree segment was not dominated by single regional buyers as in pulpwood, since export brokers and wholesalers were numerous. Finally, there was substantial potential for non-price competition, as the growers with the best-quality stock were in a position to exploit their advantage. Where a wild tree might sell for fifty cents, a sheared tree could fetch $1.50 to $2.00, which could also be compared to $1.00 for a stick of pulp requiring forty years to grow.

The matter of Christmas trees was always close to the hearts of NSWOA activists. As seen above, it formed the focus for Lord's initial Lunenburg County organization in 1967. While there was no interest in group marketing, there was considerable support for an organization that would explain techniques of how to improve the crop, review market conditions, and enhance the tree growers' stature as a significant forest segment. The first Lunenburg County field day (a September event that continues today) occurred in 1967, and the Lunenburg County Christmas Tree Growers Association was in the vanguard of the movement for cultivated trees. Throughout the 1970s, the LCCTGA consolidated its position. It took Nova Scotia growers to New England and brought American buyers to Nova Scotia. Many tree growers also maintained connections to the Forest Products Association to gain access to its private forest land access roads program.

This contrasted with the situation in eastern Nova Scotia, where cooperative marketing had been mounted in the 1950s, though with mixed results. When the NSWOA established its Christmas Tree Committee in 1970, it moved into direct marketing to serve some eastern balsam tree producers among the membership. Lord obtained a copy of the New Brunswick Christmas tree grade (the only formal one then available). He and Tom MacPhee went on a selling trip to New England, and they obtained some contacts from New Brunswick sellers and signed a few advance orders. Given the essentially unregulated situation in Nova Scotia, these first shipments were sold on the New Brunswick grade, with MacPhee doing the fieldwork in Antigonish and Guysborough, while Lord worked with perhaps a dozen small producers with similar interests in the west. Located in the Springfield area, deep in the back of Queen's, Annapolis, and Lunenburg Counties and farther from the German and Dutch individualists, the latter saw the advantage of cooperatives. The operation continued for five years, normally breaking even on overall account. One year there were several mix-ups when the grader allowed poor trees to be sent, and the settlement put the Christmas tree account into a loss. But they were able to ship upward of 100,000 trees, amounting to $200,000 worth of business by the end.

As time passed, many of the non-Christmas tree members of the provincial association grew concerned about the amount of Lord's staff time being diverted to this project and about the financial exposure of the association. One solution was to shift the tree operation out of the association entirely. By now interested and knowledgeable, Lord transformed the tree enterprise from a cooperative to a private proprietorship in 1976, with some coordination with MacPhee, who did the same in the eastern region. Lord reached agreement with the association to take two months' leave each fall during the tree season (October-November), which eased the association's always fragile finances while affording him the necessary time to direct sales and shipping. By the early 1980s, his Nova Balsam Christmas Trees was shipping some 50,000 trees per year. Today his business continues at this level, with the majority of trees being bought from local growers and about one-quarter being cut from his own 100-acre lot.

The Woodlot Owners Association showed few jealousies or suspicions in the face of other organizations. In the early 1970s, the association helped to bring together some growers from New Brunswick and Nova Scotia to form the Atlantic Christmas Tree Council. Its main achievement was to approve a joint grade modelled on the New Brunswick standard. Later, when the Maritime Lumber Bureau decided to get out of the grading business, the Atlantic body lost its principal rationale and became inactive. Meanwhile, a Nova Scotia Christmas Tree Council (NSCTC) had evolved. It was a loose federation of the preexisting regional groups, led by Lunenburg and the Antigonish-Guysborough committee, with smaller groups from Cape Breton, Cobequid, and Cumberland. Representation on the volunteer board was weighted according to membership size, and the council met several times each year to deal with matters of trade information, lobbying, and crop research.

The council membership was firmly committed to improving the industry and maximizing returns through better techniques of cultivation, consistent grading, and a focus on the premium tree market. As recently as the late 1960s, most balsam exports were wild trees. Volumes declined steadily between 1958 and 1976 (1,000,000 trees), largely due to the low quality of the uncultivated Nova Scotia product. While most tree lots are natural stands (as distinct from plantations) created by new growth after clearcuts, the value added by cultivation was significant. The basics were spelled out in the Provincial Forest Practices Improvement Board's publication *The*

Trees Around Us (1980) and later in the Nova Scotia Christmas Tree Council's *Christmas Tree Grower's Manual* (1987).[52] Early in a rotation, these natural stands required thinning and spacing and some planting to fill gaps. Then annual weeding, fertilizing, pruning, and shearing were required to produce superior trees and to meet grade standards. A managed stand might be stocked to the level of 1,000-2,000 trees per acre, facilitating an annual harvest at a rate of up to 100 trees per acre. The theory and practice of cultivation, as promoted by the council and its regional affiliates, influenced only one segment of the industry. Of the approximately 3,000 growers in the province in the early 1980s (half of them in Lunenburg County), about one-third were dues-paying council members.[53]

During the 1970s and early 1980s, the American market became increasingly stratified into premium, standard, and utility grades, with accompanying price differentials. At the same time, demand outstripped supply for much of the decade, resulting in a sellers' market in Nova Scotia. Lord recalls many growers cutting their trees and offering them at the roadside. With ten or fifteen buyers pursuing the stock, auction conditions prevailed in the open buying yards of those years. Another side of tree cultivation involved cutting to specifications. Lord recalls a major contract to Germany during the 1970s amounting to 25,000 trees per year. The clients wanted more 'open' trees, still symmetrical but less dense than the ideal balsam of the day and ideal for the continental market. This was achieved by altering the field treatments. In general, however, export markets remained strong into the mid-1980s. Growers could sell almost anything, often dictating the prices.

By this point, however, a huge increase in growing capacity loomed in both Canada and the United States. American growers were selling 35,000,000 to 39,000,000 trees annually but planting 100,000,000 to 115,000,000 in the same period.[54] This coincided with a shift in consumer demand that resulted in almost half the continental market going to artificial trees. As a result, the low-quality buyers were squeezed tightly and found their shares decline (except to some poor offshore destinations). And cultivated tree growers saw their prices begin to decline in the late 1980s in the face of the supply glut. The slumping market triggered a further spiral of oversupply, as some large American plantation operators went bankrupt, leaving their creditors holding millions of young trees. These trees were liquidated at a deep discount (in some cases twenty-five to thirty cents per tree), allowing new entrants to bring huge volumes to market over subsequent years at virtually no cost. Nova Scotia prices to growers in the mid-1990s were still lower than eight years earlier because of massive oversupply. Today the dealer must work intensely to win sufficient contracts to hold market share, visiting trade shows for garden centres, nurseries, and service clubs and contacting wholesalers and retailers throughout the eastern United States. Trees are cut to match orders according to size and grade, with dealers often dispatching staff to the growers' lots to mark trees for cutting and sometimes having to set quotas for suppliers.

Beginning in 1984, the Nova Scotia Christmas Tree Council joined the American National Christmas Tree Association in a promotional campaign known as 'Operation Real Tree.' Financed by a voluntary five cents per tree check-off by members, it has been credited with boosting demand for the product. However, Lord sees the 1986 forest products trade dispute between Canada and the United States as the time when the Nova Scotia Christmas Tree Council and the fledgling Canadian Christmas Tree Growers Association came into their own. After Washington slapped a countervailing duty on Canadian exports of cedar shakes and shingles, the Canadian government sought to retaliate by applying tough duties against American Christmas tree imports to Canada. In its desperation for a face-saving response, Ottawa shut out a 100,000-tree import stream (mainly to western Canada) while jeopardizing an export stream of 4,000,000 to 5,000,000 trees (half of this from Nova Scotia). To the Maritime producers, this came as a bolt out of the blue. Members of the council 'lived on Mike Wilson's door' for several weeks, while planes and busloads of growers travelled to Ottawa and Washington to drive home the magnitude of the error. Like no issue before, this brought the growers together while demonstrating the value of the council as a collective lobbying voice.

The council continues its lobbying and service roles to members. In addition to the grower's manual mentioned above, it has joined with government scientists in research studies of tree genetics, aiming to bring superior stock into a seed orchard program. In government relations, the council encouraged the province to amend the Lands and Forests Act to establish a Nova Scotia grade, with the department handling the grading function. Another amendment authorized the mandatory check-off of 1 percent of tree value at point of sale by the growers, subject to two-thirds majority approval by referendum. In such a market-oriented business, any 'compulsory' overtones would prove controversial. Lord notes that they watched the American association 'tear itself apart' over a proposed compulsory check-off several years earlier. By contrast, and after much debate, the Nova Scotia producers approved the levy in the summer of 1993, for application in the 1994 selling season, to fund various industry projects. Ironically, this was accepted by Christmas tree growers, while the pulpwood levy sparked such resistance, often in the same individuals.

In fact, Nova Scotia growers face a crowded agenda in the late 1990s. Intense competition from southern-grown Fraser fir trees seems now to be a permanent fixture in northeastern markets. The gypsy moth has been spreading into and across the province (carried as egg masses attached to moveable property), and the council has worked extensively since 1993 with customs authorities on both sides of the border in defining infested areas, devising a workable 'certificate of origin' system to protect access for non-infested areas, and a method of phytosanitary inspection for clean trees shipped from infested areas. Other policy issues include research into integrated pest management (a combination of natural and chemical controls), workers' compensation charges (distinguishing Christmas tree lot work from logging contracting), and assorted tax issues. The operational horizon of the council was well sketched by one prominent grower in 1993:

> This business has changed substantially during the past 10 to 12 years. The process of growing and selling trees faces more challenges today than ever before in its relatively short history. Governments regulate commerce to a much higher degree today and the industry must deal with far more competition from within. Europe prohibits our access to their market. The artificial tree industry has made substantial inroads, and many consumers are coming to believe that cutting any tree is wrong, without any understanding of the sustainable nature of our crop. The Christmas Tree Council wants to tackle these issues, but their resources are limited. Gypsy moth quarantine measures may limit our access to US markets, and this issue must be handled. The recent effort to address the countervailing duties requested under the Special Import Measures Act cost the Council and its members a great deal of money to manage and correct. European markets for balsam fir could be opened with enough effort. Training courses for growers needing pesticide applicator licensing must be further developed and funded. Growers of other tree species such as Fraser fir and Douglas fir are spending far more than us on tree improvement. We have developed further than any other group but will need core funding to continue to provide you with the best trees in the world to cultivate for Christmas trees. And our efforts to promote balsam fir and further enhance our cultural practices to maximize sustainable development remain woefully underfunded.[55]

Lord was active, of course, in all of the Christmas tree organizations since 1967. The Lunenburg County producers and the NSWOA tree-marketing cooperative came first. Yet given the overlapping connections at the county, provincial, and national levels, Christmas tree activists have tended to move easily from one level to another. Lord was also a founding member and guiding influence on the Nova Scotia Christmas Tree Council from 1974. After the establishment of his Nova Balsam in 1976, his involvement continued as one of the middle-sized growers/dealers shipping directly to the United States. When the Canadian Christmas Tree Association was founded in 1972, he was its charter president.

For Lord, the impressive development of the Nova Scotia Christmas Tree Council over the past two decades offered a welcome but in some respects

Richard Lord inspecting Christmas tree plantings on a lot in Lunenburg County, Nova Scotia.

bitter contrast to the fate of the Nova Scotia Woodlot Owners Association over three decades.[56] The two initiatives had much in common. Both sought to link local forest owners into a province-wide association to enhance member prospects with services and representation. Both were animated by a leading edge of commercially oriented owners who sensed the need and the opportunity to transform their forestry and business activities. The two associations even had a common point of origin, insofar as the 1967-70 woodlot-organizing campaign generated both the NSWOA and the LCCTPA. Yet at the same time there were a number of decisive points of contrast. It might be argued that the success of the Christmas tree growers flowed from their deliberate decision to avoid organized marketing, focusing instead on production, training, and lobbying issues and thereby avoiding the internal cleavages to which commodity-marketing schemes are prone. This also reduced both the opportunity and the need for an early compulsory levy to support the association's work, which often triggered predictable tensions between members and non-members.

While undoubtedly pertinent, this analysis overlooks the significant structural contrasts between the two types of woodlot interests. The Christmas tree export industry did not face spontaneous resistance from a concentrated pool of large domestic buyers whose provincial political leverage was substantial. Moreover, the favourable stumpage conditions and rotation period for Christmas trees permitted the early commercial recovery of silviculture costs within the growing enterprise, rather than depending on public subventions (grant or tax) to make silviculture viable on a comprehensive scale. Finally, the policy response of state officials differed critically in the two instances. The DLF was willing to treat the council as both a legitimate representative voice (despite its limited membership base in the tree-growing segment) and as a potential instrument for collaborative program delivery (particularly on questions of research, customs documentation, tree culture, and advertising). This last role was solidified by the timing of the council's emergence: immediately prior to the 1977 Forestry Development Agreement, with its unprecedented cornucopia of resources.

Conclusion

The central interests of Rick Lord's forestry career emerged early. Far more than the classroom, it was summer employment that exposed the student forester to key influences. For Lord, these had little to do with the central streams of industrial or government forestry. Instead, questions of forests for people, particularly private owners on the social margin, were captured in his early research. The St. Francis Xavier project was a natural extension of this. While few foresters in the nation would be well prepared for such a job, Lord was, by background and inclination, better suited than most. He was, as he says, 'always an organization sort of person.' On the woodlot-organizing project, his initial role was to fill a professional niche as the forester within a multidisciplinary team. Yet the key skills as organizer and later as association manager had little direct connection to his academic training, though they had much to do with his emerging professional identity. Perhaps, given his background, being a forester was less a hindrance than it might otherwise have been. Moreover, by viewing Nova Scotia woodlot owners alongside the Gloucester Acadians and the Anglophones and Francophones of the Eastern Townships, he gained a special insight into private land forestry in eastern Canada.

Much was accomplished, in an extremely short time, by the St. Francis Xavier/ARDA project. Within two years, woodlot activists from across the province were meeting together and planning an ambitious, multidimensional program. Then came the loss of innocence. Resources were exhausted, the grassroots atrophied, and momentum was lost. The association found itself marginalized in government circles and repeatedly blocked by corporate lobbying and litigation that challenged the legitimacy of government policy. Then, in the midst of guerilla warfare on the marketing front, the basic coordinates of provincial forest policy were thrown open by the prospect of the 1977 agreement. All the while, the NSWOA battled with an industrial adversary that had the resources to successfully paralyze the marketing scheme for a full decade.

By 1980 Lord's association bore the scars of this titanic battle. The need for a woodlot organization, with the constructive potential for forest policy and management, remained as pertinent as ever. This is clear from his 1977 presentation to the Nova Scotia professional foresters, which appears in the exhibit that follows. Indeed, some of this potential was being realized in New Brunswick, where provincial authorities took a somewhat more progressive line. However, the decade of struggle exacted a great toll. Exhaustion and disillusionment beset veteran members and discouraged prospective members. The prospective leadership was divided by region (east, central, and west) and by function (supply groups, group ventures, educational and lobby groups), leaving little by way of resources and mandate for the parent association. This fragmentation set back the cause of woodlot owners by at least a decade and perhaps a generation.

On the other hand, the Lunenburg group and the Nova Scotia Christmas Tree Council demonstrate the impact that progressive voluntary associations can make: harnessing scientific knowledge to practical field problems, promoting advanced silvicultural practices, providing market information to members and mounting shared promotional campaigns, and maintaining positive government relations for those issues that require external help. The difference, in large part, lay in the structural relations of the Christmas tree business. The short rotation was suited to intensive culture and cash return, and the export market removed the threat of a regional purchasing monopoly as in pulpwood.

Appendix

Forest Management Situation on Private Holdings (Richard Lord)[57]

Introduction

The maintenance and the prospects for expansion of the wood-processing industry in Nova Scotia in the future will depend substantially on what our smaller private woodland owners decide to do with their forest holdings. Unless some rather drastic changes take place in recent and current attitudes and practices, the outlook is not encouraging.

In the time available, I want to try and briefly outline why the private ownership sector will play such a decisive role, some factors which have contributed to current negative and unproductive attitudes and practices, and finally, the prospects and means of altering current trends to satisfy the needs of the owners, the industry, and our forest economy as a whole.

Future Timber Supply

Without quibbling over various inventory and other statistics, it is apparent to all that our timber resources are no longer inexhaustible. Even prior to the most recent and serious budworm infestation, specific shortages were being predicted for the near future. It is safe to project that from sometime in the late 1980s or 1990s onward, virtually all available timber will be in demand to fill even the current level of processing.

Constraints on Availability and Production

Against this coming pressure for availability and wood supplies, it is inevitable that constraints on production are going to be applied. In one arena, growing public pressure will inevitably restrict indiscriminate harvesting practices. At the same time, we can expect to see more timber land set aside for recreational and other uses. Large industrial holdings, whether Crown leases or freehold, will be the most susceptible to, and the first to feel, these pressures, if they are to have any prospect for acceptance by the private owners, [they] must first be seen to be in effect on Crown and industrial land holdings.

At the same time, there is little reason for optimism that 'new forests' will be developed in time to meet expected shortages. Although accelerated fibre production based on plantations of poplar hybrids, European larch, and other species is already beyond the experimental stage in many areas, there is little indication that Nova Scotia is prepared to move quickly enough in this direction, or that such a program could provide the amount of fibre required in the time available.

Thus, it seems clear that in the fairly near future, the demand for wood from whatever sources are available will become intense. The major potential source of wood physically available to meet needs of industry will be the collective holdings of individual, non-industrial owners. Together these people own over fifty per cent of Nova Scotia's forest resource. These properties are the most accessible and potentially productive of our forests. They are dispersed more or less evenly throughout the Province, and by their nature are less susceptible to devastation by such agents as fire, insect or disease. They are less exposed to public pressure, and to legislative restrictions; they can respond more quickly to management or other initiatives of the owner. This major sector of our forest resource could easily provide a much higher proportion of the future fibre requirements for an expanding wood-using industry. Whether this role will be assumed is a question very much in doubt.

Problems in the Private Sector

From any point of view – owner, industry, or government – the private forestry situation in Nova Scotia today is sick. The level of management and production is at an all-time low; relatively few owners have any sustained interest; mature timber is left standing; immature stands are being clearcut; and cut-overs are for all intents and purposes abandoned. A majority of owners are apathetic – or at best indifferent. Only a declining minority engage in any annual or periodic harvesting. Most operations involve the lump-sum sale of stumpage for clearcutting of the woodlot by private or company contractors, marking the termination of the owner's interest in the land. Reforestation or silviculture in timber stands is virtually non-existent. Ownership of much of the resource is being acquired by others who have no interest in commercial forestry. By various means, vast acreages are being effectively removed from future timber production.

The private forestry situation has deteriorated to this condition gradually over a considerable period of time and for a number of reasons. Until fairly recently, our private forest lands were generally very well managed, providing supplementary income to people who brought with them to this country and maintained a strong respect for the forest and a sense of stewardship. Changing social, cultural, and economic realities have weakened this sense, or in some cases perverted it, to the point where today it is reflected most often in a reluctance of the owner to permit any cutting at all.

Owners who traditionally earned their livelihood from combining a number of resource activities – farming, fishing, forestry – have come to specialize in some particular endeavour which now absorbs all their available capital and time. Others, although remaining as rural residents, have turned to various trades or employment opportunities which are more lucrative. Along the way the woodlot is neglected. Often, time has simply caught up with the owner. The property is frequently held by an elderly man – or by his widow – whose children have left. The woodlot is neglected, then eventually clearcut or sold to obtain retirement income, or to settle an estate. In either case, it is out of any sustained production.

For those who retain an active interest in the woods, either through a desire to see it properly managed or as a source of employment and income, there are substantial difficulties. Undeniably, prices paid [to] the producer for primary forest products are the lowest in Canada, and offer little margin for profit. Markets are limited and subject to fluctuation. Skilled help is virtually non-available, nor are training programs suitable for individual owners. The owner is provided with little or no information on markets, new techniques or equipment, record keeping, etc. Credit facilities to obtain equipment or undertake improvements, or to acquire additional available acreage to build a more viable unit, are non-existent.

Consequently, the more progressive woods-oriented individuals generally find it far more advantageous to neglect their own holdings and seek employment on company-operated land, or to turn to private contracting, purchasing stumpage from other owners. Economic pressures then generally force these operators into undesirable cutting practices.

These situations are not unique in Nova Scotia, other than in their impact. They create a serious problem here because so much of our total timber resource is affected, and because the economic survival of many of our rural communities is closely linked with the private resource.

The problem is compounded by what many owners see as a history of government and industry indifference, or interference, and this has created a climate of cynicism and apathy among many owners. Specifically, well-intended but impractical restrictive legislation, such as the Small Tree Act, disenchanted many private owners and operators. More generally, the objectives of the private landowner and those of government are often felt to be at cross-purposes. Take, for example, the Department's long-standing desire and various programs to increase the extent of Crown holdings by buying up private woodlots. The owner sees this, when coupled with low product prices and inadequate services from his government, as merely an insidious form of expropriation.

Unlike other provincial forestry departments – and unlike other government departments in this province – Nova Scotia has never had an active private forest land extension service. Advisory services to owners have been minimal – practically non-existent – a reflection of the lack of importance placed on the private land sector by government. In fact, in 1970, the understaffed and underbudgeted Extension Division was disbanded in a reorganization of the Department. There has been little communication between private owners and government (compared, say, to the active role undertaken by the Agriculture Extension Service), and the Department is seen by owners primarily in terms of fire protection, game laws, restrictions, and as a friendly landlord, if not a bedfellow, of industry. At the same time that field staff and services have been inadequate, the Department has had no senior staff specifically trained or experienced in, or responsible for, particular problems pertaining to private land forestry.

Similarly, industry in the Province has, with few exceptions, limited its role with respect to private landowners to that of offering a market as, and to the extent that, purchase wood was desired. Given this climate, it should not be surprising to observe that over the last decade, production has declined substantially on private lands. Where private woodlots used to provide almost sixty per cent of the wood fibre requirements of the industry, they now account for less than forty per cent. Moreover, much of the wood that is being harvested is the result of a long-term owner's decision to liquidate his holdings, and the land is unlikely to be returned to active managed forest production.

At the same time, other properties are being withdrawn from active production by owners who have acquired, or who retain, the property but have no interest in timber production, or who are simply unwilling or unable to produce wood for a variety of reasons.

Thus, we have a pattern of

(1) mature and overmature woodlands being withheld from production;
(2) final one-shot harvest operations followed by virtual abandonment of the land resource; and
(3) sustained forestry production from only a diminishing minority of owners.

The experience with regard to woodlot owner organization and cooperation can be viewed against this overall prevailing pattern. The same 'indifference and apathy' and diverse objectives, both among and between owners and government, which give rise to the need for a strong and effective collective interest group, have served to restrict the development and effectiveness of such groups.

Recruitment of members and the active involvement of owners have been less than satisfactory by some criteria, but when the above considerations are taken into account, perhaps the response is remarkably high.

The interests of private woodlot owners and of commercial logging operators can be quite diverse, but some considerable success has been achieved in bringing these two groups together – particularly in the eastern half of the Province. The efforts to obtain a workable pulpwood marketing mechanism to provide for the negotiation of pulpwood prices and condition of sale have developed into a continuing, and so far unsuccessful struggle, ... but the effort is continuing, and it will be successful. The long and frustrating delay, in addition to being extremely expensive, has undeniably discouraged many owners, and contributed to an attitude of futility and apathy.

Another major contribution to this sense of frustration has been the failure to develop a close and responsive working relationship with government officials. Far too often, owners are overly cynical and critical of government policies and objectives and assess these only from a fairly narrow and personal viewpoint. While there are many who meet that description, there are more who, more unfortunately, have been totally indifferent. Yet there still remains a substantial minority who are more open to various viewpoints and retain enough optimism to remain interested and involved.

On the other side, government officials are often too quick to exclude private sector representatives from their deliberations, to adopt a paternalistic attitude, and to attribute the shortcomings in the private sector to the very people who, by virtue of their continuing interest and involvement, have taken the lead in attempting to improve the situation. If the approach taken by individual owners or their organizations to this end is sometimes less than diplomatic, and if the panaceas called for are sometimes less than wholly practical, then reasons should be fairly evident, and the response should be more constructive and less critical. After all, as long as the patient retains enough life to respond to the pin pricks, some hope for recovery exists.

Where to from Here?

What private owners decide to do with their forest holdings will have a significant, perhaps crucial, effect on the future of our wood-processing industries. For some time, the outlook has not been promising, but it may not be too late to reverse the current trends.

It is a basic premise of this presentation that there remains a place in the forest economy of Nova Scotia for a strong, viable, and productive sector based on relatively small private forest ownerships. The tradition, the rights, and the responsibilities of private ownership are deeply instilled, and will not be surrendered easily. Nor should they be. Neither take-over by the Crown, nor absorption by the corporate multinationals, will do anything to improve the productivity of our privately held forest resources. In this regard, we can only hope, thanks for once apparently to the Federal Government, that the provincial objective of acquiring all available private holdings has been effectively squelched. Rather, the objective must be to provide a climate in which private owners will have the opportunity and incentive to retain and develop their holdings.

The problem remains, then, of how to restore these private forest lands to productivity. Given the diverse desires, interests, and objectives of the current owners, no single or simplistic responses will suffice.

Firstly, there is little prospect that many of those who have left to find trades, professions, or other means of livelihood will return to the woods. Nor will many of the older generations still holding a typical 150-acre woodlot be able to respond directly to any management opportunities or incentives which may be offered. Yet, if only one in ten or one in twenty of the current owners remain able and interested in sustained forestry production and management, and if these people can be shown some modest economic incentive, a basis for continued and greatly improved production exists. Consolidation and loan programs must be devised to respond to this need.

Owners obliged through age or necessity to surrender their holdings will have the satisfaction of knowing their resource is in good hands, and younger people, by acquiring these woodlots, will have the opportunity to expand and manage larger forest blocks, retaining individual initiative and pride of ownership.

Similarly, the group management concept, originated in Quebec in the 1960s and introduced in Nova Scotia with the West Pictou Pilot Project, may meet the needs of a variety of owners in a given locality, and foster a level of management and production which could not be undertaken individually.

Under the new Subsidiary Agreement, emphasis is to be placed on bringing private woodlots under management, and substantial incentives are being offered to encourage this. However, as has been shown, few owners have the ability, time, or interest to undertake the actual work themselves. There is a critical need for highly skilled silvicultural and harvesting crews to carry out the management plans. There is little to be gained by having owners commit their property to long-term management agreements if there is nobody available to carry out the work.

It is also critical that an equitable marketing system be developed. Whatever the merits of the arguments of any of the parties involved in the prolonged Pulpwood Marketing Board dispute, surely it can be accepted that private producers have a right to some form of collective negotiation of prices, quantities, and conditions of sale. The failure to achieve this objective has frustrated and demoralized many of the most capable and active producers, and has certainly contributed to the decline in private management and production. Few of even the most ardent advocates of a Marketing Board accept that such a mechanism will result in drastic price increases, or even that such increases would resolve many of the current problems. But the right to bilateral negotiations, and the prospects of exercising some degree of influence on the price of their product, are seen as legitimate aspirations, the denial of which generates hostility or apathy.

A related and major consideration is the unavailability of markets. In recent years, producers have been faced with reduced quotas, or have not been able to obtain markets for their wood. Obviously, this situation has done little to encourage any long-term management.

At the same time, these producers feel that government commitments to guarantee wood or stumpage to the industry at low rates and in excess of the actual requirements serve to undercut their position with respect to both supply and price. To overcome this, Crown land should, in future, be considered as the residual, not the principal, supply of wood to the industry, with priority to be given to available supply from private lands.

There must also be a change in the relationship between government and the private owners. The private sector, both to build up its self-respect and to protect its interests, must be given a greater role in influencing government policy decisions, and in implementing these decisions. For a start, greater priority must be placed on responding to the problems and needs of private owners. The attitude that government is paying the bill and must, therefore, call the shots fails to recognize the major commitment of the owner who agrees to place his land under management. Owners, through their organizations, must be allowed more direct involvement in matters that directly affect them.

As well, owners must be assisted in defining their own needs and responses, both individually and collectively. This can be accomplished through an expanded advisory service, which must go beyond technical field advice and include the dissemination of a great variety of information on such topics as markets, techniques, bookkeeping, tax considerations, and the like.

The minority of owners who demonstrate their interest and concern by becoming involved in various issues must be encouraged, and they must have the opportunity to develop educated and informed views. These are the people who, individually and collectively, by example and by leadership, can best bring about the improved management and production levels on their own lands and on private woodlots as a whole.

On the same basis, various forestry programs can best be implemented by placing the administrative and operational responsibilities in the hands of owner organizations on a fee-for-service basis. Such an approach will tend to generate a strong sense of purpose and accomplishment for these organizations, will increase their visibility and effectiveness in representing owner interests, and will develop both participation and leadership among forest owners and operators.

Finally, government and industry must be prepared to accept some risks, and some losses, in implementing various policies and programs. Too often implementation is delayed or denied because all possible loopholes or side-effects can't be safeguarded against in advance. Consequently, generally desirable programs are set aside as impossible or impractical, or are introduced with so many restrictions or conditions that prospective participants won't take part.

In summary, the factors which have led to the decline in management and production by private woodland owners must be resolved if future demand for wood fibre in Nova Scotia is to be met. The interests of all parties – government, industry, and the owners – can only be served by creating the proper conditions. Many aspects of the new Forestry Subsidiary Agreement represent a positive step in this direction, but at the same time, parts of the Agreement itself, and the manner in which it has developed, are symptomatic of some of the current, continuing malaise.

It should also be realized that many of these problems run deeper, and will not be resolved by the provisions of the Subsidiary Agreement. The Agreement will reach and assist only the minority of owners who are the most involved.

Nevertheless, it is to be hoped that in the course of implementation of the Subsidiary Agreement, substantial progress will be made in meeting at least some of the needs of many private owners and indirectly of both industry and government, and that new approaches will be evolved to respond to the needs not covered by the Agreement.

Reproduced courtesy of Richard Lord.

8
Mary Guptill: The Theory and Practice of Field Forestry

In so many respects, the decade of the 1970s was a transitional one for forestry. Mary Guptill's experiences are an apt reflection of this. Guptill was one of the first women, since Mona Roy in 1948, to complete a forestry degree at the University of New Brunswick. This makes Guptill part of the fourth generation of forestry professionals in Canada.[1] The practical difficulties of accommodating female students in the faculty, especially with respect to fieldwork, gave her early exposure to the difficulties of transferring classroom learning to the field. This played an important role in her decision to become a field forester. By the time she entered the workforce in 1978, the face of forest policy had been transformed, in the Maritimes as elsewhere, by the new generation of federal-provincial forest development agreements. Not only did this underwrite a vastly expanded program of intensive silviculture, but it also created an impressive range of new jobs for young foresters within the many new programs. Guptill spent more than a decade working in this context, as a field forester for La Forêt Acadienne, one of the newly conceived group venture enterprises unique to Nova Scotia. Her experience offers a fascinating glimpse into the challenges of planning and implementing management plans on the small private woodlots of the province's French Shore.

In both her education and her professional life, Guptill provides a telling picture of the tension between the theory and the practice of forestry. The theoretical forestry that she was exposed to in forestry school, and the stipulations and regulations of the federal-provincial forest development agreements, often came into conflict with their implementation on the ground. Guptill's career also illustrates how both forestry school and the agreements set technical parameters for field action, a condition that left Guptill both uninterested and reluctant to join the bureaucratic game. Her perspective on the Group Venture program differs from those in prior chapters. Guptill did not experience the bureaucratic tensions faced by David Dwyer as he fought for budget dollars and executive support within the DLF during the formative years. Nor did she face the frustrations of Rick Lord, who spent years advocating for woodlot owners to deliver their own publicly funded silviculture program, only to have the NSWOA bypassed in its design. What Guptill did experience was the reality of group politics at the operational level, among La Forêt Acadienne members and in dealing with wood buyers such as Mersey.

The youngest of the foresters discussed in this study, Guptill was born in 1955 and raised in the urban environment of Halifax. Her interest in forests started at summer resident camps, where she spent ten years as a child and teenager, most frequently at the YM/YWCA camp at Big Cove in Pictou County. In high school, biology was a favourite subject for her. It was after taking her exams at high school and then returning home that she first sensed a connection between science and the real world. This became evident simply by looking at the maples or oaks growing in the backyard of her parents' house.

After high school, Guptill considered going to France to learn the language, but ultimately she decided against going. She also contemplated enrolling at Dalhousie University to study biology. Although her transcript showed grades at the A level, the prospect of dealing with the large student numbers and the large classes was not at all appealing. A high school visit to the Dalhousie pathology laboratories did not help the situation. She could not see herself working with test tubes in a white coat under fluorescent lighting in a laboratory. Instead, she became increasingly intent on working in an outdoor environment.

At times, critical choices can be triggered by simple coincidence. The sight of a person wearing a University of New Brunswick forestry school jacket prompted Guptill to write to the school for information. This triggered the whole process. She applied, was offered a scholarship, and entered forestry school in 1973 at the age of eighteen. She was one of three women in an entry class of 150. The path to the bachelor of science in forestry degree was a five-year program.

In Forestry School

At the university, Mary Guptill encountered a highly academic approach to forestry. This conveyed a basic understanding of ecological patterns, plant growth, tree silvics, and forest management. The first year at forestry school covered introductory science courses such as biology, chemistry, and mathematics. The second year covered upper-year prerequisites, again in the sciences, such as biochemistry, cell biology, and computer sciences. It was at this stage that she decided to stay with forestry.

In her fourth year, Guptill began to study the economics of forestry. She also took a course in forest policy. In her fifth year, she studied intensive silviculture. She learned that trees can be managed in systems for a variety of purposes, such as wildlife habitat, recreation, and industrial use. The forestry students did tackle one major project, in the form of a group-based watershed management report. The terms of reference were spelled out at the start of the year, and a study area was assigned. Professors were available as resource people, but the class was expected to organize itself. This was an intensive exercise in teamwork and organization; everyone had to follow a bit and lead a bit.

In retrospect, Guptill wonders whether too much pure forestry was taught at the school, while applied and social science aspects of the profession were neglected. Also, much of the curriculum seemed to be uncomplicated in retrospect. 'I had learnt nothing of the double messages between the theory and the practice of forestry.' However, Guptill was exposed early to some of these double messages through forestry fieldwork during the summers. Not all of the forestry students chose to do such work. For those who did, jobs with companies were quite popular, often on cruising parties in the woods. For Guptill, this was not an option, since companies would not hire female forestry students for the camps. This began to change shortly after she graduated, as employers agreed to hire women in pairs. Given the situation, Guptill accepted summer jobs in forest research, arranged through the UNB faculty. It was in these positions that she acquired her first critical perspective on the complexities of forestry.

Three of the four jobs were based in Fredericton, normally involving a combination of fieldwork and much time in the laboratory. One season, the study centred on leaf litter. Guptill's work involved sorting samples of leaf litter taken from the forest floor, identifying different types of stems, leaves, or whatever fell to the ground. By the end of the summer, Guptill could recognize them all, even in very small proportions. Another season, she helped in a research project, based at the Nashwaak Experimental Forest. The work centred on the collection, control, analysis, and storage of various climatological data. Part of the data that she collected was used for a supervised research project, required as part of the curriculum. This involved a study of temperature gradients at various levels of the tree canopy, in the air and in the soil, to understand the impact of solar radiation within the stand. The fieldwork involved frequent readings of a measurement apparatus fixed within the canopy, which was reached by climbing a triangular radio antenna that had been erected for the purpose.

A third summer season was spent in Ontario at a research camp in Algonquin Provincial Park. Guptill's work was part of a project involving a series of spruce plantations throughout the province to test the response of spruce seedlings from all over the world when grown in different climatic zones. This was her practical introduction to tree-planting practices and growth patterns. Another project that summer in Ontario involved meteorological studies of precipitation in the forest. The instrumentation was set up in a stand of spruce, to gauge flows of rainfall. A gauge on a nearby lake measured total precipitation, while in the stand the aim was to measure patterns of rainfall landing on the surfaces of trees (leaves and trunks) and on the forest floor, the patterns of runoff, the proportions that evaporate, and the proportions absorbed by the trees.

Most of the fourth summer was spent assisting in a province-wide plantation survey in New Brunswick. The fieldwork consisted of finding old plantations, establishing stocking plots, and collecting sample seedlings. In the lab, the biomass of the sample seedlings was determined.

One advantage of these summer research jobs was that the work could carry on into the school year, offering a useful source of continuing income. More importantly, they offered firsthand experience with a variety of research problems and settings, extending far beyond the exposure available in the classroom. The work itself, combined with the many hours of discussion with forestry professors, cemented Guptill's commitment to forestry. At one point, the momentum of the research work helped to sustain her flagging interest in the program. It also led Guptill to conclude that her role within the profession was with silvicultural work, improving the conditions in the woods.

This was well illustrated by her thesis project, which examined plantation stock in northern New Brunswick. Guptill compared the techniques of container planting and bare root planting, taking samples of two-, three-, and four-year-old growing stock. Dry weights were measured to establish stem-diameter gains, and root characteristics came in for close study. The results left her with a healthy scepticism of planting for the rest of her career. The research showed what plantations were doing and what they were not doing. It was clear that, in much of the planting then being done, the root systems were not developing sufficiently well. For various reasons, the roots were experiencing serious difficulty emerging equally around the stem.

For example, pine trees were planted by the bare root technique: a special shovel was inserted into the soil to open a crack, the seedling was inserted, and the crack closed as the shovel was removed. This tended to produce a lop-sided root structure, with the trees developing greater stability on one side than on the other. On the weaker side, some trees could be pushed over with relatively little effort. Container planting involved various methods, one utilizing split plastic tubes and another paper pots. The tubes were designed to allow the roots to force their way out over time, while the peat-filled paper pots were expected to rot away in the ground. However, the paper pots were not rotting away as intended, and ultimately the chemistry of these paper containers had to be changed. As for the tubes, Guptill found sobering results where black spruce trees in tubes had gone into shock after planting. They were only ten inches tall when growth was arrested. Back at school, the techniques of planting had been thoroughly covered in course work. But the problems of implementation had barely been acknowledged, much less resolved.

A similar dilemma of theory versus practice affected the social and organizational complexities of practising forestry. These were subjects that barely arose in class. However, in the working environment, they can be critical: 'If you don't learn how to deal with people, you can't do anything.'

Guptill believes that in certain respects a change in the student body in forestry school may have accentuated the discrepancies between what was taught in the classroom and what happened in the field. Many students in her own cohort, and in the older groups preceding her class, came from rural backgrounds and had strong rapport with nature and trees. They were capable people, skilled in the outdoors, but not necessarily strong academics. The new generation seemed to be more like Guptill herself, straight out of high school, with urban backgrounds and more academic inclinations. The curriculum itself also put increasing emphasis on the theoretical rather than on the practical. The attrition rate in the program was high. Of the 150 students admitted five years earlier, only 50 graduated in 1978, and Guptill was the only female graduate. This was not unexpected, as the students had been told at the start of the program that the persons sitting to their right and left would no longer be present at the end. Not all students dropped out of the program as a result of failure. Many transferred to other programs at UNB or to forestry programs at other universities.

The group that graduated in forestry, however, was tightly knit by the end, and they knew each other well. Looking back, Guptill recalls her male colleagues as generally supportive in their own ways. They were a group of eighteen to twenty-three year olds who could be alternately kind and crude, thoughtful and careless. They shared virtually all classes together, and there was seldom a need to go outside the faculty. Overall, she concludes, 'Being a female in forestry school was a pleasure.' And in a practical vein, it was a good training ground for work in what was predominantly a male profession. (As a female student forester in the mid-1970s, Guptill was part of a path-breaking cohort within the profession in North America. The USDA Forest Service hired its first two female rangers in 1976. By 1993 more than 14 percent of the line positions within the Forest Service were filled by women.[2])

Getting a Job

By the time Mary Guptill had completed the degree, she knew that she wanted to work in the field rather than in the office or laboratory. She was willing to move to wherever opportunity dictated. There were some possibilities in Manitoba as well as in Nova Scotia. So often in such cases, the first prospect tends to cast the die. In the summer of 1978, Guptill got her first job with the forest consultant firm Timmerlinn Limited. This made her the first woman to practise professional field forestry in Nova Scotia. The work involved drawing up management plans for individual woodlots, under the Private Lands Management Program administered by the Nova Scotia Department of Lands and Forests.

In the fall, Guptill was laid off from that position. By this time, however, she was reacquainted with the province and conversant with the employment scene. She was now aware of the group ventures initiative, which ran parallel to the individual lands program and was financed similarly by the Canada-Nova Scotia forestry agreement of 1977. Unique to this province, the groups were small shareholder-owned companies or cooperatives (see Map 4). They were formed voluntarily by small private woodlot owners at the local or regional level. Their mandate was to plan and manage members' forest properties by sharing the professional and silvicultural overhead costs. In this way, it was hoped, the program would accelerate the rehabilitation and active management of members' woodlots, in anticipation of the time (estimated to arrive in as little as twenty years) when Nova Scotia would face wood scarcities.

Guptill applied for two positions as a group venture forester, one at the North Mountain group in Middleton (Annapolis County), and the other

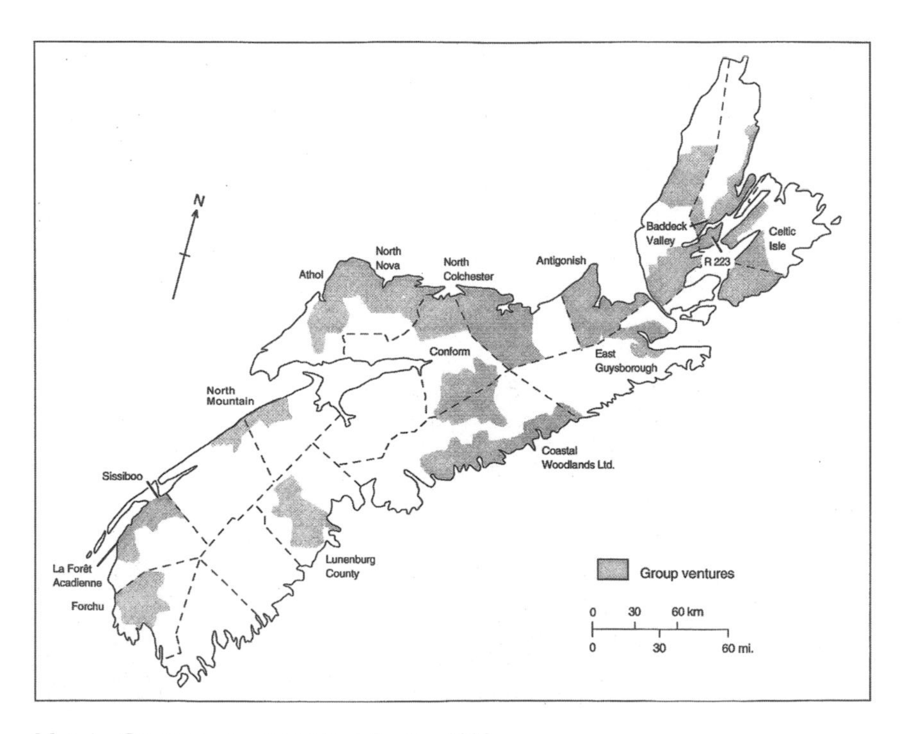

Map 4 Group ventures in Nova Scotia, 1988

with the Sissiboo group in Digby. She borrowed her mother's car and drove up the valley for two interviews on the same day, first at Middleton and then at the Hillgrove office of the DLF, where Sissiboo was set up temporarily. She had good hearings at both, but during the interview at Hillgrove she received an unexpected telephone call. It was Arcade Comeau, calling on behalf of La Forêt Acadienne, the group venture next door in the Municipality of Clare. He asked if she could come down to their office in Concession for a third interview. She agreed and met Comeau and the manager, Howard Ferguson, later that day. Guptill was offered the jobs at both Sissiboo and La Forêt. After taking time to think about both, she accepted the latter.

Her decision was based partly on the fact that La Forêt seemed already to be well established, with a strong manager and secretary in place. (This was at a time when there was much demand for the establishment of new groups across the province, and many were still at the formative stage. Sissiboo was not so far along, and it needed staff who could aid in the overall organization of a business.) Interested chiefly in the field aspects of forestry work, Guptill found La Forêt the more attractive prospect. In the back of her mind, once again, was her interest in learning French, and here again Clare offered the best prospect. It was an intuitive decision, based on her wish to take care of the forestry side of the operations while others would take care of the business aspects. She joined La Forêt in the fall of 1978, renting a huge house (with eight rooms) in Petit Ruisseau/Little Brook and setting about her work. Her expectations were high. Now she would be able not only to develop management plans but also to implement them.

Terms of Employment

A key objective of the Canada-Nova Scotia Subsidiary Agreement on Forestry, concluded in June 1977 and extending over a five-year period, was to promote increased timber productivity and commercial involvement on small private holdings. The problem was defined as follows:

> Privately-owned productive forest land in the Province constitutes approximately three-quarters of the resource with control distributed among an estimated 30,000 owners. It is estimated that up to one-half of this total, mainly in holdings of less than 1,000 acres, may not be available for forest production in the future. Therefore if a concerted Forest Management Program is not undertaken, as much as 2.5 million acres will be effectively unproductive while the remaining lands are being overexploited. In addition, these lands are now supporting far less timber than they are capable of growing, due to poor management practices in the past. A strong thrust to increase growth and production from these lands and to reduce the effects of fragmented ownership is essential.[3]

To this end, a strong emphasis was 'placed on the grouping of contiguous small blocks and the creation of group management ventures on a formal basis.'[4] Just over $1 million of the total $25 million (five-year) appropriation was allocated to the group program. The intention was to establish eight such groups across the province, modelled on the West Pictou Forest Owners Group, a prototype supported by the Nova Scotia DLF since 1975.

The core of each group was the membership, those woodlot owners who agreed to place their lands into management under group auspices. This relationship was the key to the entire structure: eligibility for assistance flowed from shareholder status in the group, which conveyed both privileges and obligations. Once a management plan was prepared for the woodlot and accepted by a shareholder, he or she agreed to accept the group as an authority on forest management. Shareholders also agreed to participate in management for ten years and to pay the group a 5 percent commission on the roadside price of all woodlot products sold.

Organizationally, each group was funded to support a core staff, usually including a manager, a bookkeeper-secretary, and a forester. The manager provided overall direction, handled shareholder business, undertook liaison with government agencies, and supervised forest operations. The secretary handled the office business and maintained records, while the forester dealt with the members' properties in cruising woodlots and preparing management plans. The operations budget also covered the costs of maintaining the office.

Forest improvement work approved in management plans was eligible for financial support under the authority of the agreement, under terms designed to cover the full costs of the treatments. The actual work could be done by crews employed by the group, by contractors engaged by the group, or by the landowners themselves. A crucial element in this silvicultural effort was an operational manual devised by federal and provincial authorities to define eligible activities, performance standards, and financial contributions. The range of eligible treatments was defined in the *Manual of Procedures and Standards* issued by the Forest Development Agreement. This detailed directory, known to the staff as the 'bible,' constituted the central parameter of woodlot planning: 'Only certain silvicultural treatments are eligible for subsidy, and these have been determined by the provincial government. The complete list and definitions have changed over the years as knowledge of the resource increases and as budgets tighten. However, the core treatments remain and include various stand improvement practices such as regeneration cuts and thinnings in merchantable and non-merchantable stands. Stand establishment is also funded, as are boundary line renewal and road construction. The provincial government also sets the quality standards and carries out periodic inspections to ensure that the standards are being met.'[5]

In all, there were four forestry agreements running almost consecutively for eighteen years. The initial 1977-82 deal was followed by a nine-month hiatus before the 1982-9 sequel was signed. It was succeeded in turn by a two year (1989-91) deal and a final 1991-5 agreement. In 1993 the federal government decided to terminate all its federal-provincial forestry agreements. In an effort to lobby Ottawa for a renewal, the government of Nova Scotia created the Coalition of Nova Scotia Forest Interests. However, when it became clear that the federal will was set, the coalition was reassigned to explore postagreement policy options.[6] In the spring of 1995, the provincial minister, Don Downe, declared that any future private lands funds would be restricted to the (individual) private landowner stream, leaving the eighteen groups to determine their own future as private organizations in the commercial marketplace.

As forest policies, these complicated agreements operated at several different levels and changed over time. Over the life of the four agreements, policy objectives, as well as budget suballocations and eligibility criteria, varied considerably. For example, the initial 1977 subagreement gave considerable emphasis to promoting new markets for (underutilized) hardwood species.[7] This meant that silvicultural improvements for hardwood stands received considerable support. However, the subsequent subagreement of 1984-9 gave far less emphasis to hardwoods, with the result that the detailed terms were adjusted accordingly.[8] For the Clare region, where hardwood and mixed-wood stands are prominent and the potential was considerable, this had first a positive and then a negative effect. At one stage, the forest was to be managed for the maximum number of jobs. At another point, pulpwood production was the most important. And at yet another point, the environment was supposed to be stressed.[9]

To complicate matters further, there were constant personnel changes, especially among the foresters, within the coordinating government agencies.[10] Such changes of personnel could themselves alter the priorities among policy objectives as well as affect the program's continuity and stability as people had to be trained and retrained. People would move up the career ladder or change agencies completely, and sometimes the 'bible' would change accordingly.

Ultimately, however, the individual treatment definitions and the shifting rationales and the procedures behind such treatments were a result of the changes in the objectives of the federal-provincial agreements. As they filtered down the program ladder, these varying objectives had a definite impact on the group ventures. The groups were always made aware of the prevailing goals and objectives, and they had to submit progress reports to the Department of Lands and Forests on a regular basis and then explain how the new objectives had been met.

These complicated yet changing agreements set the parameters and terms of employment for Guptill as the field forester for La Forêt. They could be enabling as well as constraining, and she had to negotiate a fine line to cope with them. The process illustrates once again, as it had in forestry school, the tension between the theory and the practice of forestry.

History of the Municipality of Clare and Its Forests

The Municipality of Clare, the home of La Forêt Acadienne, comprises some 227,000 acres of land, of which 65 percent is held by small owners. The municipality takes the shape of a triangle, with one side running along the shore and the other two municipal lines converging to a point in the interior (see Map 5). The population is predominantly Acadian, numbering some 9,675 residents according to the 1993 census. Theirs is a deeply rooted

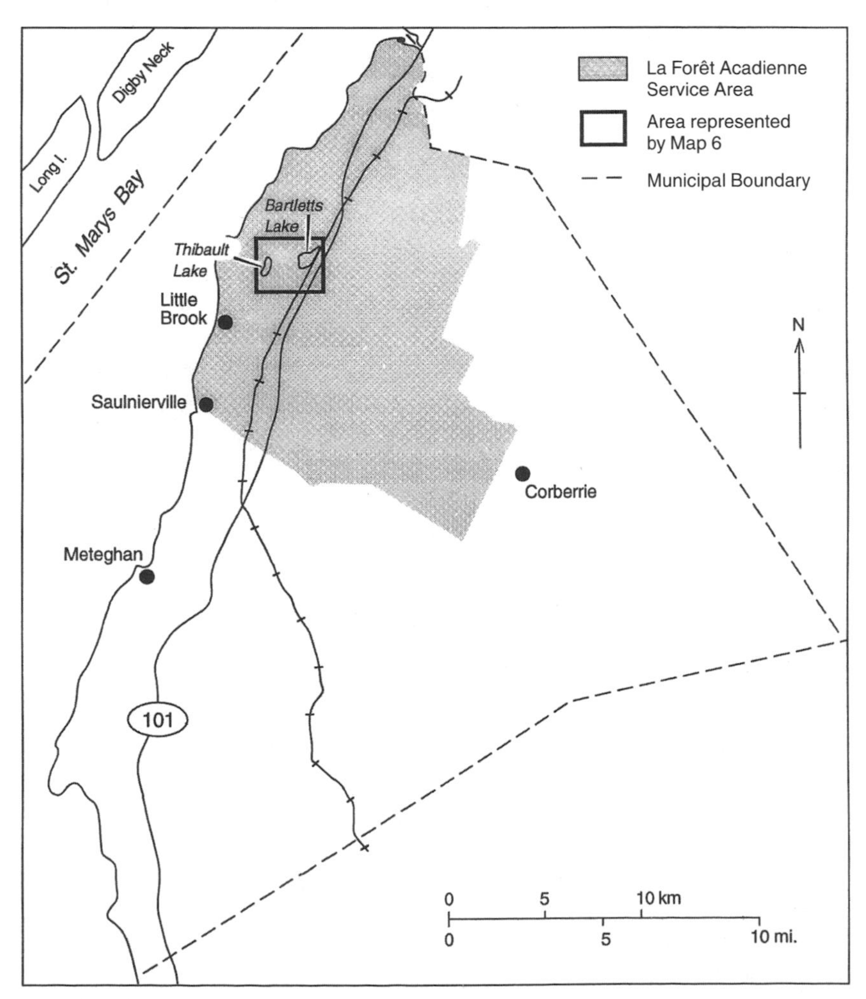

Map 5 Municipality of Clare and La Forêt Acadienne service area

society, with a network of local institutions reinforcing the Acadian identity. The provincial Crown owns relatively little forest land in Clare, and what it does is found in the 'back' or interior. Private land titles were first granted to settlers in the mid-eighteenth century, and most of the coastal belt had passed into private ownership by the end of that century. Some of the land was then cleared for farming and pastures, while the woods were relied upon for firewood and building materials.

The pattern for land survey followed the French colonial system. Properties began at the coast and extended inland, in long, relatively narrow strips. After all the shore tracts had been disposed, the system was repeated farther inland beyond a boundary road or trail. The result is well illustrated by the network of secondary roads that continues today. Over time, these original grants were divided length-wise, at right angles to the coast, the main road, and many of the main landscape features. This was to ensure that each farmer's son had access to all the positive and negative features of these highly varied lots, 'to the shore or the main road, [and] to the fields behind the house, as well as equal claim to the orchards, spruce stands, hardwood knolls, bogs and swamps.'[11] Generations of further subdivision led to ever more narrow woodlot properties. Today they are often less than five chains (330 feet) wide, while stretching three miles in length, as shown by the woodlot map forming part of the exhibit at the close of this chapter.[12] Significantly, this land tenure and use system has directly shaped the nature of the modern forest in Clare. Individual owners were confined by the dimensions of their properties to rather narrow strip cuts. Even if desired, it was next to impossible to undertake the larger rectangular block cuts common to squared-off properties. Equally, the strip cuts tended to regenerate rapidly by natural seeding from adjacent lots, thereby 'encouraging and maintaining the tolerant species while preserving the gene pools.'[13]

Despite its settler history, much of the land of Clare offers poor prospects for farming. The soils are alternately infertile, shallow, wet, or stony. As a result, much of the land cleared for agriculture has reverted to forest over time. It is for the most part a mixed-wood forest (i.e., combining softwoods and hardwoods) whose predominant species are red spruce, hemlock, yellow birch, sugar maple, red maple, and beech. These species are mostly shade-tolerant, capable of surviving in conditions of very limited direct sunlight but holding growth potential that can be released should sunlight later be restored. As Guptill herself points out,

> By Maritime standards, these are desirable trees; sought for either fibre strength, timber quality, longevity, heat production, easy regeneration, or rot resistance. They produce a variety of marketable products, and utilization quality varies from low grade pulpwood and fuelwood to clear timber and furniture stock. They are mostly shade tolerant, capable of surviving

> adverse conditions while maintaining the ability to grow rapidly once conditions improve ... Within each stand it is not unusual to find scattered 150-year-old spruce, a main canopy level of 45- to 60-year-old yellow birch, beech, maple and spruce plus an understorey of 1- to 15-year-old spruce regeneration.[14]

In effect, a set of physical and social factors combined to permit a highly effective system of 'natural' reforestation in Clare.

Appearing before the Nova Scotia Royal Commission on Forestry in 1983, Guptill argued that the Municipality of Clare offered one of the most promising settings both for intensive silvicultural management and for the group approach. Collectively, these small woodlot owners possessed a considerable wood inventory. Moreover, the pattern of small-scale but ecologically variable properties could, in principle, be more effectively planned and treated on a coordinated than an individual basis. As Guptill put it, 'The private land forests of Clare are diverse, are inherently flexible and thus are well suited to the practice of intensive forest management. In addition, the ground itself has a gentle topography with most of the boulders below or incorporated into the topsoil, facilitating the use of small, low-cost equipment.'[15] These lands were very accessible as well, with an extensive secondary road grid together with an off-road system of tractor trails that followed the land forms, across the knolls (high ground) and around the swamps and bogs. Furthermore, the low-volume (but diverse species and products) output from improvement and commercial treatments could be more effectively consolidated and marketed under the auspices of the group enterprise. Finally, there were dozens of local industries to purchase local roundwood, offsetting the dependence on export sales. When Guptill arrived as the field forester for the fledgling La Forêt, the prospects must have been both exciting and formidable.

The Operations of La Forêt Acadienne

One of the original eight groups, La Forêt Acadienne was organized as a limited company in 1978. Its initial service area extended from the municipal line at Weymouth to Saulnierville and from the shoreline almost to Corberrie. This territory was later increased twofold by extending the original boundary (see Map 5). La Forêt was effectively resident-controlled, since only three voting shares could be held by any one owner. By 1984 La Forêt had seventy shareholders who had committed approximately 5,000 acres to forest management. Membership grew steadily but slowly over the years, surpassing 100 by 1987 and mounting. The program became increasingly popular across Clare as plans and improvement work and returns began to flow. Although the group never advertised its presence or tried to actively recruit members, there was always a waiting list of potential members.

Admittance was on a 'first come, first served' basis, though other criteria were applied in some cases. If, for example, an applicant did not know where his or her woodlot was located, other applicants might have been given priority. Or, if the tree stands of a particular lot did not qualify for treatment under the agreement, a woodlot owner might have been told, 'We cannot help, beyond writing the management plan.' Overall, La Forêt's rate of progress was limited largely by the scale of funding and the work that could be done by a manager, a forester, and a forest technician. Later the group was able to hire a field supervisor as well.

The diverse forest land ownership patterns in Clare presented La Forêt with a unique set of problems but also opportunities in practising forest management. Normally, the hauling of wood along the length of the narrow lots was uneconomic. Stands were small and tended to vary considerably along the lengths of woodlots while extending across many contiguous lots. Thus, there were definite advantages to being able to treat neighbouring lots at the same time. Another factor was the large number of woodlot owners per acre. This meant that a diversity of perspectives and values would likely figure in rather small forest areas, requiring more time to be spent in explaining the program. The high proportion of boundary lines per forest acre was also a factor. Given all this, it is not surprising that the typical small woodlot yielded only small amounts of wood for small returns.

Guptill thus describes the work of La Forêt as 'miniature scale with miniature funding.' The three-person staff was in charge of most activities. They included conducting land searches; providing services to members and associated public relations activities; dealing with harvesting and selling wood, and communicating and explaining the process to the woodlot owner; hiring and firing crews and supervising their work; and bookkeeping. The clerical and record-keeping load was intense, as all forest management plans, silvicultural work done, and market transactions of wood had to be kept separate for each member of the venture. And there was no backup!

As forester for La Forêt, Guptill was involved in all aspects of ground-level delivery of intensive forest management on small private woodlots. Primarily responsible for writing forest management plans for each individual woodlot and for prescribing silvicultural treatments, she also helped to supervise and assess the work as it was done, to answer technical inquiries from members, and to do extension work on behalf of La Forêt.

A crucial step in the program was the initial contact made with woodlot owners and the locating and cruising of their properties. Locating boundary lines was often not easy. It is inherently difficult to maintain the property lines on long, narrow strips of land, and some of the lines in Clare have been seriously neglected over time. On square or rectangular properties, where two or more broad sides may face a water break or a road, it is much easier to maintain the property lines. But on a seventy-acre lot measuring three miles by 200 feet, there are six miles of boundary lines to be maintained every ten to fifteen years, compared to only 1.25 miles of boundary lines on a square block of the same acreage. Furthermore, given the age-old tenure of many properties in Clare, owners frequently did not know the boundaries of their woodlots at the outset. The first step toward management involved boundary renewal.

In this task, Guptill often had to start at a point several woodlots over (where the boundaries were known) and then work her way toward the woodlot in question. She would then try to identify signs of old boundaries and flag them for renewal. This was intricate and time-consuming work. There were sometimes jogs in the property lines, and Guptill recalls some properties that were, at their narrowest points, no more than eighteen feet wide. Whenever an adjacent boundary line had already been renewed, this facilitated the job considerably. Over time, as more and more woodlots were located and lines renewed properly, the task became easier. Accurate boundary work was a prerequisite to ensure that work would be done on the proper woodlot. In any case, it was particularly critical in such a 'crowded' woodlot context. Often it was done during slack times in winter and spring. This was the time when the ground was soft, and it was difficult to move equipment without damaging the surface. It was sometimes easier to see

Mary Guptill working in the woods on the French Shore, Nova Scotia.

old boundary lines then, though the difficulty of walking in the deep snow sometimes took away from that advantage.

The next stage involved cruising the woodlot to establish stand types, age classes, and wood volumes along with topographical features that would affect the silvicultural approach. Here again the local conditions affected the nature of the work. Since the long, narrow woodlots ran at right angles to the topographical features, the pattern of tree stands was unusually complex and diverse. This presented problems for taking inventories. Normally, a 200-acre square woodlot would have up to about ten large stands. However, the narrow, rectangular lots in Clare might contain up to twenty different forest stands yet be only forty acres in size. The much smaller stands would be separated by landscape features such as swamps, knolls, streams, or rivers. Since each stand required a set number of field samples, cruising these smaller lots required many more samples, more time, and more detailed notes to aid in the mapmaking. Guptill noted that one can cruise many 200-acre properties faster than a forty-acre Clare woodlot. Moreover, each forty acres may necessitate dealing with a separate owner.

From the cruise results, the management plans were devised. Each plan set out the results of the cruise, the maps of the property, and recommendations for silvicultural work. Obviously, not all of the potential work was needed, certainly not all at once. There were priorities in terms of the land base, the degree of accessibility, and so on. The forester then presented the finished plan to the woodlot owner, who would decide whether to accept the plan and whether to formally join La Forêt and commit to a decade of management. Here the group staff began to work with the landowner-member to achieve the necessary treatments. Coordinating these operations for scores of members simultaneously was an extraordinary challenge.

Guptill designed a system for classifying and organizing the huge number of specific recommendations that flowed from the management plans. She compiled a set of loose-leaf binders according to the different types of silvicultural treatments proposed. For example, one section would list 'merchantable thinnings,' with an entry for any thinning being proposed in any management plan, according to owner, size, and other relevant information. Each entry was also accompanied by notes unique to that setting. Thus, a particular thinning site might be qualified as 'inaccessible except for the driest years.' Similarly, the list of 'cleaning' treatments might note that a designated site 'can be done at any time,' with an indicator of the owner's priorities attached.

Organized in this way, the binders became a tool that La Forêt's manager, Arcade Comeau, could rely on to plan the work of his woods crews. They could be adjusted according to locale, adjacent properties, available access, crew training, necessary equipment, climate and season, and so on. In this way, the group's service plan was prepared and modified as necessary. La

Forêt treated about 300 acres per year at that time. Wherever possible, it tried to work adjacent lots together (this was a goal of the group program as originally conceived). However, neighbouring properties were not always available at the right time. Consequently, a glimpse at today's map of managed properties can be misleading. More often than not, four to five years might separate the sign-up decisions of neighbouring owners, and this in turn became a factor in the timing of work. Sometimes a single isolated property simply had to wait.[16]

On-the-ground forest improvement work was done in one of three ways: by group venture crews, by private silviculture contractors, or by the landowners themselves. Provided that the job was done to designated standards, it was financed and the workers paid according to fee schedules under the agreement. Since there were initially no private contractors in the area served by La Forêt, the group organized its own crews. Most years, this amounted to nine regular workers and additional help as needed. In the beginning, these crews had only the most rudimentary equipment, and their work was extremely labour intensive. Since the group did not own any equipment, whatever was available was provided by the work crews themselves. It was no reflection on the quality of the workers to note the limitations of the equipment. Often it consisted of rundown tractors and worn-out saws. Some of the early tractors did not even have winches available to drag the logs to the trail for cutting and loading. In such cases, the crews felled at the stump and cut the lengths into cordwood (pulpwood and fuelwood), which was then manually moved to the tractor on the trail and loaded onto the wagon. The heavier sawlogs were pulled out by the tractor.

Later, when the tractor was equipped with a winch, the operator could pull the logs out to the trail instead of having the tractor go off the trail to the logs. With the upgraded tractor, one man and a helper could handle the job. There was now less need to cut as many secondary trails to get the tractor closer to the stump. The most advanced arrangement for working on the small Clare woodlots consisted of a tractor with a remote-control winch, towing a wagon with a grapple loader. With this rig, the operator had more choice as to where to fell trees and was better able to avoid dropping them on the regenerating stock left on site. In difficult situations, the operator felled against standing trees. He then attached the winch and, standing well away, used the remote control and the winch line to tug the tree to the ground. It was then limbed, hauled to the trail side, and cut into products. In this arrangement, the grapple loaded both the cordwood and the logs onto the wagon. Since the operator routinely worked from the deepest part of the stand toward the front, he used the trees marked for later cutting as bumper trees. Once he had a load, he pulled the wagon to the roadside, where he separated the cordwood and the sawlogs. The advances here came on several tracks: safety, productivity, flexibility, labour saving, and less ground damage. By 1987 only one of the La Forêt employees had amassed sufficient capital to acquire such a rig.

While road building was an eligible stand improvement expense, La Forêt chose not to get involved until a significant proportion of adjoining members' most productive lands were under management. Even then, the road would be sited according to the location of the best stands of all the owners in the area (not just those under management). In the meantime, crews would work with the available trails, bequeathed by generations of previous woodlot work.

As a field forester, Guptill drew her greatest satisfaction from working close to the land and its owners. Each property was different and presented its own challenges. For her, the art of field forestry lay in assessment and implementation. The client was the La Forêt shareholder who owned the land. The target was the woodlot itself.

Due to highly variable stands and often small forest areas, woods workers had to make difficult choices in deciding which trees to cut. It was Guptill's job to keep pace with and answer their questions. Applied, intensive silviculture was almost a hobby for Guptill, and she found coaching and encouraging those who carried the saws highly satisfying. She liked her role as field forester, and she had little interest in formal administrative or policy issues and the type of career advancement that would take her away from the forest.

The Challenges of Woodlot Forestry in Clare

Mary Guptill's decade of work with La Forêt Acadienne gave her an appreciation for the political and technical difficulties in achieving intensive management. At the group level, there were certain limits to the small organization. La Forêt's operating staff was limited to the three core positions, and they were always heavily burdened with work. There was also the question of tools and equipment for the woods workers, as detailed above. On another level, there was the financing question. La Forêt could always have done more work, more quickly, if the levels of funding under the agreement had permitted it. The state of the wood markets also affected the group's development, since commercial sales offered a major source of funds (in addition to the incentives supplied by governments) both for shareholders and for La Forêt.

These were all external constraints not directly within the control of the group venture. There was one further variable that affected their forest work from a distance. This was the detailed terms of the Canada-Nova Scotia agreement as spelled out by the *Manual of Procedures and Standards*. Guptill accepted these regulations and the red tape that went with them for two reasons. On the one hand, as she herself put it, 'My job as a field forester was to implement. Somebody else decided what kind of treatments were to be authorized. My job was to work within them.' In the day-to-day work, this was a given. It was not the field forester's role either to complain about the regulations to the landowner or to make excuses or blame the regulations for inconsistencies in program delivery. On the other hand, Guptill believed that there had to be standards to apply across the province, since large sums of public money were being used to underwrite a great variety of silvicultural programs by groups, individuals, and companies. There was always the danger that some people might financially abuse the system, as well as the danger of treatments being poorly designed or delivered by the burgeoning community of government and private foresters. For these reasons, regulations and standards were necessary, and the *Manual of Procedures and Standards* set them out in detail.

But while Guptill thought that the administrative structure of the federal-provincial forest agreements was basically sound, there were also complications that arose from time to time. In retrospect, it started out quite simply. In 1978 there were few rules or regulations on how to execute the forest management program. The first claim written by Guptill was a letter to the effect that 'We have completed ten acres of thinning at the rate of $100 per acre. Could you please forward the funds for the expense.' La Forêt could design its own procedures, and many decisions were delegated to the foresters and managers on the ground.

'It took some time for the bureaucracy to catch up with us,' Guptill recalls. In a sense, perhaps, the groups were permitted to make a good-faith start, with the elaborate rules coming later. But over time, the regulations and procedures increased by leaps and bounds, filling a binder with ever more loose-leaf sheets. Guptill dealt with these challenges in a practical way – by knowing the 'bible' to the best of her ability and knowing the lands on which she worked. Through this procedure, La Forêt tried to keep a smooth and consistent relationship with its clients, contractors, and workers, while at the same time abiding by the rules of the time. Obviously, a forest cannot be managed on a five-year cycle (the maximum length of the agreements), Guptill argues, but some continuity had to be maintained despite the changing objectives.

They thus tried to maintain the work on the ground as steadily as possible while managing, as best as they could, the rocky ground of bureaucratic rules and personnel changes. One of the commitments that La Forêt tried to follow throughout this period was to maintain work for the woods workers on a regular and consistent basis. In fact, Arcade Comeau, the manager, even succeeded in maintaining the momentum during the nine-month break in funding in 1982 (and, even more astoundingly, in the 1988-9 situation).

Given the unique forest conditions in the Municipality of Clare, Guptill was concerned about one major bias in the forest management procedures: the lack of funding available to maintain forests in relatively good condition.

According to the agreement, there was only money available for a one-time treatment of any given area. Once a treatment was applied to a particular stand, no funding was available to manage that stand for the future. Future work had to pay for itself. In other words, if a plan called for selection logging in uneven-aged stands, to be repeated every ten years as a technique of stand improvement, it was expected that the proceeds from this logging would cover its costs and more. In one sense, this 'one-shot' rule was understandable, since there was so much land to cover and such a demand for the scarce improvement funds. Guptill agrees that the funds had to be used economically. However, the one-shot treatment of tree stands was not always the most effective approach to silviculture. Particularly in La Forêt's service area, it often took a small amount of money applied on a number of occasions to maintain the good quality in a stand – that is, to maintain it in relatively good health, constantly producing high-quality products. The continued but partial approach is very labour-intensive and puts the greater level of wages and commercial returns into the hands of local people.

The one-time-only treatment provision, by contrast, influenced both the type of sites given treatment and the types of treatments applied. Owners who wished to gain the greatest level of benefits from the program were inclined to clearcut their merchantable stands (i.e., maximizing the dollar return) and to opt for the full set of renewal treatments to be paid for by the program: site preparation, planting, weeding, and spraying, which together made the most expensive treatments per acre. In effect, the structure of the program encouraged the spending of a lot of money and time to redeem degraded sites that had been logged destructively or before their time. Guptill recalls one forest property within the service area of La Forêt: a piece of former pasture land thick with immature spruce.[17] It had been inherited within a family, and the descendants were not in agreement on what to do with it. With the desire to realize and maximize their immediate assets, they decided to clearcut the land. La Forêt was then charged with the task of 'managing it back to life' with money available from the program.

This sort of case created many complications. Due to the clearcut, no seed trees were left. Nor were there tree seeds left in the ground, since the trees cut were immature. Cutting immature trees also left a heavier burden of bush on the site – that is, a larger percentage of tops, branches, and small wood than if mature trees had been harvested. Cherry and raspberry and blackberry seeds that had lain dormant since the land had been a pasture were now coming up instead of tree species. In mature forests, enough time has elapsed so that such seeds die. In addition, the transformation from a pasture soil into a forest soil had not been completed, a transformation that includes the evolution from one bacteriological/fungal community to another. Pasture soils are not forest soils and are thus not as likely to support trees as well.

In such a case, the site can regenerate either in so-called weed trees or in merchantable stock, depending on how it is handled. If tree growth that is on its way to producing quality wood products is wanted, one is forced to prepare the site, plant artificially, and then apply herbicides. There might still be problems. The seedlings planted might not come from the region being planted, and their success is rarely as good as natural regeneration. The red spruce, for example, interbreeds with black spruce and differs from place to place. In addition, planted trees are seldom as good as trees that are naturally seeded. Their roots are often misshaped and may never sufficiently anchor or nourish.

Thus, the situation of the land, and the presence of the program, could combine to require decisions that might not have been taken otherwise. Guptill believes that such monies could have been employed in maintaining and improving already productive forests on a continual basis. The ultimate objective in forestry, of course, is to prevent situations such as the above from surfacing in the first place. But the reality is a long way from that state in Nova Scotia.

There might have been reasons other than silvicultural concerns for the principle that a forest area only be treated once. The success of the federal-provincial agreements was measured by the number of acres treated. This biased the funding formulas in favour of clearcutting and herbicide applications, which were conducted on large acreages and showed quick and sometimes very impressive results. It was much more difficult and time-consuming to perform selection cuts and thinnings. The results were less spectacular in the short term, but the long-term benefits might have been that much greater. At La Forêt, the group rarely had to justify to the funding agencies the expensive clearcut, plant, and herbicide scenario but frequently had to defend a woods worker's choice as to why this or that particular tree was cut in the relatively inexpensive merchantable thinning or shelterwood treatment.

The difficulty of rationalizing money spent on selection cuts was apparent on a couple of occasions. Selection cuts on the woodlots under management by La Forêt typically involved stands of mixed ages and species: some old-growth red spruce, a main canopy of hardwood, and an understorey of red spruce. At one point, Guptill wanted to have some of these stands thinned. However, the regulations at the time stated that thinnings could only be made in stands that could withstand a delay in cutting and in those that would respond to a thinning (i.e., increase in diameter). Inspectors from the Nova Scotia Department of Lands and Forests and from Forestry Canada were regularly sent to all groups to ensure that proper procedures were followed. In this case, the inspector thought that this stand, given the old age of the spruce, would not respond to a thinning. The trees were mature, so it was stated, and they were not supposed to grow any more. Following stipulated procedures, the disagreement was resolved, and Guptill had her plan approved. Time has since proved her right. She based her claim on the fact that the trees on the French Shore are different. They not only live longer but also maintain their ability to increase in diameter after release longer. The soils are good, and the moisture from the ocean provides an excellent growing climate. Even abandoned pasture lands, in contrast to other areas of the province, reforest in red spruce rather than in white spruce. Competition from the balsam fir is uncommon because the balsam aphid survives the warm winters in the area.

This was a minor technical difference in the assessment of treatments to particular stands between La Forêt and the federal and provincial inspectors. Its successful resolution was a credit to the procedures for resolving conflicts contained within the program. The conflict nevertheless illustrates that a definition of maturity depends on the overall purpose for forest management and the site-specific conditions of a tree stand. The troubling implication of the above is that it is harder to justify spending money on maintaining those forests that are fine, whereas there is unquestioned money available to redeem poorer forests.

La Forêt had other problems with the bureaucracy. When La Forêt was formed, it set a minimum size of five acres for membership. This was not controversial, though when visitors toured the area the group was called upon to demonstrate the suitability of its small-scale holdings for management. Shortly before Guptill left La Forêt, a new rule was implemented to require a fifty-acre minimum. This was not appropriate for the Municipality of Clare. Most woodlot owners in Clare own fewer than fifty acres, sometimes as little as five acres of prime forest lands, but as a collective they sit on one of the most important contiguous tracts of forest in the province, with good soils, good accessibility through road systems, and little practice of clearcutting in the past. It was left to her successor to once again fight for the five-acre provision.

When it came to formulating treatment plans, the staff of La Forêt tried to be guided by the natural progression of succession on their lands. At times, this led to conclusions at odds with prevailing orthodoxy, as shown by the debate over herbicide sprays. La Forêt used herbicides when necessary but generally tried to minimize their application. When faced with a forest mess, a choice was always available. When appropriate to the landowner's goals and under certain conditions, herbicides offered an alternative to waiting for nature to reestablish a desired crop over 50 to 150 years. Guptill's approach to herbicides was practical: they constituted one tool among many. As well, Guptill could see the danger of elevating one particular approach above all others. In a presentation to the Nova Scotia Royal Commission on Forestry, in 1983, she made the point succinctly:

> La Forêt is greatly concerned whether cutting methods and herbicides are being used properly. The cutting method chosen can lock a forest manager into a set of decisions that result in the area being site prepared, planted and then sprayed with herbicides. Often, however, on the same piece of land, a different cutting procedure would bypass the need for site preparation, planting and spraying and yet result in an acceptable forest cover. The problem here lies not in whether the treatment is safe, but whether it could be avoided through better foresight and better choice of prior treatments. Forest managers should be required to spend more time analyzing and scheduling the sequence of treatments on a particular piece of land instead of viewing each stage as an isolated incident. By thus viewing the forest as a continuum, more treatments would be used in the proper time and place. Treatments would be designed to avoid problems instead of solving problems, hence fewer clearcuts and less need for herbicides. As a result, the public might be more willing to accept the use of clearcutting and herbicides where they are needed and where they are appropriate.[18]

Yet there are unquestioned monies available for weeding by herbicides and for the management of messy and degraded lands. This might be a good thing, especially when tree plantations meet with success in growing wood fibre rapidly. But it is less justifiable when it takes away money from the use and maintenance of forests that are still in good shape.[19]

The Socioeconomic Impact of La Forêt Acadienne

The Group Venture program was never without its critics. However, based on her own experience, Mary Guptill has no doubts about the positive impacts that La Forêt Acadienne was able to exert on the Municipality of Clare. They can be summed up as the impacts on the land itself, on landowners, and on the community. There is no question that the forest land itself was improved as a result of the cumulative treatments of the group program. As argued above, the joint shareholder enterprise was particularly effective in achieving returns of scale in organizing improvements to small properties.

For the small landowners who formed the active and potential membership, La Forêt was similarly influential. These members were a varied lot. Guptill dealt with widows, mechanics, pig farmers, cattle raisers, retired people, mill workers, mill owners, shipyard workers, teachers, highway workers, painters, and even some non-resident Americans. Most of them were forty-five or older, since few young people owned forest lands. Bearing in mind that the Canada-Nova Scotia agreements aimed to promote forest management and harvesting on small private properties, the group provided an organizational vehicle that combined extension work and professional management expertise. Guptill has suggested that many landowners were able, for the first time, to understand their woodlots as an appreciating rather than a depreciating asset. This injected a long-term outlook quite new to many. At the same time, the group was able to demonstrate links between silvicultural improvement and commercial returns. Perhaps most important was the element of choice, and in this sense control, offered to woodlot owners:

> The private land program has also given woodlot owners an alternative when harvesting wood. Prior to the program, owners who could not do the work themselves had to hire a local logger. And generally, the logger cut where and when he wanted to cut, and the job varied according to the quality of the wood, according to how much money the owner wanted and according to how careful the logger was. Unfortunately, the price of wood is not conducive to carefully picking out trees here and there. Nor are the economics favourable when working on ten hectares stretched over four kilometers. The result is usually a high-grade, with no consideration for regeneration. By participating in the program, owners can have the mortality wood removed, while leaving the better, healthier wood to continue growing. Thus they have an alternative: if they do not need or want to harvest their entire crop, they can have it managed so that it will live longer while improving in quality.[20]

In this simple coupling of owner involvement and ecological choice lies a challenging political statement, particularly in the forest context of modern Nova Scotia. The same strengths may help to explain the hostility often directed toward the groups by the larger industrial contractors and processors.

It may be at the community level that an organization such as La Forêt has its greatest influence. This was probably magnified in Clare given the close ties within the Acadian community. The group venture was 'of' this community as well as 'in' it. It was locally based, owned, and staffed, and it was accountable to local shareholder members. Not only did it provide continuing employment for up to one dozen local people, but it also facilitated the flow of wood products to local processors, thereby accelerating the network of local commerce. The huge cultural support system in the community helped to disseminate the ideas and practices promoted by La Forêt.

Meanwhile, the monies funnelled through La Forêt were distributed among truckers, haulers, woods workers, and owners (stumpage). The spin-off benefits were many. For every dollar pumped into the program by governments, another would be created by the sale of the wood. In addition, most of the monies paid out by La Forêt were wages, which meant that most monies stayed in the region. On top of that, the local forest was improved and brought back into productive use. Few government programs can boast of such results.

La Forêt employed workers because there were no contractors in the area. In the latter years of Guptill's employment, three to four two-men crews were employed. These forest workers were provided with regular jobs at relatively good wages by local standards. Some temporary workers were used as well. The 'bible' set maximum rates for various treatments, and La Forêt paid these rates. As silvicultural workers, they could have been paid at even higher rates.

Most of the work was done by these crews. Arcade Comeau tried to team up inexperienced with experienced workers and to assign work that was appropriate to a particular crew's skills. Some workers started to work during their summer vacations from high school and then became full-time woods workers after graduating. Others were attracted to the work because it provided an alternative to commuting the longer distances outside the community to make a living.

When planning a particular job, Guptill tried to line up several cuts in a close geographical area to minimize movement of workers from place to place. This saved workers travel and helped the truckers who later picked up the wood at the roadside.

While the group handled most work through hired workers, there were a few woodlot owners who performed their own work. Those who did were usually interested in their woodlot and performed the work in their spare time or when on vacation. Guptill would then come in and ribbon the trees to be cut or help the woodlot owner to do it.

Reflections on Politics and Markets

It is impressive to note what the Canada-Nova Scotia forestry agreements did for the Municipality of Clare and the woodlot owners of La Forêt Acadienne. But it is also interesting to speculate about what the program could have done had its mandate been broader. Put simply, the agreements were technical, aimed at offering technical services to provide more fibre for the forest industry. The welfare of forest communities was only a secondary objective, and no provisions were made to improve the economic status of small woodlot owners in the wood market.[21] It is therefore not strange to find that Guptill feels uneasy about the politics and markets in the forest. There was only one basic and stable market for pulpwood in this part of the province. 'They were good to us,' she says in one breath. 'Arcade called them up on Friday mornings and asked how many loads we could have. They answered one, two, or whatever.' In another breath, however, she argues that 'They were clearly a monopsony, and you had to play the game. They played it dead seriously.'[22]

The best way to counter the monopsony, Guptill argues, was to try to grow quality wood and to find alternative markets. There were other markets: two large sawmills in the vicinity bought sawlogs, and some smaller sawmills were also competitive. One mill sawed hardwood for lobster crates, but it closed. There were also, from time to time, export markets for pulpwood to Europe and New Brunswick (Irving) that paid higher prices. These lots were shipped by boat from Weymouth. In the early fall of 1987, the price of pulpwood for the overseas market was 13 percent higher than it was locally. Small niche markets occasionally surfaced, such as a local manufacturer of wooden crafts for the tourist trade, needing a load of pine logs. At one time, La Forêt sold pulpwood in long lengths as wharf poles and realized a 69 percent increase in price. In the winter of 1986-7, fuelwood sold as lathwood for a 56 percent increase in price. Overall, La Forêt tried to be as well acquainted with the local market as possible.

The market for hardwood changed over the years. The local outlets for hardwood sawlogs declined as lobster traps for the fishing industry changed from wood to wire. Meteghan Woodworkers folded just before or after Guptill's appointment. N. Melanson's hardwood mill burned down a few years later. The demand for hardwood cordwood increased over time as the so-called oil crisis raised the price of domestic oil. Some was sold to the Canexel hardboard mill in Lunenburg County, but most of it was sold as fuelwood locally.

Guptill refers to the battle over marketing as too confrontational for her liking. 'I am in it for the long run,' she states, and 'I prefer to work with the situation by trying to find compromises ... If you choose sides, you are quickly labelled for all time, and you cease to have credibility. The weak lose in confrontations. The pulp companies have paid people who fight their battles very well, while the weak have volunteers who, once defeated, fall to the wayside.' Guptill sees the group ventures as a means to empower small woodlot owners. And, though it was relatively easy to convince sufficient (for the amount of funding available) small woodlot owners in the Municipality of Clare that the Group Venture program was something for them, some locals remained suspicious about the government's intentions, believing, as in the past, that it was out to take away their lands.

'We changed things slowly,' Guptill recalls. 'People began to trust after they saw you, after you had been around for a while.' Initially, they thought that she merely followed in the wake of Arcade, who greased the wheel of the whole process and who, Guptill believes, deserves a lot of credit for the success of La Forêt. They started out to work lands acre by acre in slow fashion, which lent itself to the small crews whom they employed. They could not be too political. Instead, they went about the business of taking care of the forest, a concept for which the groups were good, but it was very hard to implement.

Progress was tenuous. It was difficult to see a future in which the forest would yield enough money for intensive management to be widespread. Perhaps it would be possible if the general level of knowledge was higher – that is, if there was a greater social conscience about what was acceptable cutting and what was not. Perhaps, as well, more woodlot owners would be needed to do their own work. Guptill recalls that they were far from these ideals, if indeed they could ever have been reached. It costs either money or someone's time to maintain and upgrade the forest, and there was (and remains) little cultural or legal pressure to do so. The woodlot has always been the woodlot owner's bankbook. Withdrawals are made with little concern for the future, either because of ignorance or through lack of choice.

Progress will never be made unless people are provided with an equitable, or nearly equitable, alternative to destructive forest practices. La Forêt tried to let the managed lots convince woodlot owners that this alternative was a necessity. If somebody needed money and had to cut wood, then Guptill would plan a best scenario with alternatives to clearcutting or high-grading. She would say 'Cut that stuff; it is old and will die or rot anyway' and then recommend that immature stands be left alone or thinned. Perhaps this approach was not aggressive enough.

Mary Guptill in her greenhouse at Meteghan, Nova Scotia, 1998.

In 1987 Guptill became pregnant with her first child. She stopped working a couple of months before her delivery, feeling tired when walking in the woods. After her daughter was born, she worked off and on part time as child care arrangements allowed. In 1988 she resigned her position with the group venture. Today she operates a small tree nursery, shears a few Christmas trees, manages 400 acres of woodland spread over twelve woodlots, and takes care of her daughter, Rosiane.

Conclusion

Mary Guptill's career illustrates well the tensions, contradictions, benefits, and limits of field forestry under the Canada-Nova Scotia forestry agreements that ran from 1977 to 1995. Working as a field forester for La Forêt Acadienne group venture in the Municipality of Clare, she struggled to implement intensive silviculture in the region. The assistance that she provided was technical. It was facilitated by the funds that paid for her salary and the silvicultural measures taken on the ground. It was constrained by the financial limitations of the agreements, their ever-changing goals and objectives, and their bias toward type A forestry (or industrial) prescriptions. In this climate, Guptill struggled to implement a forest management regime that tended more toward type B forestry – that is, more sensitive toward the forest ecology of the region and the maintenance of the fine forests existing there already.

The statistics on plans approved and treatments accomplished are important indicators of the success of La Forêt. But Guptill also underlines La Forêt's less tangible but no less significant role in shaping community attitudes:

> By nature of its cooperative approach, a Group involves a vast number of people, many who have not yet gained an appreciation of the concepts and benefits of intensive forest management. A great deal of emphasis is therefore placed on education of the public. This extension work is required of the two- or three-member staff over and above their daily operational work. Beside being able to communicate with the public, the staff must deal with the politicians who direct the program and the two government bureaucracies which fund and administer the program. They must be able to translate the results of research into guidelines for woods workers, plan with the owners who control the land, and deal with the buyers and shippers of the wood products.[23]

It would be hard to find a more concise description of the complex social and political vortex in which the groups were situated. In the course of performing woodlot management, La Forêt was called upon to act as advocate, broker, advisor, regulator, auditor, employer, and servant to many distinct interest groups. This represented an extraordinary level of functional demand to impose on such a modest organization. Fortunately, Guptill found her particular role to be both challenging and rewarding.

Yet the nature of the Group Venture program, and Guptill's position in it, posed severe limitations. Its technical focus meant that the social and economic position of small woodlot owners in the wood fibre market was never addressed squarely. The program thus did little to address the historical inequities that had built up in the forest industry. Guptill herself was reluctant and resigned to the fact that activism on the part of La Forêt in the wood market, lobbying for a better share of the market and better prices, would be counterproductive. Such action might have served as a good complement to the government funding for technical services and left a stronger legacy in the empowerment of Clare woodlot owners.

Appendix

Typical Woodlot Management Plan (developed by La Forêt Acadienne Ltée)

Some of the difficulties confronted by La Forêt Acadienne in forest management in the Municipality of Clare are illustrated by a typical woodlot management plan. The following woodlot management plan, containing three lots and dated 31 March 1990, contains descriptions of the tree stands and recommendations for treatment (Map 6).

Three complicating features stand out. The first is the long, narrow, and small sizes of the woodlots. Lot 1 is 15.4 hectares; lot 2 is 2 hectares (this lot is only 5.5 metres wide); and lot 3 is 1.3 hectares. The second problem is the tendency of topographical and forest features to run in the opposite direction to the lots. There is thus a huge variety among the various lots, calling for different treatments for each stand. The stand variety is indicated for lots 1 and 3. The variety in lot 2 is similar though not shown because of the narrowness of the lot. The third complicating factor is the poor access to the lots. Mary Guptill wrote at the time, 'Access to these lots and others in the area is terrible. A suitable road system to service the land does not seem evident. There are, at present, too many landowners, too many swamps, too little land under management and too little money.' Finally, the picture is complicated by the scattered and often isolated lots under management with La Forêt. Map 6 shows other lots under management by La Forêt in grey shade. These lots, while most likely owned by different individuals, at least have the potential to be managed collectively. But those lots not under management by La Forêt cannot be part of a larger collective management plan; the overall management picture is thus made very difficult.

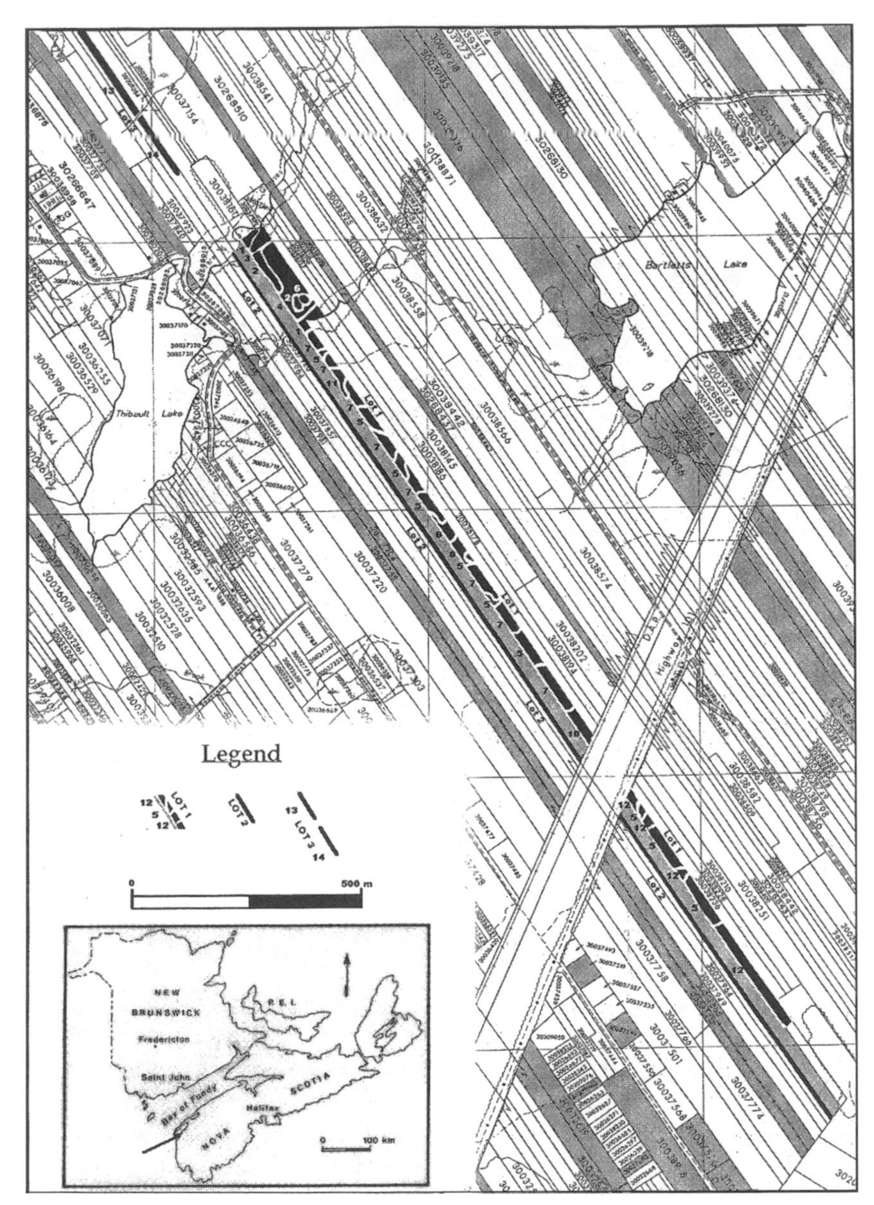

Map 6 Sample woodlot and shareholder property pattern

On La Forêt-managed forest lands, Guptill recommended to her clients a very different type of forest management than clearcutting and establishing forest plantations to produce a maximum amount of fibre for the pulp and paper mills (Table 5). Guptill based the recommendations on sound forest management practices for silviculture and harvests, with considerations

Table 5

Treatment for sample woodlot management plan

Stand(s)	Stand description	Recommendation(s)
1	Meadow flooded by beaver dams.	Maintain as wildlife habitat through protective strips and vehicle-exclusion zone.
2	Cut-over.	Leave to develop on its own.
3	Alder patch.	Leave to develop on its own.
4	Dense stand of mature red spruce.	Merchantable thinning.[1]
5, 14	Poorly drained cut-over and wooded areas.	Leave as wildlife habitat.
6	Old field with regeneration of alders, white spruce, and shrubs.	Leave to develop on its own.
7, 13	Well-drained areas cut over entirely 12 years earlier. Regeneration mostly in red spruce, though sometimes out-numbered by beech, maple, and fir seedlings.	Selected cleanings.[2]
8	Tall, healthy yellow birch, beech, and sugar maple, with scattering of spruce sawlogs.	Merchantable light thinning and removal of diseased beech.
9	A mature red spruce stand. Fifty years earlier, the poorer trees were cut, now leaving well-spaced, tall, large trees. The holes have been filled by beech and yellow birch, which have not grown well.	Shelterwood cut recommended, but timing is not important.[3] The cut will remove the last old growth on the lot, but more such growth will remain on the adjacent lot.
10	Equal proportion of 40-year-old yellow birch, red spruce, and balsam fir. The spruce trees are understorey, and time is running out for the fir.	Merchantable thinning.
11	Old orchard trees that have died. Regeneration of 40-year-old red spruce and some hardwood.	Merchantable thinning.
12	Poorly drained cut-over with many remnants.	Leave to develop on its own.

[1] A merchantable thinning refers to the selective harvesting of a stand for commercial purposes, but it also improves the growing conditions for the remaining trees.

[2] A cleaning refers to the removal, usually of hardwoods and shrubs, in a stand to allow the more ready growth of softwood seedlings. This can be done manually or by the application of herbicides.

[3] A shelterwood cut refers to several successive cuts in a stand over a number of years, leaving the remaining trees to reseed and shelter the regeneration.

made for both the environment and wildlife. Her recommendations tried to preserve or increase diversity in species and age structure of the stands. The methods included favouring valuable habitat species such as apple, alder, and serviceberry when they occurred in stands of commercial timber. Cavity trees were left where it was safe to do so in order to provide nesting habitat for various species of birds and mammals. Some areas were left to develop on their own. Thus, the implementation of management plans was to improve the quality of not only the timber but also the wildlife habitat.

Reproduced courtesy of Mary Guptill.

9
Conclusion

The foregoing accounts of Nova Scotia foresters show the unique political forces shaping professional practice in a small Canadian province as part of the wider picture of twentieth-century forestry development. The principal message of the stories is that forestry is as much about politics as it is about trees. Foresters are coloured by the organizations for which they work, the place of these organizations in larger institutional frameworks, and the positions that individual foresters hold within their organizations. As such, there is bound to be some tension between forestry practice on the ground and the standards set by the profession and the science taught in school. Any type of forestry is, in fact, ideological and coloured by certain biases and assumptions.

In making this argument, our book differs from those that portray professional foresters principally as experts, struggling to bring enlightened practices to forest management in the face of limited visions by politicians, businesspeople, farmers, landowners, and, most recently, environmentalists. The field programs advanced by such experts have displayed a certain similarity and conventionality in twentieth-century Canada. Following the sociologist Bruno Latour, we may describe these as 'immutable mobiles' – that is, uniform, standardized, and unchanging methods that can be transferred and applied to any geographic context.[1] At the macro-analytical level, the sustained yield concept (the idea that forests yield an annual increment of fibre that can be calculated and harvested) qualifies as an immutable mobile. At the micro-analytical level, a generalized preference for clearcut harvesting (justified on grounds of logging efficiency) represents another. One of the overriding premises of scientific forestry is an acute sensitivity to site. On a theoretical level at least, all foresters are aware of this. However, institutional priorities tend to bend or even override the art of site-based management. Often this is forcefully expressed in the shorthand slogans of daily work: 'We are in the softwood business here,' 'Uneven-aged management is not commercially viable,' or 'Cheap stumpage rates keep mills competitive.'

We also differ from studies claiming that all organized professions, including forestry, function as 'conspiracies against the laity,' deploying professional mandate, vocabulary, and tradecraft as insulators against public involvement or accountability. As such, professional practice may be either self-reinforcing or co-opted by the leading elements in society.[2] One of the most powerful antidotes to this is an alternative knowledge domain accessible to, or even the product of, social communities. The forest practices that flow from this idea might be referred to as 'mutable immobiles,' highly flexible methods restricted to local application in a specific place. Such methods might be devised by local people familiar with the history and idiosyncrasies of their environment. Alternatively, immobile methods may be fashioned by experts in strict response to field conditions.

It seems to us that there is considerable creative space between these extreme views. This book shows that detailed historical studies of individual foresters in specific situations can shed light on both the strengths and the weaknesses of the profession, its views on certain scientific and social issues, and its ability and potential to tread new and unexplored ground. The narratives, we hope, show that there is much to be learned from exploring the careers of key professionals who have cut against the grain of conventional wisdom and orthodoxies. They provide an insider's view of the profession, and they identify not only the prevailing currents but also the countercurrents that bid for influence in their times. Many of these elements might have generated (or might yet generate) a more environmentally sound and socially equitable kind of forestry. Indeed, some of the elements advocated by the dissenting foresters in this volume pointed the way toward alternative forms of forestry altogether. Ultimately, the history of professional dissent gives a deeper perspective to the more recent public and environmental challenges of forestry. The foresters in this volume show that professional forestry might be capable of generating answers with a wider application (as 'mutable mobiles') to the present human and environmental crisis. Billie DeWalt refers to such mutable mobiles as contextualized, holistic knowledge developed in specific locations but adaptable and applicable to similar phenomena in other circumstances.[3]

It would be foolhardy, however, not to recognize that there have been formidable obstacles to the reform of forestry. Such obstacles are well documented in the careers of the foresters reviewed here. In summing up this book, we therefore begin by outlining the conventional grain in Nova Scotia prior to the mid-1960s. This time represents a landmark in Nova Scotia forest policy, a point at which the provincial government committed most of the province's Crown wood supply to the pulp and paper industry. Thereafter, forest policy became more rigidly and militantly committed to Aldo Leopold's type A forestry and the creation of a secure industrial fibre supply. We then outline how this ingrained pattern was challenged by the foresters of this book and how they cut against the grain in articulating elements of an alternative forest policy. The mid-1960s constituted an especially fluid period when various dissenting initiatives cut against the grain of the current orthodoxies. Although they failed to be implemented, they represent constructive lessons for an alternative forest policy. We conclude by emphasizing that forestry is a thoroughly ideological concept, constructed to serve certain interests for certain purposes. The identification of these interests, and their biases, is essential to build a forestry that is more fully mindful of the forest environment and the people who live in and depend on it. The narratives here suggest that there is among foresters a rich tradition of dissent. The profession has been ill served by neglecting or even suppressing this fact. In the current debate of absolute morals between industrial and environmental advocates (cut it all versus preserve it all), this tradition is well worth exploring for constructive alternatives in between.

Identifying the Grain

In the previous seven chapters, we have explored experiences of foresters who cut against the grain in various ways. It seems to be appropriate, before reviewing the lessons one last time, to consider the orthodoxy against which their efforts often strained. Put another way, we need to identify the grain of convention that both provoked and constrained the voices of dissent. This should be grasped at two connected levels: the national, and the provincial.

Some of the guiding principles of the Canadian (and in many respects North American) professional outlook can be summed up quickly. From the turn of the twentieth century onward, there was a general consensus that the prevailing generation of the great Canadian forest had a limitless potential. The succession of commercial timber products broadened the range of commercially valuable products, from deals to lumber to pulpwood, opening vast new stands to exploitation in the process. This is the basis of an important corollary, that industrial demand would set the pace and the terms of forest harvesting. In the few cases when government authorities intervened to shape these terms, with manufacturing conditions, log-exporting restrictions, or land ownership rules, the results were highly imperfect.

There was also a variety of logging strategies for cutting and removing timber, depending on the species and the specifications being sought. But all relied upon natural regrowth as the standard method of regenerating a future crop. In addition, the standing crop (both Crown and private) became an object of state attention. For the first half of the twentieth century, 'forest protection' meant fire protection, and provincial forest agencies organized comprehensive programs to monitor and suppress fire that typically combined corporate and state resources in cooperative relationships. Significantly, this first major conservation measure was aimed not at the activity of commercial operators at the cutting edge but at curtailing external (natural and social) sources of resource depletion.

Following the Second World War, a series of engineering advances in bush machinery began to transform the logging process. Ever more powerful wheeled vehicles were introduced into the bush. Year-round operations replaced seasonal patterns of winter logging and spring removal. With the rate of extraction accelerating, and the unit logging cost declining, the capacity for wholesale removal or clearcutting became the strategy of industrial choice for the first time. The postwar period also saw the application of new approaches to the balancing of harvest and regeneration. The regulation of annual allowable cut offered the possibility of sustained yield harvesting, and for postwar foresters this passed from being a theoretical to a practical possibility. By the mid-1960s, wholesale clearcutting opened the need for new approaches to regeneration. The planting of nursery stock was certainly not a new technique, but the scale of application grew exponentially from this point. As a host of successional complications emerged in diverse ecological settings, 'intensive silviculture' became a more pressing matter if annual allowable cuts were to be increased to accommodate new industrial plants.

Given the natural, commercial, and state policy differences across the country, each jurisdiction adapted to these general trends in its own way. For Nova Scotia, the political and economic context of forestry adds nuance and distinction to this general portrait. Some of the highlights of this history include the late entry and ambivalent welcome of professional forestry to the province; the marginal position of the small woodlot owner segment in local wood markets; the colonial legacy in the province's sawmilling industry; and the terms and conditions under which the pulp and paper industry was established.

Professional forestry came somewhat late to Nova Scotia (relative to provinces farther west). It was only in the 1920s that 'forests' came to rival 'lands' in the provincial resource department and only in 1926 that the first trained forester was hired. This inaugural program was soon suffocated, and it was more than a generation before systematic forestry policy was squarely addressed. In the meantime, the Department of Lands and Forests retreated into a focus on Crown forest improvement, to the virtual neglect of private land forestry. Only with the Bélanger-Bourget inventory of the 1950s did modern management become a possibility. Crucially, the state-led transition from a sawmill-based to a pulp-based industry coincided with the transformation of management prospects. By the mid-1960s, the province's forests were overcommitted to industrial production. The province had by this time invited two additional pulp and paper corporations: American

Scott Paper, and Stora Forest Industries. From this point forward, DLF District and Extension Foresters found their roles subject to change. In time, the intense utilization of the fibre supply demanded compliance with and adjustment to the massive entry of new harvesting machinery and reforestation techniques.

Forest science was redefined as well. In particular, it was enlisted in the silvicultural campaign both to boost regeneration rates and to perfect resource supply models. As one corporate leader wrote,

> To support the growth of the forest industry that can be foreseen, we need more trees, we need better trees, and we need trees that grow faster. Science has been asked to show the way to these ends. To find out how much wood can be safely cut each year, we need to know the volume of standing timber in the Province, and the rate of annual growth – net growth, which means the actual growth of new wood less the normal ravages of wind, fire, insect, and disease. It is the computer that has come to the rescue in finding the answer to these problems.[4]

Finally, in the 1970s, a new generation of federal-provincial forestry agreements tied together the goals of increased harvests and intensive regeneration. This called for stepped-up ecosystem manipulation on a scale not seen before in Nova Scotia. While these forces shaped the 'grain' of convention in Nova Scotia forestry, the subjects in this book reacted, in varying degrees, against it. In the following sections, their initiatives are reviewed under the categories of forest economics, forest science, and social forestry.

Working against the Grain in Forest Economics

Otto Schierbeck shed telling light on the terms on which government and industry interests embraced professional forestry in the 1920s and early 1930s. Schierbeck advocated a more efficient resource use along conservationist lines of maximum sustained fibre yield, with the fullest domestic utilization of the resulting harvests.[5] He set about performing his duties without compromise, but he quickly ran into problems. For the majority of rural residents, lumbermen and settlers alike, forest management was irrelevant. Cut-and-run logging was the order of the day, fire protection was ignored by settlers or deemed overly expensive by lumbermen, and Crown forest properties were routinely subjected to trespass. Common to all social strata was a political and economic culture premised on partisanship, clientelism, and crude material exploitation. This left little room for rational resource use guided by professional expertise.

Schierbeck nevertheless satisfied the emerging needs of pulp and paper enterprises for accurate forest inventories and effective suppression of fire. He also organized campaigns of popular education, the most aggressive and telling of which involved the demonization of forest fires. His efforts were part of the wider set of corporate services offered by an emerging bureaucratic state. In some respects, however, the Nova Scotia forest industry was not ready for the new bureaucratic order. While Schierbeck was prepared to provide services to industry, he also demanded and expected something in return. He thereby exposed some of the problems in the politics and forest sector of Nova Scotia. At the time of his appointment, the forest industry was characterized by crude lumber exports sawn by small portable sawmills, while alongside it an emerging pulp and paper industry sought major concessions and incentives to locate in the province. Schierbeck criticized many aspects of these dealings. He observed that forest management stipulations of Crown leases were routinely ignored; land and fire taxes were routinely unpaid; forests were high-graded (taking the best, leaving the rest); the province's hardwoods (an important ingredient in the Acadian forest) were neglected and abused; lumber was sawn crudely and marketed poorly; and the pulp and paper industry was accommodated on exceedingly generous terms.

The Nova Scotia phase of Schierbeck's career ended abruptly with his firing in 1933. This aptly illustrates the ambivalent position of professionalism in state circles. Schierbeck was an uncompromising technocrat in the conservationist tradition, and that proved to be his undoing in a system in which patronage and clientelism remained vibrant. Politicians and industrialists expected services, not demands, from the profession, while Schierbeck expected both. He was thus both hired and dismissed by Conservative governments, an unusual event during a time of patronage appointments. Schierbeck was clearly a public forester. He aspired to a position of authority and independence from both government and industry, at a time when the profession was still weak and his department remained highly traditional even by the standards of the provincial bureaucracy.

John Bigelow developed his own critique of the Maritime forest industry, and his career provides another example of the tenuous position of early professional forestry in Nova Scotia. It was while working for a Newfoundland pulp company from 1935 to 1938 that Bigelow became disillusioned with his profession, its engineering focus (cutting trees as fast and cheaply as possible), and particularly the industry's treatment of pulpwood cutters. It was also there that he became convinced that forest management depended on socioeconomic reforms rather than on field practices alone. This conviction remained with him throughout his career.

Returning to Nova Scotia to join the Marketing Division of the Department of Agriculture, Bigelow found an outlet to confront the monopoly situation in pulpwood buying. As in Newfoundland, the small woodlot owners were paid rock-bottom prices for their pulpwood and were cheated at the scale by powerful buyers. In contrast to Schierbeck, Bigelow worked in a department more concerned with the welfare of rural (principally farm) producers. This involved a separate set of social assumptions from the Department of Lands and Forests, which was preoccupied with protecting forest fibre for the mills. Furthermore, enjoying strong support from his superiors, Bigelow managed to organize successful pulpwood producer cooperatives for the export market. Although these cooperatives were short-lived, as the Second World War closed access to these markets, they showed that a higher price for pulpwood was a strong incentive for forest management.

The Department of Agriculture also allowed Bigelow to address the old colonial system of predatory middlemen who profited from the crude exploitation of the forest resource under the merchant-dominated lumber trade. He did so by providing secretarial services to the Nova Scotia Forest Products Association, an embryonic trade organization of progressive lumbermen, and later by leaving government to serve as the secretary of the Maritime Lumber Bureau. Given the early and extensive settlement pattern in Nova Scotia, the pioneer forest had been extensively degraded by the Second World War. Bigelow was appalled by the ways in which an agent-exporter-lumber-producer system operated to furnish cheap crude export staples at the expense of the forest and the smaller lumbermen. Agents ordered so-called merchantable grades from the exporters, who then distributed orders to a few large lumbermen and hundreds of smaller ones. With the terms of trade set by agents and exporters, the lumbermen were at the mercy of the merchant element. Similarly, grading practices were open to abuse, as agents exaggerated the cull and the proportion of poor lumber and as exporters passed the expense on to the producers. The traders and agents, in short, as Bigelow wrote, were 'happy with things as they are, content with the continual deterioration of Nova Scotia's forest.'[6]

The Nova Scotia lumber trade was here different from that in other jurisdictions. The most accessible forests for large stationary mills were gone. So-called portable sawmills were therefore used to reach the least accessible areas, where the lumber was sawn on the spot and then hauled to roads, railway sidings, and ports for export.[7] The supply of sawlogs and the scale of operations were greater elsewhere, which meant larger operations that had more clout in negotiating and even bypassed the agents and the exporters. Many larger sawmillers in the other provinces cut lumber to specification for the domestic or US market.

Bigelow's experiences also demonstrated the thoroughly political nature of the lumber trade. The reformist elements in the industry were well outnumbered by those still comfortably rooted in the practices of colonial times, reacting with suspicion to the reforms promoted by Bigelow and others. Through the NSFPA and the MLB, Bigelow provides a colourful portrait of these lumbermen, who were also thoroughly political in the partisan sense.

Some still expected and used political office to advance their own economic or political careers. Others were incompetent hangers-on from the colonial era, when political office and party loyalty were key ingredients for personal emolument.

The degraded state of the forest, and the continuing primitiveness of the lesser sawmilling industry, help to explain the state's growing interest by midcentury in a pulp and paper-centred strategy of economic development. Donald Eldridge offers one perspective on this transition from his years as a large lumber operator. While the smaller operators cut over the woodlots and then surrendered them to tax sales, the larger operators cut, held on to, and bought more forest lands. Doing so helped not only their businesses but also the pulp and paper companies to which many of these lands were sold in subsequent years.

Bigelow was also a key figure in the transition of industrial leadership from sawmilling to pulp and paper. Through the frustration of reform battles in the sawmilling industry, at the provincial and the regional levels, he was pushed toward greater consideration of the pulp and paper industry. As he wrote at the time, 'It will only be economically feasible to manage a woodlot in Nova Scotia when we have more demand for more kinds of the wood now growing on these woodlots. The most direct way to get more demand is to get more wood-consuming industries started. Wood-consuming industries of economic size, by today's standards, will not start in any area until they foresee a 25-50 years or preferably a perpetual supply of wood.'[8]

The terms under which the pulp and paper companies were established in the province provide an equally telling example of the political forces shaping the context of forestry. Bigelow provides us with an interesting perspective. His first insights stemmed from the temper of the time. When the Mersey Paper Company came to Nova Scotia, it was treated to major concessions by the province, among them the provision of cheap hydroelectric power, fire protection, and pulpwood from Crown lands. But Bigelow's other revelations came from firsthand experiences. He was a principal figure in contacting and negotiating the entry of Nova Scotia Pulp Limited (NSPL) in the late 1950s. He was also the official who drafted the Nova Scotia Pulp Limited Agreement, the contract under whose terms the company acquired its Crown timber lease. Bigelow worked hard to make the point, to both Cabinet and the bureaucracy, that Nova Scotia's forest resources should not be sold cheaply. He advised a stumpage rate of $4.40 per cord. The premier at the time, however, revised the stumpage to one dollar per cord, on the grounds that it had provided a stable operating environment for the Mersey Paper Company since 1929.

Bigelow's career testifies to the broader political and economic significance of forest management and to the reluctance of provincial authorities in acknowledging this. By coincidence and conviction, Bigelow found himself in positions to promote forest management by shaping markets. These efforts focused on the economics of forest use and the relations among commercial stakeholders in the industry. He addressed crucial issues relating to pulpwood production, lumber marketing, and paper investment – from bureaucratic positions in the Departments of Agriculture and Trade and Industry. His experiences also underlined the limited mandate, the narrow perspectives, and the restricted set of policy instruments enjoyed by the Department of Lands and Forests.

Yet the coming of Nova Scotia Pulp Limited illustrates the limits of technical advising. Decades of concessionary treatment proved impossible to resist, and, after years of fruitless effort, provincial authorities were unwilling to risk losing the NSPL project over the stipulation of detailed terms. A few years later, the issue arose again, in the context of negotiations toward a third pulp mill. Lloyd Hawboldt experienced the tensions firsthand when the DLF was called upon to advise on Crown wood supply. At issue were the findings of the 1958 forest inventory and the interpretation of these data for long-term timber use. The Hotel Nova Scotia meetings marked a crucial juncture in the relations between the government and its forest technocrats. Not only did ministers decide (against initial professional advice) to commit Crown forest resources to the Scott Maritimes project, but they also did so by diminishing and humbling their own professional experts. This was another crucial moment in shaping the mandate and the practice of foresters in modern Nova Scotia.[9]

In sum, the experiences of the early foresters illustrate some defining and constraining features for the profession, industry, and state. They include the late entry and ambivalent welcome of professional forestry to Nova Scotia; the continuing marginality of small timber owners in the primary wood market; the colonial legacies in the lumber trade (with its merchant domination, its poor product, and its political embeddedness); and the eventual commitment of the Crown forest estate to pulp interests on highly concessionary terms. These constitute pivotal circumstances that illuminate all subsequent events and struggles. Together they define the 'grain' with which professional foresters, in industry or government, had to contend.

Working against the Grain in Forest Science

Politics is a crucial element in explaining various science debates and conflicts in the Nova Scotia forest community. Three controversies stand out in the foresters' accounts in this book. These deal with the science underlying methods of taking forest inventories and estimating forest yield; the science of understanding and combatting the spruce budworm infestations of eastern Nova Scotia softwoods; and the science underlying forest management systems more generally. Superimposed on these issues is the question of whether forest science offers a uniform and transferable knowledge base, an 'immutable mobile' in Latour's terms, or whether it offers a unique and contextual knowledge base, a mutable immobile whose application is adjusted according to place and time. The stories told here suggest that the research outputs of the federal forest service, and the theoretical disciplines taught at forestry school, have tended toward the first model, while provincial forest research and field forestry have tended toward the latter.

The first issue of interest is the way in which forest science was shaped by political and economic forces. It is striking how frequently and thoroughly the executive powers of government overlooked the advice and information provided by their own foresters on the state of the resource and the terms of management. The times when pulp and paper industry investment were solicited and eventually established in the late 1920s, late 1950s, and 1960s were defining moments.

Both generally and at specific moments, the inventory data and stumpage payments advocated and recommended by foresters to government were challenged, questioned, and modified under ministerial pressure to accommodate pulp and paper industry investment. Chief Forester Otto Schierbeck dared to challenge the forest cruises of the Crown lands claimed by the Mersey Paper Company, as part of the terms of its coming to Nova Scotia in the late 1920s. Schierbeck argued, reasonably and professionally, that Mersey had underestimated the forest resources on large areas of Crown lands, thereby overestimating the amount of land that the company needed. However, his advice was not taken seriously by the government, and the dispute played an important role in his dismissal in 1933.

About thirty years later, DLF Division of Extension Director Lloyd Hawboldt faced a very similar situation. He and his colleague Dick Bulmer saw inventory estimates and permissible harvest levels revised upward on at least two occasions. The most crucial moment was in 1964 when a Crown lease was under negotiation for Scott Maritimes. For a full week, their inventory models, assumptions, and data were questioned in-camera by a senior minister with trial law training. By the end, Ike Smith had constructed his own forest estimate, whose fibre content proved just enough to accommodate the third pulp mill. This was not a unique event in Nova Scotia (though the fact that department foresters were forced to accept the new forest estimates may have been). The methods of calculating annual allowable cuts (AACs) were thoroughly flexible and manipulable and typically yielded results, well into the 1970s, corresponding closely to industrial needs.[10]

The second aspect of the politics of science refers to the spruce budworm spray debate. Controversy has raged over the treatment of this insect since the 1920s, when it began to affect the fibre supply of the forest industry. Industrial foresters have argued in favour of spraying chemical insecticides and biocides, and this has become the standard view within forest science.[11]

Advocates of a silvicultural tradition contend that wasteful and exploitative harvesting practices have resulted in more frequent and serious damage by the spruce budworm and that the severity of the infestations can be minimized through appropriate forest management and silvicultural measures. In a recent review of the debate, Allan Miller and Paul Rusnock have argued that, in spite of industrial foresters having spent massive efforts at refuting the silvicultural hypothesis, the balance of the debate suggests that there is a lot to commend in that hypothesis.[12]

Lloyd Hawboldt subscribed to the silvicultural hypothesis from the first day that he confronted the spruce budworm. Writing in the mid-1950s and again in the 1970s, he vigorously opposed the pressures from the forest companies, the Canadian Forest Service, and the Faculty of Forestry at the University of New Brunswick. In fact, he provided the intellectual foundation for the environmental movement that succeeded in pressuring the provincial government to implement a spray ban in 1976.[13] He was not, however, successful in implementing silvicultural methods to deal with the spruce budworm. This has no doubt contributed to the discrediting of the silvicultural hypothesis and the claims that the extent of budworm infestations cannot be affected by such measures.

The third science issue relevant here is the more general difference between what Aldo Leopold has called types A and B forestry. Type A forestry is the agronomic kind, which has been increasingly utilized by industry and adopted and rationalized by the profession.[14] Here the focus is technocratic control of the forest environment. The use of exotic species, monoculture, heavy machinery, and chemical and biological agents is the order of the day. Type A foresters even claim the ability to accommodate ecosystems management and sustainable forestry, seeing themselves as planners who can impose ecological order on the landscape. Their techniques remain 'immutable mobiles,' ensembles of concepts and techniques that can be transferred and applied universally.[15] Immutable forest mobiles are not without value. One familiar mobile of the early twentieth century was the diameter cut limit. The Nova Scotia version was enacted in 1943 by the Small Tree Act. Despite the inevitable controversies, it was somewhat successful in saving some of the province's sawlog supply during a period of intense exploitation. Overall, though, most foresters were critical of the STA for taking no account of the varying properties of site and stand. In many cases, densely stocked stands of small-diameter softwood needed thinning to release chronologically 'old' trees for growth. Donald Eldridge noted the sawmillers' complaint that some tree stands opened up by the application of the diameter limit were made weaker and vulnerable to windthrow.

But while the STA showed the weaknesses of the concept of the immutable mobile, and while most foresters recognized them, a more sophisticated version of it was clearly becoming dominant in Nova Scotia forestry.

Various federal-provincial cost-shared forest management arrangements represent a long line of triumphs of type A forestry and the immutable mobile model in forest science. The first triumph is illustrated by the demise of Hawboldt's studies of the birch dieback in the 1950s. The team of scientists was able to identify the immediate cause of the disease and advance the knowledge on some of its deeper causes. Yet the unit was closed on the assumption that the centralization of scientific efforts under the federal Canadian Forest Service would yield better results. The Canada Forestry Act of 1950 allowed for such centralization of research in the hands of a federal agency.

Mary Guptill provides us with an apt picture of the discrepancy between the immutable afforestation mobiles developed by the professionals of a centralized bureaucracy and the needs of local forest conditions on the French Shore. While generous in her assessment of the federal-provincial forestry agreements that ran from 1977 to 1995, her account shows some of the contradictions and 'inefficiencies' in trying to force a general program and set of recommendations on a local situation. The agreements not only favoured type A forestry methods at the expense of other measures, but they also provided an uncertain source of funds and changing objectives and goals within the parameters of type A forestry. And in 1995 they were terminated. The program was biased toward clearcutting, which paved the way for woodlot owners to take advantage of the full set of one-time renewal treatments paid for by the program, such as site preparation, planting, and spraying. This encouraged the treatment of degraded and poor stands and yielded high figures of the number of hectares treated, an important condition trumpeted by the program developers. But the program was plagued with inconsistencies between the way in which certain trees were supposed to grow (according to the experts) and the way in which they grew in the field.

Guptill would rather have seen monies to maintain and improve good stands on a continual basis. Against the odds, she managed to promote the implementation of some aspects of type B forestry, which is more sensitive to local ecological and cultural conditions. The natural conditions provided for fast-growing, insect-resistant, commercially desirable, and shade-tolerant tree stands. The cultural situation yielded long, narrow woodlots, which, when cut, were naturally reseeded by the trees on adjacent lots. These circumstances were highly suitable for a selection cutting system. Overall, Guptill shows the relevance of local knowledge and local concerns for long-term forest management.

Working against the Grain in Social Forestry

Social forestry flows from the recognition that resource use affects people's life prospects, and forest relations should be designed and applied in this light. In contemporary times, social forest planning would be geared to communities of users or stakeholders. In earlier generations, it was mediated by commercial, administrative, and legal processes to reconcile conflicting claims. Mainstream North American forestry was not entirely unaware of societal context, for the turn-of-the-century conservation movement asserted a public interest as the rationale for controlled use and renewal. Yet this early posture declined with the formalization of technique and vocation in the decades following the Second World War. Otto Schierbeck is a fascinating example of this process at work, and his predicament is highly revealing. An apt expression of the founding generation of foresters, his efforts and writings show an impressive awareness of the tensions besetting good management. His undoing, by the combined forces of tradition and profit, sent a strong message to subsequent foresters about the range of permissible behaviours. Indeed, the memoirs of his successor, Wilfrid Creighton, take on a fascinating significance when read in this light.

Professional forestry came to Nova Scotia more slowly than it did elsewhere. Yet the private land base and the variety of commercial use created a far more diverse forest constituency than is typically found in Canada. Even if professional thinking in general tended to ignore social context, the Nova Scotia forest outlook seemed positively agnostic on the social question. This did not apply to all resource matters, however, as a comparison with the wildlife side of the DLF mandate attests. As early as Schierbeck's time, private stakeholders were recognized in the form of landowners' and users' groups (fish and game clubs and federations), and consultative channels provided third-party input.[16]

In making this point, it is important that we respect historical context. From the outset, professional forestry was geared to servicing resource use. The question was how far this would be equated with industrial interest. Nova Scotia foresters should not be expected to challenge or defy the foundational values of business life. But neither should they be excused from silence on or ignorance of redistributive economic claims. In the early twentieth century, these were raised repeatedly over at least three common problems: the land market, forest taxation, and scaling practices in logging.

From the start, the emphasis on servicing industry and developing forestry and forests as magnets for industry compromised the social services provided by the Department of Lands and Forests. The department focused on protecting and developing the resource, not on the people who relied on it. Forest businesses were treated with deference and permitted to control operational procedures in their working areas, Crown as well as private. These standards and practices, including marketing practices, were also extended, through the contract logging system, to the large proportion of forest lands held by small woodlot owners (mostly farmers) incorporated into the same system of concessions. The small owners faced different marketing obstacles than did the corporate landowner or the Crown, yet no allowance was made for this fact in policy. Despite a debate stretching back to the 1930s, farm wood was not incorporated into the organized marketing regime that governed grain, livestock, and dairy products. Instead, the industrial buyers were free to forge rural mercantile chains of brokers and dealers who perpetuated a form of backwater capitalism as late as the 1960s. This allowed the pulp companies to dictate minimal prices for private pulpwood, while their agents further tightened the screws on woodlot owners through arbitrary scaling (volume measurement) and grading (quality measurement) of wood.

But this process did not go uncontested. The Department of Agriculture was the government agency that took up the cause of woodlot owners as an extension of its farm constituency. The friction that then developed between the Departments of Lands and Forests and Agriculture is illustrated in several of the foregoing accounts. John Bigelow criticized Lands and Forests in the 1930s and 1940s, when there was no sign that marketing issues in general, or the woodlot interest in particular, seemed of any relevance to its program. This contrasted sharply with the organized constituencies and flexible regulatory outlooks prevailing in Agriculture. The latter certainly recognized the importance of paradigm choice. This lay behind Waldo Walsh's proposal that Bigelow and Creighton be named joint deputy ministers of the DLF for one year, after which time their performances would determine the permanent appointee. Although the two departments operated at the time under a common minister (who expressed support for the plan), it was never carried out.

Bigelow's moves from Agriculture to the Maritime Lumber Bureau and later to the Department of Trade and Industry provide a microcosm of the frustration of woodlot owners and lumbermen in having their needs met by the Department of Lands and Forests. While these were not necessarily conscious moves on Bigelow's part to find alternative ways of addressing the economics of small woodlot management, they did afford Bigelow the opportunity of doing so. Thus, it was an ironic moment in the mid-1950s when he addressed a group of woodlot owners and identified the Department of Trade and Industry as a champion of their cause: 'This organization of woodlot owners has, I am happy to say, now been formed in this area on a voluntary and informal basis. It has the wholehearted support of the Department of Trade and Industry and will be aided by this Department in every way possible to develop into a formal organization with a membership of the majority of the woodlot owners of the area, authorized to do business for its members and to acquire knowledge and business experience as rapidly as possible.'[17]

Immediate attempts to form a woodlot organization were, however, frustrated and compromised by the opposition of some of the major actors in the government and by the indifference of the Department of Lands and Forests. Here again it is clear that, in its failure to recognize the relevance of social relations, the state forest agency assimilated a set of unexamined social assumptions. The micro-tracts of 200 to 400 acres, which typified the woodlot sector in Nova Scotia, bore little resemblance to the extensive corporate or Crown licence forests for which North American foresters were trained to work. In itself a source of regret to Nova Scotia foresters, this spatial 'irregularity' was reinforced by a series of regressive cultural stereotypes that served to further distance small owners from professional managers. Woodlot people were portrayed as part-time harvesters who responded to short-term pricing conditions in search of supplementary income. As such, they were viewed as unfit for forest management. Where some anthropologists would see woodlot behaviour as a rational response to material scarcity, and a positive hedge against overspecialization, foresters viewed it as evidence of marginality, primitiveness, and even ignorance. From here it was only a short step to the question of enterprise and management. Again owners were portrayed as people clinging to primitive practices such as field burning and collaborating in the degradation of their properties by selling stumpage to 'woodpecker' sawmills. Entirely lost in such assessments was the impact of uncertain and (in some cases) punitive wood purchasing, the impact of municipal taxation, and the unpredictable demands for labour and cash return in a mixed enterprise (bricolage) livelihood combining subsistence and commercial activities.

In sum, neither academic training nor professional norms presented the woodlot as a legitimate enterprise of consequence to the forest system. If David Dwyer is correct that foresters go to the woods to escape people, then it is a short step to denigrate the small holding as insufficiently 'natural' to merit serious attention. Only in this way can we explain the studied disinterest that state foresters displayed toward small holdings. The first organized survey of woodlot attitudes and analysis of woodlot economics were not conducted until the 1970s.

Against this backdrop, the prospect of an organized woodlot movement was seen as similarly foreign. Although the Nova Scotia government welcomed and supported associations representing the forest industry (NSFPA) and the forest profession (NSS-CIF), its indifference and later hostility toward the NSWOA are striking. Perhaps the perceived passivity of the tens of thousands of woodlot owners was preferred to the uncertainties and complications of facing a collective voice. Ever since the 1930s, two pillars of DLF policy were to enhance the management of Crown forest lands and to expand the Crown forest estate by a vigorous purchase program aimed at doubling the 25 percent in Crown hands. As Donald Eldridge pointed out, this placed the DLF formally in competition with industrial buyers. In practice, however, it was the small ownership segment that found the Crown chasing their lands, and the suspicion of government takeover was deeply rooted in the rural outlook by the 1960s.

In any event, there is little evidence that the DLF ever recognized the positive potential of a densely membered, province-wide woodlot association as either a consultative body for policy formulation or an instrument for the delivery of programs to private lands. Excluding the six extension forest specialists within the DLF, the professional outlook of Nova Scotia foresters remained rooted in simple stereotypes. In comments dripping with condescension, the deputy minister of lands and forests, Bob Burgess, told a pulp industry meeting in 1970 that 'Joe Farmer may not know what silviculture *means*, let alone believe it's any good.'[18] This attitude carried over to the forest group ventures, established under the 1977-82 agreement. Despite their unexpected rapid growth at the community level, the groups came under criticism from industrial quarters as artificial creations of public subsidy. At one point, the Nova Scotia Section of the Canadian Institute of Forestry endorsed a resolution critical of the group concept, and David Dwyer was prompted to resign his lifelong membership in protest. For Mary Guptill, who spent her career in group forestry work, its suitability was self-evident, and the work of undoing political stereotypes was a continuing concern.

Against this backdrop, efforts to advance the interests of small woodlot owners, through organization, marketing reform, and management planning, continued through the 1960s, 1970s, and 1980s. The accounts of Donald Eldridge and Rick Lord tell the story of these attempts from two sharply differing points of view. As a forest industry lobbyist and then as deputy minister of lands and forests, Eldridge waged ideological warfare against the woodlot movement, painting it as a threat to the industry rather than as an integral part of it. Lord spearheaded the woodlot-organizing campaign despite an evident lack of legitimacy in DLF circles. However, his initial success was frustrated in the later fight for a marketing act by legal obstruction from the pulp companies and a limp implementation by the government.

Both Lloyd Hawboldt and David Dwyer also supported the aims of the Forest Improvement Act, which attempted to formulate private and Crown land management standards. This entailed multistakeholder forums in district boards, embracing industry (both pulp mill and sawmill segments), professional foresters, wildlife interests, and small woodlot owners. Through a parallel process of industrial lobbying at executive levels, and administrative indifference on the operational plane, this scheme was similarly scuttled. Industrial opposition and governmental indifference to improvement of the woodlot sector have continued unmitigated in the 1980s and 1990s despite occasional recognition in official circles that reforms have been overdue. Indeed, when such reforms have been taken seriously by committed bureaucrats, they have revived staunch opposition by the pulp and paper sector and its supporters in government.[19]

In the Realm of Possibilities

With so many qualifiers, it is perhaps difficult to conceive of any outcomes other than the ones in place at present.[20] However, this book has demonstrated that the prevailing tendencies in Nova Scotia forestry have been challenged repeatedly along the way. Authorities were not without choices as dissenting voices advanced constructive proposals on behalf of more sensitive ecological and social development (i.e., mutable mobiles). We believe that these are well worth exploring and that they may provide important lessons on which to build future strategies.

In the hands of individuals such as those profiled above, the techniques of professional forestry cannot be dismissed as irrelevant to reformist concerns. Indeed, many of their ideas pointed in powerful new directions for forest management. They also nursed social concerns that extended well beyond commercial profitability through maximum industrial production. From time to time, the dissenters generated mutable mobiles, ideas, and practices based on modern science and techniques yet malleable to specific local situations. These were not, in other words, fixed tools and concepts to be slavishly applied to any forest environment. This underlying impulse, as much as the particulars of dissenting programs, holds significance for forestry debates today.

The urge for immutable mobiles is not confined to mainstream technical forestry. In fact, it could be argued that its generalizing and totalizing impulse applies to both of the chief protagonists in contemporary forest debate, and in this respect both are inherently ideological.[21] Professional forest specialists based in government and industry consider themselves scientific and economic realists. At the same time, they sense ideology in the simple misconceptions and emotive beliefs of their middle-class urban critics and environmental activists. The critics are ideological to the extent that they lose touch with practical reality, neglecting the economic importance of primary forest production and dismissing the sophisticated technical foundations of modern forest maintenance. Furthermore, the critics see an undifferentiated 'nature' as the wellspring of an alternative forestry (implying that commercially inspired forestry is somehow 'unnatural'), calling for the preservation of old-growth or virgin forest stands (with little apparent awareness that today's stands are the product of complex successional processes), and equating ecological sensitivity with non-economic values (suggesting that foresters are ignorant or dismissive of nature in the round).

For 'green advocates' convinced of the material excesses of an overconsumptive society, ideology refers to the canon of professional forester beliefs that unites state and business technocrats in a self-legitimating elite. They also seem to lose touch with reality, becoming ideological in the conviction that processes of planning, management, and renewal confer an institutional capacity to effectively direct natural processes. Furthermore, they confuse quantitative accuracy in measurement and mapping trees and stands with qualitative knowledge of the underlying stand ecologies. They generate a false confidence in the transformative power of logging and silvicultural systems to replace and even enhance the disappearing forests, and they confuse the renewal of a 'resource' with the sustainability of nature.

It is into this overly simplistic dichotomy that we have tried to introduce a continuum of intermediate dissenting experiences. Significantly, they were generated within the domain of institutional forestry without losing their intellectual vitality. This should caution the critics of professional forestry against prematurely dismissing alternatives strictly because they come from industrial or government forestry. At the same time, these accounts offer trenchant lessons for those who defend institutional forestry. The urge for continuity and legitimacy constantly risks suppressing the creative and progressive edges of the vocation, often in harsh and arbitrary ways. Losing memory of the pluralistic foundations of forestry practice is also to be deplored. The fact that these narratives of dissent are not even mentioned in Wilfrid Creighton's history of the Department of Lands and Forests, or Ralph Johnson's history of the Nova Scotia forests, is telling.

There is some evidence today of a convergence between preservationist and industrial foresters. This convergence may seem paradoxical since many non-fibre perspectives on forest management still firmly reject the engineering premise of total managerial control (as embodied in plantation monocultures, herbicides and insecticides, clearcutting, and planting). Yet a powerful group of forest scientists contends that forest management can be achieved by a landscape-planning and maintenance regime that combines fibre plantations and non-industrial areas. What is sometimes called 'new' or 'sustainable' forestry, or forest ecosystem management, claims the ability to manage such a spatialized system. Based on the principle of integrating ecological and environmental values in forest commodities production, this form of forestry, like its sustained yield predecessor, aspires to control the forest ecosystem. First forest ecosystem dynamics are modelled for a fuller understanding, and then uses are planned accordingly. This involves unbundling the forest into roving ecosystems in which the forester can plan functions in minute detail, in which harvests mimic natural forest disturbances, and in which old-growth sections may move from place to place as they mature and die. New forestry thus claims to take account of all the human requests of the forest, working toward compromise and balanced solutions between the extremes of utilitarian and preservationist ideologies.[22]

Certainly, there are problems with this spatialized ideology of forest use. The preserved areas can provide a screen (a 'green lie') for the continued environmental and social costs incurred by industrial forestry. The non-industrial areas may become resorts for the enjoyment of the affluent, the same people who may also profit from intensive forestry. In relation to these concerns, James Proctor argues that the assumption that the forest can be everything to everyone is fallacious.[23] Judging from the debate on setting aside old-growth or ancient forests in North America, a resolution seems to be more distant than ever. It is not possible to set aside large tracts without affecting the fibre supply. Similarly, it appears impossible to agree on the extent of forest to be set aside; the environmentalists contend that too little is preserved, while the forest industry claims the opposite.[24] Proctor thus argues for an approach based less on policy and management and more on values, ethics, and ideas (all components of a different ideology). Such values should contain three key guiding notions: a realist form of moral pluralism that takes into account the places of both humans and animals in the forest; an anthropogenic moral basis that recognizes that there are both utilitarian and intrinsic values in nature; and a sense of ethics rooted in, but not limited to, place, which shows respect for the environments in which one resides and makes sure that resource use in one place does not result in unsustainable resource use elsewhere.

The views of the dissenting foresters in this volume speak to the development of such values. Their views and values are not only radically different from the dominant grain in Nova Scotia forest politics, but they also deviate from the extreme views expressed by industrial and environmentalist interests generally. Collectively, their ideas amount to a conservationist ideology centred on social equity and ecological concerns. From the 1920s to the 1950s, Schierbeck and Bigelow advocated a broad 'nationalist vision' wherein sustained yield forestry, value-added industries, and a fair return from Crown stumpage and private lands were to constitute the focus of a Nova Scotia forest policy. Somewhat later, Bigelow, Lord, and Dwyer called for a stronger position for the tens of thousands of small woodlot owners as partners in the forest estate. More recently, Hawboldt and Guptill provided strong statements in favour of a forestry that begins by preserving forest integrity rather than by meeting industrial demand. Indeed, the dissenting accounts above speak to the contemporary efforts to transcend the environmentalist/industrialist divide.[25] All elements, we contend, will have to be incorporated into an alternative ideology and ethics to serve human, non-human, and environmental ends.

Notes

Chapter 1: Introduction

1 For organizational histories, see Ken Fensom, *Expanding Forestry Horizons: A History of the Canadian Institute of Forestry: Institut Forestier du Canada, 1908-1969* (Montreal: CIF, 1972); Kenneth Johnstone, *Timber and Trauma: 75 Years with the Federal Forestry Service, 1899-1974* (Ottawa: Supply and Services, 1991); Wilfrid Creighton, *Forestkeeping: A History of the Department of Lands and Forests in Nova Scotia, 1926-1969* (Halifax: Province of Nova Scotia, 1988).

2 H. Vivian Nelles, *The Politics of Development: Forests, Mines, and Hydro-Electric Power in Ontario, 1849-1941* (Toronto: Macmillan, 1974), 182-214.

3 Michael Howlett and Jeremy Rayner, 'Do Ideas Matter? Policy Network Configurations and Resistance to Policy Change in the Canadian Forest Sector,' *Canadian Public Administration* 38 (1995): 382-410.

4 Richard Rajala, *Clearcutting the Pacific Rain Forest: Production, Science, and Regulation* (Vancouver: UBC Press, 1998); Ian Radforth, *Bush Workers and Bosses: Logging in Northern Ontario, 1900-1980* (Toronto: University of Toronto Press, 1987).

5 Thomas Dunk, 'Talking about Trees: Environment and Society in Forest Workers' Culture,' *Canadian Review of Sociology and Anthropology* 31, 1 (1994): 14-34, and '"Is It Only Forest Fires that Are Natural?" Boundaries of Nature and Culture in White Working Class Culture,' in L. Anders Sandberg and Sverker Sörlin, eds., *Sustainability – The Challenge: People, Power, and the Environment* (Montreal: Black Rose Books, 1998), 157-66; Russ Janzen, 'Hegemony and Genealogy: Managerialist Discourse in the Forests of British Columbia,' in Sandberg and Sörlin, *Sustainability*, 149-56.

6 It is impossible to provide an exhaustive list of such references. For a recent statement, see Elizabeth May, *At the Cutting Edge: The Crisis in Canada's Forests* (Toronto: Key Porter, 1998).

7 See the contributions by Sandberg, Clancy, and Bissix and Sandberg in L. Anders Sandberg, ed., *Trouble in the Woods: Forest Policy and Social Conflict in Nova Scotia and New Brunswick* (Fredericton: Acadiensis Press, 1992). See also Elizabeth May, *Budworm Battles: The Fight to Stop the Aerial Insecticide Spraying of the Forests of Eastern Canada* (Halifax: Four East Publications, 1982), and the contributions by McMahon, Majka, Livesey, and Schneider in Gary Burrill and Ian McKay, eds., *People, Resources, and Power: Critical Perspectives on Underdevelopment and Primary Industries in the Atlantic Region* (Fredericton: Acadiensis Press, 1987).

8 The literature on the modern professions is extensive. One helpful survey of professions and professionalization is Magali Sarfatti Larson, *The Rise of Professionalism: A Sociological Analysis* (Berkeley: University of California Press, 1977).

9 The leaders include British Columbia (43 percent), Quebec (20 percent), and Ontario (14 percent), followed by Alberta (8 percent), New Brunswick (5 percent), Nova Scotia (2.5 percent), Saskatchewan (2.4 percent), and the remainder. *Compendium of Canadian Forestry Statistics, 1994* (Ottawa: Canadian Council of Forest Ministers, 1995), 56.

10 L. Anders Sandberg and Peter Clancy, 'Small Woodlot Owners, Property Rights, and Forest Management in Nova Scotia,' *Journal of Canadian Studies* 31, 1 (1996): 25-47.

11 O.L. Loucks, *A Forest Classification for the Maritime Provinces* (Fredericton: Canada, Department of Forestry, 1962). For another helpful account of the Nova Scotia forest environment, see Nova Scotia Museum, *Topics and Habitats* and *Theme Regions,* volumes 1 and 2 of *The Natural History of Nova Scotia* (Halifax: Nimbus, 1996).

12 Indeed, some ecologists argue that the notions of an orderly and balanced succession of stages for ecosystems is unrealistic. The new paradigm of ecology stresses that flux and unpredictability are integral parts of successional patterns. See S.T.A. Pickett and Richard S. Ostfeld, 'The Shifting Paradigm in Ecology,' in Richard Knight and Sarah Bates, eds., *A New Century for Natural Resources Management* (Washington, DC: Island Press, 1995), 261-78.

13 William Wicken, '26 August 1726: A Case Study in Mi'kmaq-New England Relations in the Early 18th Century,' *Acadiensis* 28, 1 (1993): 5-22; and Stephen Patterson, 'Indian-White Relations in Nova Scotia, 1749-61,' *Acadiensis* 28, 1 (1993): 23-59.

14 Robert M. Leavitt, *Maliseet Micmac: First Nations of the Maritimes* (Fredericton: New Ireland Press, 1995).

15 Ellice Gonzalez, *Changing Roles for Micmac Men and Women: An Ethnocultural Analysis,* National Museum of Man Mercury Series, Canadian Ethnology Service Paper 72 (Ottawa: National Museums of Canada, 1981).

16 William Wicken, 'Mi'kmaq and Waustukwiuk Treaties,' *UNB Law Journal* 43 (1994): 241-53, and 'Heard It from Our Grandfathers: Mi'kmaq Treaty Traditions and the Syliboy Case of 1928,' *UNB Law Journal* 44 (1995): 145-61; Alf Hornborg, 'Environmentalism, Ethnicity, and Sacred Places: Reflections on Modernity, Discourse, and Power,' *Canadian Review of Sociology and Anthropology* 31 (1994): 245-67, and 'Mi'kmaq Environmentalism: Local Initiatives and Global Projections,' in Sandberg and Sörlin, eds., *Sustainability,* 202-11.

17 Ralph S. Johnson, *Forests of Nova Scotia* (Halifax: Four East Publications, 1986).

18 Creighton, *Forestkeeping.*

19 Mike Parker, *Woodchips and Beans: Life in the Early Lumberwoods of Nova Scotia* (Halifax: Nimbus, 1992).

20 Eric Mullen and Millie Evans, *In the Mersey Woods* (Liverpool, NS: Bowater Mersey Paper Company, 1989).

21 Ray Raphael, *Tree Talk: The People and Politics of Timber* (Covelo, CA: Island Press, 1981), and *More Tree Talk: The People, Politics, and Economics of Timber* (Washington, DC: Island Press, 1994).

22 Raphael, *More Tree Talk,* xv.

23 For a recent example, see May, *At the Cutting Edge.* May emphasizes the heavy overcuts of timber, relative to AAC, in most Canadian jurisdictions. For an earlier powerful statement, see Jamie Swift, *Cut and Run: The Assault on Canada's Forests* (Toronto: Between the Lines, 1983). For a summary critique drawing on the historical and international literature, see James C. Scott, *Seeing like a State: How Certain Schemes to Improve the Human Condition Have Failed* (New Haven: Yale University Press, 1999), Chapter 1.

24 For an early statement, see J.F. Franklin, 'Toward a New Forestry,' *American Forests* 95 (1989): 37-44. For a variety of views, see Gregory H. Aplet et al., eds., *Defining Sustainable Forestry* (Washington, DC: Island Press, 1993). For some initiatives in Canada, see Canadian Council of Forest Ministers, *Sustainable Forests: A Canadian Commitment* (Hull: Canadian Council of Forest Ministers, 1992), and Ontario Forest Policy Panel, Diversity, *Forests, People, Communities: A Comprehensive Forest Policy for Ontario* (Toronto: Ministry of Natural Resources, 1993). For an academic account, see Hamish Kimmins, *Balancing Act: Environmental Issues in Forestry* (Vancouver: UBC Press, 1992).

25 For a good example, see Gordon Baskerville, 'The Forestry Problem: Adaptive Lurches of Renewal,' in Lance H. Gunderson, Crawford S. Holling, and Stephen S. Light, eds., *Barriers and Bridges to the Renewal of Ecosystems and Institutions* (New York: Columbia University Press, 1995), 37-102.

26 Chris Maser, *The Redesigned Forest* (Toronto: Stoddard, 1990), 55. This is the point made by so-called ecoforesters. For their critique of traditional forestry, including its more recent 'ecological versions,' see Alan Drengson and Duncan Taylor, eds., *Ecoforestry: The Art of Sustainable Forest Use* (Gabriola Island, BC: New Society Publishers, 1997).

27 Some have called for politics to become a more central focus of forestry students' curricula; see Paul E. Ellefson, 'Politics and Policymaking: A Teaching Challenge in Forestry,' *Journal of Forestry* 91, 3 (1993): 24-7.
28 For an example of this response, see the series of articles by retired New Brunswick forester Edward S. Fellows, which appeared in the Fredericton *Daily Gleaner* in 1993 (6, 13, 20, 27 February; 7, 13, 27 March; 3 April).
29 This is evident in the series of federal 'State of the Forest' reports produced annually by the Department of Natural Resources since 1991. The charter document of this perspective is *Sustainable Forests, a Canadian Commitment.*
30 Witness the 1990s debate over old-growth cutting practices in British Columbia. Labelled 'the Brazil of the North,' it was featured in the Sierra Club publication, Bill Devall, ed., *Clearcut: The Tragedy of Industrial Forestry* (San Francisco: Earth Island Press, 1993). For a recent riposte, see Patrick Moore, *Pacific Spirit* (Vancouver: Terra Bella, 1995).
31 Herbert Kaufman, *The Forest Ranger: A Study in Administrative Behaviour* (Baltimore: Johns Hopkins University Press, 1960), 65.
32 Alexander A. MacDonald, 'Exploitation of the Private Owners of Forest Resources in Nova Scotia,' unpublished paper, 1975.
33 Edward S. Fellows, 'Popular, Professional Views Often Clash,' *Daily Gleaner* (6 February 1993).
34 See J. Murray Beck, *Politics of Nova Scotia*, vol. 2., *1896-1988* (Tantallon, NS: Four East Publications, 1988), 171-2.
35 For details on the evolution of two leading programs, see *The Fiftieth Anniversary of the Faculty of Forestry at the University of New Brunswick, 1908-1958* (Fredericton: UNB Forestry Association, 1958); and John W.B. Sisam, *Forestry Education at Toronto* (Toronto: University of Toronto Press, 1961).
36 For more detail on Fernow's 'Toronto' period, see Andrew Rodgers III, *Bernhard Eduard Fernow: A Story of North American Forestry* (Princeton: Princeton University Press, 1951).
37 Of the dozen charter members, Sisam notes, 'Three of these men were employed with the Dominion Government, three with industry, four with provincial governments and two with the university.' John W.B. Sisam, 'Historical Highlights, Canadian Institute of Forestry,' *Forestry Chronicle* (April 1983): 55.
38 Fensom, *Expanding Forestry Horizons.*
39 Ralph S. Johnson, 'Forest Legislation,' Panel Discussion for Nova Scotia Section of CIF, New Glasgow, NS, 17 September 1964.
40 Johnson, *Forests of Nova Scotia*, 300.
41 This would be the forestry equivalent to the chartered accountant (CA) credential for accountants, the provincial bar admission certificate for lawyers, or the College of Physicians and Surgeons membership for medical doctors.
42 Mike Fleming, 'Professional Foresters Group Formation Urged,' *Chronicle-Herald* (22 November 1975): 21.
43 Nelles describes it as the 'central agency of the conservation movement.' Nelles, *The Politics of Development*, 184, 188, 193, 198, 200-1.
44 The commission published reports and proceedings throughout its mandate. One of them was a forest inventory of Nova Scotia, produced by Fernow, who thought that its publication would give credibility to the Faculty of Forestry at the University of Toronto. Bernhard Fernow, *The Forest Conditions of Nova Scotia* (Ottawa: Commission of Conservation, 1912). For two accounts of the commission, see C. Ray Smith and David R. Witty, 'Conservation, Resources, and Environment: An Exposition and Critical Evaluation of the Commission of Conservation, Canada,' *Plan 11* (1970, 1972): 55-71, 199-216; and Stewart Renfrew, 'The Commission of Conservation,' *Douglas Library Notes* 19, 3-4 (1971): 17-26.
45 See, for example, Nova Scotia Members of the Canadian Society of Forest Engineers, 'Forestry, Economy, and Post-War Reconstruction,' Department of Lands and Forests annual report, Halifax, 1944.
46 Most CIF-NSS members of the time were captured in a photograph from the 1959 annual meeting in Liverpool, Nova Scotia, reprinted as 'The Way We Were,' *Forestry Chronicle* 71, 6 (1995): 792.

47 Johnson, *Forests of Nova Scotia*, 300.
48 Canadian Institute of Forestry, Nova Scotia Section, *A Forest Policy for Nova Scotia* (Halifax: CIF, 1971).
49 For example, within the forest sector consultative committee of the Nova Scotia Voluntary Economic Planning organization, the NSS-CIF was recognized as a stakeholder entitled to representation at policy deliberations since the early 1970s. This committee addressed most of the explosive policy questions of the day, including organized pulpwood marketing by small producers and budworm spray authorization. See Peter Clancy, 'The Politics of Pulpwood Marketing in Nova Scotia, 1960-1985,' in Sandberg, ed., *Trouble in the Woods*; and Anthony Lamport, *Common Ground: 25 Years of Voluntary Planning in Nova Scotia* (Halifax: Government Services, 1988), 104-16.
50 Ian Mahood and Ken Drushka, *Three Men and a Forester* (Vancouver: Harbour Publishing, 1990).
51 Fred Stevens, 'The Section's First Sixty Years,' *Pulp and Paper Canada* 78, 11 (1977): 22.
52 Nova Scotia author Thomas Raddall captures the contrast dramatically in his story 'The Lower Learning,' in which an elderly timberman encounters two young forestry students on a job in the bush. See *The Dreamers* (Porter's Lake, NS: Pottersfield Press, 1986).
53 A. Paul Pross, 'The Development of Professions in the Public Service: The Foresters of Ontario,' *Canadian Public Administration* 10 (1969): 376-404.
54 R. Peter Gillis and Thomas R. Roach, *Lost Initiatives: Canada's Forest Industries, Forest Policy, and Forest Conservation* (New York: Greenwood Press, 1986), 80-7.
55 Creighton, *Forestkeeping*, 43. On the impact of another senior government forester, Frank A. MacDougall, see Richard S. Lambert and A. Paul Pross, *Renewing Nature's Wealth* (Toronto: Government of Ontario, 1967), Chapter 18 and passim.
56 The phrase was coined by Graham Allison in *The Essence of Decision* (Boston: Little Brown, 1971).
57 The US Forest Service has been treated extensively in the academic literature. See, for example, Michael Frome, *The Forest Service* (New York: Praeger Publishers, 1971); Harold T. Pinkett, *Gifford Pinchot, Private and Public Forester* (Urbana: University of Illinois Press, 1970); Harold Steen, *The U.S. Forest Service: A History* (Seattle: University of Washington Press, 1976); Christopher McGrory Klyza, *Who Controls the Public Lands? Mining, Forestry, and Grazing Policies, 1879-1990* (Chapel Hill: University of North Carolina Press, 1996); and Paul Hirt, *A Conspiracy of Optimism: Management of the National Forests since World War Two* (Lincoln: University of Nebraska Press, 1994).
58 One survey of the federal service is Johnstone, *Timber and Trauma*.
59 The philosophical tradition of Crown management has been explored in the case of Ontario by Nelles, *The Politics of Development*. The evolution of the Ontario DLF is comprehensively explored in Lambert and Pross, *Renewing Nature's Wealth*.
60 One account of the history of the Department is provided by longtime Provincial Forester and Deputy Minister Creighton in *Forestkeeping*. Within the DLF, the wildlife division is examined by Donald Dodds, *Challenge and Response: A History of Wildlife and Wildlife Management in Nova Scotia* (Halifax: Department of Natural Resources, 1993). In 1992 the Department of Lands and Forests became part of the new Department of Natural Resources.
61 John Kenneth Galbraith, *The New Industrial State* (New York: Signet, 1978).
62 Johnson, *Forests of Nova Scotia*, 243.
63 In his 1960 study, Kaufman argued that the US Forest Service reflected a strong ethic of 'voluntary conformity' produced by 'post-entry training, both inside and outside, and by placement, transfer and promotion policies.' *The Forest Ranger*, 198.
64 Jamie Swift, 'Cut and Run: The MacAlpine Case,' *Harrowsmith* October-November 1983: 38-45, 107, 112, and Swift, *Cut and Run*, Chapter 8. MacAlpine was subsequently rehired following legal action. For the United States, see the interesting case of US Forest Service forester Jeff DeBonis's well-known protests against the failure of his employer to meet environmental standards in forest management. DeBonis is the founder of the Association of Forest Service Employees for Environmental Ethics. Jeff DeBonis, 'Natural Resource Agencies: Questioning the Paradigm,' in Knight and Bates, eds., *A New Century*, 159-70.
65 Donald F. George, 'The View from the Finish Line,' unpublished paper, Antigonish, NS, n.d.

66 For the decline of birds as agents of insect control, see Matthew Evenden, 'The Laborers of Nature: Economic Ornithology and the Role of Birds as Agents of Biological Pest Control in North American Agriculture, ca. 1880-1930,' *Forest and Conservation History* 39 (1995): 172-83. For various statements in support of silvicultural methods in controlling the spruce budworm, see Alan Miller and Phil Rusnock, 'The Rise and Fall of the Silvicultural Hypothesis in Spruce Budworm (Choristoneura fumiferana) Management in Eastern Canada,' *Forest Ecology Management* 61 (1993): 171-89. See also, by the same authors, 'The Ironical Role of Science in Policy-Making: The Case of the Spruce Budworm,' *International Journal of Environmental Studies* 43 (1993): 239-51; and Alan Miller, 'The Role of Citizen-Scientist in Nature Resource Decision-Making: Lessons from the Spruce Budworm Problem in Canada,' *The Environmentalist* 13, 1 (1993): 47-59.

67 R. Van den Bosch, *The Pesticide Conspiracy* (Garden City, NY: Doubleday, 1978).

68 For various perspectives on this issue, see Thomas R. Dunlap, *DDT: Science, Citizens, and Public Policy* (Princeton: Princeton University Press, 1981); and John H. Perkins, *Insects, Experts, and the Insecticide Crisis: The Quest for New Pest Management Strategies* (New York: Plenum, 1982).

69 One of the most striking of these is the 1993 folio published by the Sierra Club in Devall, ed., *Clearcut.*

70 In the context of Swedish forestry, Anders Öckerman has shown how thoroughly fluid the various harvesting techniques of the clearcut and selection cut methods may be. Anders Öckerman, 'Culture versus Nature in the History of Swedish Forestry,' in Sandberg and Sörlin, eds., *Sustainability*, 72-9.

71 John Smith, *The Acquisition of Forest Reserves* (Halifax: DLF, 1975), 6. The subsequent Nova Scotia proposal for substantial federal assistance for the purchase of private forests was denied the following year.

72 This is a theme treated in 'Introduction: Dependent Development and Client States: Forest Policy and Social Conflict in Nova Scotia and New Brunswick,' in Sandberg, ed., *Trouble in the Woods*, 20-1; Sandberg and Clancy, 'Small Woodlot Owners,' 25-47; Land Research Group, 'Whither Our Land? Who Owns Nova Scotia? And What Are They Doing with It?' *New Maritimes* 8, 6 (1990): 14-25; Bill Parenteau and L. Anders Sandberg, 'Conservation and the Gospel of Economic Nationalism: The Canadian Pulpwood Question in Nova Scotia and New Brunswick, 1918-1925,' *Environmental History Review* 19, 2 (1995): 55-83; and L. Anders Sandberg and William Parenteau, 'From Weapons to Symbols of Privilege: Political Cartoons and the Rise and Fall of the Pulpwood Embargo Debate in Nova Scotia, 1923-1933,' *Acadiensis* 26, 2 (1997): 31-58.

73 Clancy, 'The Politics of Pulpwood Marketing'; Bissix and Sandberg, 'The Political Economy of Nova Scotia's Forest Improvement Act.'

74 Edward Grumbine, 'Policy in the Woods,' in Devall, ed., *Clearcut,* 259.

75 Jack Westoby, *The Purpose of Forests: Follies of Development* (Oxford: Basil Blackwell, 1987), 328.

Chapter 2: Otto Schierbeck

1 An earlier version of this chapter appeared as L. Anders Sandberg and Peter Clancy, 'Forestry in a Staples Economy: The Checkered Career of Otto Schierbeck, Chief Forester, Nova Scotia, Canada, 1925-1933,' *Environmental History* 2, 1 (1997): 74-95.

2 Vivian Nelles, *The Politics of Development: Forests, Mines, and Hydro-Electric Power in Ontario, 1849-1941* (Toronto: Macmillan, 1974), 183.

3 Schierbeck was born on 13 December 1881 in Struer on Jutland in Denmark, the son of a section engineer of the Danish State Railways. After finishing his primary education at Sorø Academy in 1898, he entered forestry school in Copenhagen and graduated in 1905. He moved to North America the same year and worked from 1905 to 1908 as a topographer with an engineering party for the railway firm of MacKenzie and Mann and Company. Among other projects, Schierbeck worked on railway construction in Quebec, Ontario, and Cuba. In 1909 he returned to Denmark to take up the position of *Overbanemester* at the Danish State Railways, a position that put him in charge of planting trees and shrubs along one section of the Danish railway network. From 1917 to 1920, he was the owner of

Schierbeck's Forest Tree Nursery in Svebølle. In 1920 he returned to Canada, this time as forester for Price Brothers and Company in Quebec, where he served until 1923. He then took up the position of forester for Frank John Dixie Barnjum Limited in Nova Scotia, where he stayed until 1926. From that year until 1933, he held the position of Chief Forester for the Department of Lands and Forests in Nova Scotia. Dismissed from that position, he worked as a lumber operator, consultant, and magazine editor for the next three years. For a few months in 1936 and 1937, he performed research for the International Paper Company at its operations in Maniwaki, Quebec. Schierbeck ended his career with the Dominion Forest Service Experimental Station in New Brunswick. In early April 1941, he contracted pneumonia; he died at Lincoln, New Brunswick, on 22 June 1941. *Forestry Chronicle* 17 (1941): 134; Otto Schierbeck, 'Fra Amerika,' *Tidskrift for Skovvaesen* 18 (1906): 63-7; Aksel Thyssen, *Danske Forstkandidater* (København: Danske Forstkandidaters Forening, 1956), 60. We are indebted to Jette Pedersen of the National Forest and Nature Agency for providing the Danish material on Schierbeck.

Schierbeck described his training as follows: 'The Foresters all receive their technical training at the Agricultural College in Copenhagen, whose standing is similar to that of MacDonald College [of McGill University in Montreal]. The course is from six to seven years' duration, two years of which is spent in the woods with a graduate Forester. During this latter time the Forestry students have to take part in all of the work on the district and submit a survey, an independent cruise, and prepare at least two yield tables. They have three examinations; the first is elementary, including botany, zoology, chemistry, etc.; the second gives scientific forestry; and the third and last is held in the woods where the graduating students have to mark several stands for thinnings, report on logging, sawmills, plantations, etc.' Otto Schierbeck to Georges Maheux, 23 February 1931, RG 20, vol. 734, no. 18, Public Archives of Nova Scotia (hereafter PANS).

4 There were other Canadian foresters who expressed and then 'paid' for criticism of Canadian forest management. Few of them, however, rendered a critique as broad and as deep as Schierbeck's. For an account of Judson Clark in Toronto, see Nelles, *The Politics of Development*, 209-11; for the tragic fate of the director of the Dominion Forest Service, Ernest H. Finlayson, see Peter Gillis, 'Limits of Federal-Provincial Cooperation, 1920-1936: Ernest Herbert Finlayson and Canadian Forestry,' in David G. Brand, ed., *Canada's Timber Supply* (Chalk River, ON: Forestry Canada, 1991), 4-12; and Kenneth Johnstone, *Timber and Trauma: 75 Years with the Federal Forestry Service, 1899-1974* (Ottawa: Forestry Canada, 1991), 64-75.

5 Nelles, *The Politics of Development*, 182-214.

6 For more on Barnjum, see Thomas Roach and Richard Judd, 'A Man for All Seasons: Frank John Dixie Barnjum, Conservationist, Pulpwood Embargoist, and Speculator!' *Acadiensis* 20 (1991): 129-44; L. Anders Sandberg, 'Forest Policy in Nova Scotia: The Big Lease, Cape Breton Island, 1899-1960,' in Sandberg, ed., *Trouble in the Woods: Forest Policy and Social Conflict in Nova Scotia and New Brunswick* (Fredericton: Acadiensis Press, 1992), 65-89; William Parenteau and L. Anders Sandberg, 'Conservation and the Gospel of Economic Nationalism: The Canadian Pulpwood Question in Nova Scotia and New Brunswick, 1918-1925,' *Environmental History Review* 19 (1995): 57-84; and L. Anders Sandberg and William Parenteau, 'From Weapons to Symbols of Privilege: Political Cartoons and the Rise and Fall of the Pulpwood Embargo Debate in Nova Scotia, 1925-1933,' *Acadiensis* 26 (1997): 31-58.

7 Barnjum's career is described in Roach and Judd, 'A Man for All Seasons,' and Sandberg, 'Forest Policy in Nova Scotia,' 68-73. Contemporary articles describe Schierbeck's early career in Nova Scotia as public forester for the province, a position paid for by Frank Barnjum, but Schierbeck was really a forester for Barnjum's private holdings, cruising the forest holdings and monitoring insect infestations. According to his son, Schierbeck considered himself an employee of Barnjum until his appointment as Chief Forester in 1926. John Schierbeck, interview with the authors, Ashcroft, BC, 4 November 1993.

In 1926 Barnjum made the following 'suggestion' to Attorney General John C. Douglas: 'I cannot urge you too strongly to be advised in these forestry matters as much as possible by Mr. Schierbeck. I can assure you that he is the ablest man in his profession on this continent. Personally I have not made a move in forestry matters without first consulting him. If he is given a fairly free hand the responsibility is all on his shoulders, while if his advice is not followed he cannot have the same heart in his work and would always have a perfect alibi if any serious calamity happened to our forests.' Frank J.D. Barnjum to John C. Douglas, 2 May 1926, RG 20, vol. 729, file 1, PANS.

8 Quotation from Otto Schierbeck to John Price, 7 March 1927, RG 20, vol. 729, file 1, PANS. Otto Schierbeck to Frank J.D. Barnjum, 3 December 1926, RG 20, vol. 718, no. 15, PANS; Otto Schierbeck to Frank J.D. Barnjum, 12 July 1927, RG 20, vol. 742, no. 2, PANS.

9 The magazine had a short life, with half a dozen issues published in 1934. These are contained in the Public Archives of Nova Scotia. The issues came out in April, July, August, September-October, and November.

10 Otto Schierbeck to Frank J.D. Barnjum, 17 September 1929, RG 20, vol. 726, no. 9, PANS.

11 This arguably manifested itself even more strongly in the conquered colonies. In India, for example, British foresters steered forestry in commercial directions and 'carefully restricted peasant access, by restricting it to areas of forest not deemed commercially profitable.' Ramachandra Guha, 'The Malign Encounter: The Chipko Movement and Competing Visions of Nature,' in Tariq Banuri and Frédérique Apffel-Marglin, eds., *Who Will Save the Forests? Knowledge, Power, and Environmental Destruction* (London: Zed, 1993), 85; Curt Meine, 'The Oldest Task in Human History,' in Richard Knight and Sarah Bates, eds., *A New Century for Natural Resources Management* (Washington, DC: Island Press, 1995), 12.

12 'Is Reforestation Bunk?' *Forest Crusader* April 1934: 18.

13 See, for example, 'A Business Meeting in Hell,' *Forest Crusader* July 1934: 8-9, 18, 21, and the correspondence in RG 20, vol. 746, no. 31, PANS; 'Sweeping Victory Expected over the Insect Pests,' *Morning Chronicle* 23 July 1927. For an example of Schierbeck's supreme confidence, see the document 'Working Plan and Experimental Forest' in this book. On the importance of fires in ecosystem dynamics and the negative effects of fire suppression, see Stephen Pyne, *Fire in America: A Cultural History of Wildland and Rural Fire* (Princeton: Princeton University Press, 1982), and 'Flame and Fortune,' *New Republic* 8 August 1994: 19-20.

14 Department of Lands and Forests, *Annual Report, 1925-26* (Halifax: Queen's Printer, 1926), 6.

15 Ibid.

16 Wilfrid Creighton, interview with the authors, 4 August 1994; Otto Schierbeck to Chief Rangers, 22 September 1930, RG 20, vol. 730, file 4, PANS; Otto Schierbeck to Chief Rangers, 18 May 1929, RG 20, vol. 730, file 4, PANS. Schierbeck described the partisan system aptly: 'How important the chairman of one of the political wards becomes when his party gets into power! He holds regular audiences for all humble applicants. He is flattered, patted on the back, browbeaten. His relatives and close friends are courted and brought into the fray of jobs. The brethren of the lodge to which he belongs are certainly entitled to some preference. Should poor Johnny, who once called him a bad name, but otherwise might be qualified for the job, try for it, he of course has no chance; but foremost of all, of course, only the stalwarts of the party are entitled to a showing.' See 'Politics and Conservation: Is Our Political System Faulty?' *Forest Crusader* April 1934: 26.

17 For another example, see Melanie Dupuis, 'In the Name of Nature: Ecology, Marginality, and Rural Land Use Planning during the New Deal,' in Melanie Dupuis and Peter Vandergeest, eds., *Creating the Countryside: The Politics of Rural and Environmental Discourse* (Philadelphia: Temple University Press, 1996), 93-113; Nelles, *The Politics of Development,* 188; Sandberg and Clancy, 'Small Woodlot Owners,' 25-47.

18 Otto Schierbeck to Chief Rangers, 18 May 1929, RG 20, vol. 730, file 4, PANS; *Halifax Chronicle*, 25 July 1927 and 26 December 1927; Wilfrid Creighton, *Forestkeeping: A History of the Department of Lands and Forests in Nova Scotia, 1926-1969* (Halifax: DLF, 1988), 44, 81.

19 *Forest Crusader* November 1934: 14. For an earlier account expounding the values of thinning in the Nova Scotia forest, see George A. Mulloy, 'Nova Scotia's Great Opportunity: A Province that Grows Prolific Crops of Forests and Is Rapidly Re-Establishing Lost Areas,' *Illustrated Canadian Forest and Outdoors* 25 (1929): 93-4.

20 Quotation from 'New Mill Boon to Nova Scotia: Forest Chief Hails Mersey Plant as Big Asset to Province,' *Halifax Chronicle* 16 December 1929; Schierbeck, 'Nova Scotia Has Fine Young

Tree to Replace Forest Primeval of Evangeline's Time,' *Canada Lumberman* 15 September 1928: 43.

21 In 1942 one official from the Dominion Forest Service wrote that 'This form of demonstration has not had much results except ... that the Kentville demonstration had been beneficial and other owners in the district had attempted to do something with similar woodlots. As I said in my farmers woodlot report recently, I believe that this work ... has had somewhat the opposite effect to the one intended. It was looked upon as a handout from the Government – the farmer getting a lot of useful wood cut for him at no cost, and still retaining his woodlot in better condition. A different reaction might have been obtained by merely marking a stand for thinning and letting the farmer do his own cutting.' Comments on H.D. Long's Report on Nova Scotia Plots, George A. Mulloy, November 1943, RG 39, vol. 148, file 47361, National Archives of Canada (hereafter NAC).

22 Some, for example, were sawmillers who might have been affected by Schierbeck's reforms in tax collection and Crown land leasing. For the concept of capture, see Herbert Kaufman, *The Forest Ranger: A Study in Administrative Behaviour* (Baltimore: Johns Hopkins University Press, 1960), 75-80. The concept of capture was also highly relevant with respect to the province's game management, where Schierbeck was embroiled in additional controversies that are beyond the scope of this chapter. The Russian pulpwood affair is detailed in Creighton, *Forestkeeping*, 38.

23 Creighton assumed the role of Provincial Forester, a much-watered-down version of Chief Forester, under a new Liberal administration after the Conservative government was defeated in 1933. 'Will Specialize in Forestry Activities,' *Canada Lumberman* 1 July 1933: 19. The salary component on the department's expenditures dropped from $75,561 to $40,256 from 1933 to 1934. Nova Scotia, *Public Accounts, 1933-34* (Halifax: Queen's Printer, 1934).

24 As Schierbeck explained, 'The only solution to the difficult lumber market in the Maritime provinces is, to my mind, the introduction of better machinery, which can produce a better grade of lumber. If Canada wishes to make a bid for the European market, she must produce a better grade of lumber. There is always overproduction of poorly graded and poorly sawn lumber; well graded and well sawn lumber will always find a market at a reasonable price. The only way to crowd out competitors is to excel in the product offered. Canada has the raw material – she only needs improved workmanship to control the European market.' 'Many Gang Saws Sold,' *Canada Lumberman* 1 August 1930: 138.

25 Nova Scotia Department of Lands and Forests, *The Imperial Economic Conference Report on Labour Conditions in the Woods, Forest Inventory, Duration of Supply, Forest Industries, and the English Market* (Halifax: Queen's Printer, 1932), 31; 'Hardwood Stands of Nova Scotia: Recent Controversy Regarding Methods of Manufacturing Sheds Light on Hardwood Resources of Maritime Province,' *Canada Lumberman* 15 April 1929: 34. There was some support for Schierbeck's position. The hardwood question was debated in New Brunswick forestry circles. In 1934 Harold Innis lent his support. See Harold Innis, 'Complementary Report,' in *Report of the Royal Commission Provincial Economic Inquiry* (Halifax: King's Printer, 1934), 131-230.

26 Quotation from Otto Schierbeck, 'Some Handicaps that Must Be Overcome in Canadian Lumber Export,' *Canada Lumberman* 1 October 1933: 45. 'Nova Scotia Will Send Lumber Envoy,' *Canada Lumberman* 15 March 1933: 24.

27 'Some Handicaps,' 44.

28 Bonnycastle Dale to G.H. Prince, 4 April 1929, RG 106, box 63, file D, 1929-32, Provincial Archives of New Brunswick. William Parenteau generously provided this citation.

29 *Halifax Chronicle* 20 March 1929. For Schierbeck's response, see *Halifax Chronicle* 2 April 1929.

30 Sandberg, 'The Big Lease,' 65; *Canada Lumberman* 1 October 1929: 40; Nova Scotia Department of Lands and Forests, *Annual Report, 1930-31* (Halifax: Queen's Printer, 1931), 62; Otto Schierbeck to Scotia Lumber and Shipping Company, 29 December 1930, RG 20, vol. 742, no. 21, PANS.

31 Nova Scotia Department of Lands and Forests, *Annual Report, 1925-26* (Halifax: Queen's Printer, 1926), 8; Otto Schierbeck to William L. Hall, 21 May 1931, RG 20, vol. 734, no. 13, PANS; Otto Schierbeck to William L. Hall, 21 February 1931, RG 20, vol. 734, no. 15, PANS.

As a sign of the political clout of the business segment of Nova Scotia, Alexander S. MacMillan went on to become premier of the province from 1940 to 1945.

32 The fight against monopoly was a characteristic of the broader conservationist movement. More often than not, however, and in contrast to Schierbeck's efforts, the fight was more rhetorical than real.

33 Quoted in Sandberg, 'The Big Lease,' 78.

34 For more on this, see Sandberg, 'The Big Lease,' and Parenteau and Sandberg, 'Conservation and the Gospel of Economic Nationalism'; Otto Schierbeck, 'The Sins of the Fathers Shall Be Visited upon the Children unto the Third and Fourth Generation,' RG 20, vol. 751, no. 3, PANS, 2.

35 See, for example, the different wood volumes assigned to the so-called 'Big Lease' on Cape Breton Island. Sandberg, 'The Big Lease,' 70.

36 Otto Schierbeck to William L. Hall, 9 April 1930, RG 20, vol. 749, no. 5, PANS.

37 Otto Schierbeck to William L. Hall, 22 July 1930, RG 20, vol. 749, no. 21, PANS; Otto Schierbeck to William L. Hall, 7 October 1930, RG 20, vol. 749, no. 21, PANS.

38 Doull wrote, 'I had a conference with Colonel Jones [the general manager of Mersey] yesterday and he will immediately instruct Seaborne [the woodlands manager] to have another attempt to settle with you the technical matters that are outstanding in regard to the Mersey contract. Kindly give these matters your best attention and if possible have them disposed of. If it is not possible for you and Mr. Seaborne to agree they will have to be cleared in some other way.' John Doull to Otto Schierbeck, 20 January 1932, RG 20, vol. 7399, no. 9, file 4-6, PANS.

See also Nova Scotia Department of Lands and Forests, *Annual Report, 1931-32* (Halifax: Queen's Printer, 1932), 61-2; Creighton, *Forestkeeping,* 38. The 'less generous' agreement might in fact have saved Mersey money. Surveying the lands was expensive. According to Johnson, Mersey's forester, 'Mersey would have saved millions of dollars had there been no million-cord agreement.' He nevertheless claimed that it was necessary to hold these lands for financing purposes. Ralph Johnson, *Forests of Nova Scotia* (Halifax: Four East Publications, 1986), 148. Johnson's statements are clearly suspect. The holdings of large tracts of land also constituted a weapon in the setting of price and delivery quotas from private lands. It also shut out competitors.

39 This was a political appointment that reneged on a previous appointment made within the Department of Lands and Forests. Creighton, *Forestkeeping,* 112.

40 This was in contrast to Scandinavia, where forestry and woodlands departments were integrated. See M. Juhlin-Dannfelt, 'Forestry Conditions in Canada and Sweden from a Swedish Forester's View-Point,' *Illustrated Canadian Forest and Outdoors* 20 (1924): 531-2.

41 Nova Scotia Department of Lands and Forests, *Annual Report, 1925-26* (Halifax: Queen's Printer, 1926), 46. For an elaboration of the sustained yield concept, see the contributions in Knight and Bates, eds., *A New Century.*

42 Schierbeck ceaselessly promoted this aspect of the forest of his adopted home province. See, for example, 'Nova Scotia Has Fine Young Tree,' 41-3. He also argued that Nova Scotia possessed excellent transportation facilities and a good source of rural labour. 'Nova Scotia Is Looking Forward to Greater Overseas Trade in Timber,' *Canada Lumberman* 15 July 1932: 20.

43 By *selective thinning* Schierbeck meant *selection thinning,* a method used to thin the poorer elements in a tree stand to improve the growth and quality of the remaining stand. In many other contexts, selective thinning or cutting means high-grading, cutting the best and leaving the rest in a tree stand. Otto Schierbeck, 'Selective Thinning,' *Forestry Chronicle* 12 (1936): 368, 372-3.

44 'Where the woods department is operating now approximately 2,000 square miles at heavy transportation costs, it would be able to confine its operations to 55 square miles, the operator would be able to construct *permanent* automobile main roads, *permanent* well graded side roads and *permanent* rough snake roads. His logging costs would be reduced many fold, fire protection, insect control etc. would be reduced to a minimum, and he would be assured of a continuous supply to his mill.' Schierbeck, 'Selective Thinning,' 369.

45 Nova Scotia Department of Lands and Forests, *Forests and Forestry in Nova Scotia* (Halifax: Queen's Printer, 1930), 7.

46 Subsequent research has called these claims into doubt. Gordon Baskerville suggests that during winter logging the snow provided seedlings with protection that allowed for successful regeneration. Gordon Baskerville, 'The Forestry Problem: Adaptive Lurches of Renewal,' in Lance H. Gunderson, Crawford S. Holling, and Stephen S. Light, eds., *Barriers and Bridges to the Renewal of Ecosystems and Institutions* (New York: Columbia University Press, 1995), 63.
47 Canadian Pulp and Paper Association, Woodlands Section, *Proceedings, 1927-1938*, 131-2.
48 Ibid.
49 Interestingly, Fernow, the first chief of the Division of Forestry of the US Department of Agriculture and then dean of the forestry school at the University of Toronto, started his investigations on the balsam fir as a pulpwood species when he was involved in a speculative venture on Cape Breton Island. Fernow held an interest in a Crown lease, the 'Big Lease,' whose 251,200 hectares were dominated by balsam fir stands. See Sandberg, 'The Big Lease,' and Andrew D. Rodgers III, *Bernhard Eduard Fernow: A Story of North American Forestry* (Princeton: Princeton University Press, 1951), 455.
50 Schierbeck, 'Selective Thinning,' 372-3.
51 Canadian Pulp and Paper Association, Woodlands Section, *Proceedings, 1927-1938*, 131-32.
52 See Nova Scotia Department of Lands and Forests, *Forests and Forestry*, and *The Imperial Economic Conference Report*; Schierbeck, 'Selective Thinning,' 366-74 (in this instance, he was addressing the readers of *Forest Chronicle*, the professional journal of the Canadian Institute of Forest Engineers).
53 Meine, 'The Oldest Task in Human History,' 29.
54 For a powerful argument in favour of such a venture, through the use of an experimental forest, see the document 'Working Plan and Experimental Forest' in this book.
55 The premier of the province provided the following letter for Schierbeck's supposed resignation: 'The duties of this office have been performed by Mr. Schierbeck faithfully and honestly and the Government has a high regard for his ability. During Mr. Schierbeck's term of office there has been a re-organization of the Fire Ranger Service which has been satisfactory. There has also been during that time established a reforestation program which has been successful in demonstrating the different kinds of trees that can be made useful in the Province. In regard to the technical ability and knowledge of surveying and cruising the Government regards Mr. Schierbeck's work as having been of high quality. Mr. Schierbeck's resignation is accepted owing to the necessity of economy in the public service.' The Premier of Nova Scotia, To Whom It May Concern, 12 June 1933, RG 3, vol. 12, 13 June (2), PANS.
56 Quotation from Schierbeck, 'Selective Thinning,' 366. For the more general decline of Canadian forestry in the 1930s, see Peter Gillis and Thomas Roach, *Lost Initiatives: Canada's Forest Industries, Forest Policy, and Forest Conservation* (New York: Greenwood Press, 1986).
57 RG 20, vol. 746, PANS.
58 Rudyard Kipling, 'Road-Song of the Bandar-Log,' *Jungle Book* (London: Macmillan, 1983 [1894]), 84.

Chapter 3: John Bigelow

1 Frank John Dixie Barnjum, Canada's 'foremost forest' conservationist and speculator, made Nova Scotia his home. Thomas Roach and Richard Judd, 'A Man for All Seasons: Frank John Dixie Barnjum, Conservationist, Pulpwood Embargoist, and Speculator!' *Acadiensis* 20 (1991): 129-44.
2 See Chapter 2 on Chief Forester Otto Schierbeck.
3 Peter Clancy, 'The Politics of Pulpwood Marketing in Nova Scotia, 1960-1985,' in L. Anders Sandberg, ed., *Trouble in the Woods: Forest Policy and Social Conflict in Nova Scotia and New Brunswick* (Fredericton: Acadiensis Press, 1992), 65-89.
4 James Kenny, 'A New Dependency: State, Local Capital, and the Development of New Brunswick's Base Metal Industry, 1960-1970,' *Canadian Historical Review* 78 (1997): 1-39.
5 This is a relative statement. There were progressive elements in the Department of Lands and Forests as well, but their voices were repressed. The more progressive developments in the department, however, lay in the innovative ideas on how to understand and restore the

Acadian forest to a healthier state rather than in promoting the economic status of small woodlot owners and sawmillers. See Chapter 4 on Lloyd Hawboldt.

6 This account is based on numerous interviews with John Bigelow between 1991 and 1997. Bigelow died on 16 December 1997. For an obituary, see *Chronicle Herald* 18 December 1997.

7 The war represented a last revival of the era of wood, wind, and sail, when ships were scarce and shippers looked in all directions to move cargo. Stanley T. Spicer, *Maritimers Ashore and Afloat*, vol. 2 (Hantsport, NS: Lancelot,' 1994), 71-83.

8 All subsequent comments on these magazines come from Bruce A. White, *Elbert Hubbard's 'The Philistine: A Periodical of Protest' (1895-1915)* (Lanham, MD: University Press of America, 1989), 48.

9 These included operator of a motor boat on pier construction work; electrician and plumber's helper; carpenter; construction worker; labourer; pile driver operator; blacksmith's helper; logger and shipper of timber for pier construction; tugboat operator for Canadian Gypsum Company Limited; foreman; and person in charge of construction of new ferry dock at East Ferry.

10 H. Vivian Nelles, *The Politics of Development: Forests, Mines, and Hydro-Electric Power in Ontario, 1849-1941* (Toronto: Macmillan, 1974); Peter Gillis and Thomas Roach, *Lost Initiatives: Canada's Forest Industries, Forest Policy, and Forest Conservation* (New York: Greenwood Press, 1986).

11 For more on Barnjum, see Roach and Judd, 'A Man for All Seasons.'

12 See Peter Neary, 'The Bradley Report on Logging Operations in Newfoundland, 1934,' *Labour/Le Travail* 16 (1985): 193-232.

13 See also James Hiller, 'The Origin of the Pulp and Paper Industry in Newfoundland,' *Acadiensis* 11 (1982): 42-68, and 'The Politics of Newsprint: The Newfoundland Pulp and Paper Industry, 1915-1939,' *Acadiensis* 19 (1990): 3-39.

14 This was not unusual in the North American context. See Kenneth Finegold and Theda Skocpol, *State and Party in America's New Deal* (Madison: University of Wisconsin Press, 1995).

15 F. Waldo Walsh, *We Fought for the Little Man* (Moncton: Co-op Atlantic, 1978).

16 Alexander A. Macdonald, 'Policy Formulation Process: Nova Scotia Dairy Marketing, 1933-1978,' unpublished paper, 1980.

17 For more details on these exploitative relations, see John Bigelow to Dr. R. McGregor, 14 February 1944, RG 20, vol. 871, no. 2, PANS.

18 For an account of Killam, see Douglas How, *A Very Private Person: The Story of Izaak Walton Killam and His Wife Dorothy* (n.p.: n.p., 1976).

19 Even Thomas Raddall, Liverpool's nationally renowned novelist, was employed by Mersey for some time, and Jones supported his writings financially. This is perhaps why the Mersey mill is idealized in Raddall's novels, while its early competitors, among them Frank Barnjum, are portrayed in less favourable light. L. Anders Sandberg, 'The Forest Landscape in Maritime Canadian and Swedish Literature,' in Paul Simpson-Housley and Glen Norcliffe, eds., *A Few Acres of Snow* (Toronto: Dundurn, 1992), 109-21.

20 The consequences were dire for sawmillers. See Nancy Colpitts, 'From Sawmilling to Park,' in Sandberg, ed., *Trouble in the Woods*, 90-109; and L. Anders Sandberg and Peter Clancy, 'Property Rights, Small Woodlot Owners, and Forest Management in Nova Scotia,' *Journal of Canadian Studies* 31 (1996): 25-47.

21 Useful but less critical accounts of these relationships include Eric Mullen and Millie Evans, *In the Mersey Woods* (Liverpool, NS: Bowater Mersey, 1989), and Mike Parker, *Woodchips and Beans: Life in the Early Lumberwoods of Nova Scotia* (Halifax: Nimbus, 1992).

22 The program of action is outlined in the document 'The Utilization and Marketing Problem of Farm Woodlot Products in Nova Scotia' in this book.

23 See, for example, Alexander F. Laidlaw, *The Campus and the Community: The Global Impact of the Antigonish Movement* (Montreal: Harvest House, 1961), and James Sacouman, 'Underdevelopment and the Structural Origins of Antigonish Movement Co-operatives in Eastern Nova Scotia,' in Robert Brym and James Sacouman, eds., *Underdevelopment and Social Movements in Atlantic Canada* (Toronto: New Hogtown Press, 1979), 109-26.

24 For a more detailed description, see John Bigelow, 'Cooperative Marketing of Pulpwood,' RG 20, vol. 870, no. 12, PANS, and John Bigelow to Waldo Walsh, 22 December 1975, MG 1, vol. 2490, no. 6, PANS.

25 L. Anders Sandberg, 'Forest Policy in Nova Scotia: The Big Lease, Cape Breton Island, 1899-1960,' in Sandberg, ed., *Trouble in the Woods*, 65-89.

26 Bigelow gave a detailed account of some of the success to a commission enquiring into the price of pulpwood sold by settlers in Quebec on 7 October 1941. RG 20, vol. 868, no. 1, PANS. For a personal assessment of the trip, and the relative success of the Nova Scotia effort compared with the situation in Quebec, see John Bigelow, 'Report to Mr. F.W. Walsh on the Meeting with the Quebec Forestry Commission re Pulpwood Prices and Cooperative Marketing,' RG 30-3/14/1032, St. Francis Xavier University Archives (hereafter SFXUA).

27 This is in spite of being well known for taking low-quality pulpwood. The quality of the pulpwood cut for the Europeans was well above that delivered to Mersey and the two groundwood pulp mills in Nova Scotia. The latter took rough wood, did not worry about the balsam fir, and even accepted a certain amount of redheart in the bolts. When one European visitor to Mersey's operations commented on the redheart of pulpwood, his local host asked if he had a chance to observe the girls in the local town. He was then told that their freckles did not take away from their attractiveness. Perhaps as a consequence, Mersey's newsprint was not known as the best on the continent. Bigelow recalls one technician who inspected the company's product and described it as a 'lousy sheet.' Mersey did not make it any easier for those pioneer producers who began to produce chips from wood waste as a raw material for the pulp process. Mersey was approached to finance chippers, available in Sweden, but declined, stating that they would not fit in with its system of operations. Some sawmillers got their own chippers, but when they approached Mersey and Minas Basin they could not sell the woodchips.

28 This is confirmed by Johnson himself. See Ralph Johnson, *The Forests of Nova Scotia* (Halifax: Four East Publications, 1986), 243. Yet Johnson is referred to by the Department of Lands and Forests as the 'father of forestry in Nova Scotia.'

29 And while it is true that local municipal assessors, with the backing of local politicians, assessed the large landowners at a higher rate than small woodlot owners, the amounts were relatively small. Ralph Johnson nevertheless spent a good part of his professional career trying to change the tax system, but he had little success.

30 McLelan was content with the volume of his business but once tried to set up a trade with the West Indies. He was given some names of people to write to, but he decided to go there himself.

31 See Chapter 16, 'Timber Control,' in John de Navarre Kennedy, *History of the Department of Munitions and Supply* (Ottawa: King's Printer, 1950), 243-69.

32 He dealt primarily in fir and spruce. The trade was brisk. Transport from the rest of Canada was difficult. At one stage, the Panama Canal route to Europe for British Columbia lumber was considered too long and risky, so BC lumber was railed to a depot on the St. Lawrence River and then shipped to England.

33 This saw had finer teeth and would cut the ends square and smooth.

34 Many lumbermen, even though supportive, would not come through in the end. One was W.S. Wilbur of Elmsdale, Halifax County, who operated a lumber business on a rational and sincere basis. He was always supportive of the Maritime Lumber Bureau's efforts; however, when it was ready to put its grade stamps on the lumber, he would not consent. 'I got along without it until now, so I can do without it in the future' was his rationale. The Wilburs continued to operate successfully for years after.

35 The Eddy Lumber Company had a similar reputation. It was smaller but put equal stress on quality. It always settled disputes and claims fairly.

36 In Sweden, where no grading rules were in force, the same system was in operation.

37 Some of the broad-minded lumbermen whom Bigelow dealt with in the Nova Scotia Forest Products Association and the Maritime Lumber Bureau included Prescott Blanchard from Truro, a retired lumberman who came to meetings, had good basic views, and was a valuable ally; Peter Boutilier from Yarmouth, a man on the ground, who operated a large sawmill himself, financed many more smaller operators, and could tell the pedigree of 99

percent of the people in his area; H.W. Brady from Bridgewater, who operated a large planing mill and built sashes and doors and may even have exported some lumber; Lee Carter from Springhill, who was small, solid, and knew his business; Simon Fraser of Fraser and Grant in Pictou and Antigonish Counties, a medium-sized operator who might have run one or more mills, sold through the Halifax exporters, and was well aware of the need for quality improvement; Ernst Harrison from Halfway River, Parrsboro, and Southampton in Cumberland County, who operated on a small scale but was reliable, worked for quality, had little education, but was honest; Laurie Isenor, who produced sawn lumber and box shooks, was reliable, interested in, and supportive of the efforts to reform the industry; and Ignatius Jesus Soy of the Maple Leaf Lumber Company in Londonderry Station, Colchester, who was supportive of the bureau's efforts. His son Jim was versatile and able. He supervised tree harvests in Scotland during the war and bought and cut stumpage from Hollingsworth and Whitney lands near Parrsboro after the war. He exported it through Halifax.

38 Wilfrid Creighton, *Forestkeeping* (Halifax: DLF, 1988), 52.
39 MacMillan also routinely neglected to pay his land tax. See Chapter 2 on Otto Schierbeck.
40 This is the election in which the Conservatives tried to capitalize on Anderson's activities but still lost. See J. Murray Beck, *The Politics of Nova Scotia*, vol. 2 (Tantallon, NS: Four East Publications, 1988).
41 Ibid.
42 At one point, for example, when a Quebec pulp and paper consortium was interested in Nova Scotia as a production site, Premier Angus L. MacDonald had to instruct the department 'not to be hostile to it.'
43 This is confirmed by Creighton himself in *Forestkeeping*, 43.
44 Some of these lands were carefully nursed back to productive forests by the department's foresters but then 'given' away to the pulp and paper companies as Crown leases in the 1960s.
45 In a telegram, never shown to Bigelow, the minister wrote to Mersey, 'In consideration of your request to me that no obstacle be placed in the way of Mersey Paper Company in negotiations for the purchase of the Tusket River property we are today with great reluctance wiring A.J. LaCroix that we are relinquishing our option on the property provided it is sold to a Nova Scotia industry.' J.H. MacQuarrie to R.L. Seaborne, 4 June 1946, RG 20, vol. 873, no. 6, PANS.
46 Waldo Walsh Papers, MG 1, vol. 2490, no. 6, PANS.
47 John Bigelow to Art MacKenzie, 23 January 1948, Waldo Walsh Papers, MG 1, vol. 2490, no. 6, PANS. Bigelow believed that the branch should pay immediate attention 'to a closer utilization of ... raw material and the attainment of full economic value from our remaining forest stands.' The principal means by which to achieve this goal was to provide 'a marketing service and necessary assistance to primary producers to enable them to organize and centralize the selling of forest products.'
48 See the correspondences in MG 2, vol. 970, folder 21-4, PANS.
49 The Nova Scotia Members of the Canadian Society of Forest Engineers, *Forestry, Economy, and Post-War Reconstruction in Nova Scotia* (Halifax: DLF, 1944). The signatories of the report give a good indication of the state of forestry in the province. Johnson was the chairman and principal writer. H.P. Webb was a retired forester who had left the University of New Brunswick forestry school as professor over a wage dispute. He went under the nickname Horrors Webb and sold life insurance in Halifax for the Standard Life Insurance Company, a Scottish firm. His sales pitch was mild-mannered, and he stressed the stability and longevity of the company. Brian Alexander was an operations man and employee of Minas Basin. G.W.I. Creighton was the Provincial Forester for the Department of Lands and Forests. R.S. Cumming was the first secretary for the Maritime Lumber Bureau and later professor of economics at Dalhousie University. E.S. Fellows was a private consultant. D.W. Hudson was a forester with the Department of Lands and Forests. He was a poor organizer and boss. Bigelow once worked for him on surveying, and during that summer he hurt his kneecap badly, and another boy of the crew, a Fraser from Yarmouth, drowned. Otto Schierbeck, the Chief Forester of the department at the time, put Hudson on other duties after

that. He later worked for the Fraser Companies in New Brunswick. S.J. (Sid) Patterson was a classmate of Bigelow's who worked in a forest nursery. R.L. Seaborne, Mersey's woodlands manager, and R.K. MacKinnon, a temporary employee at Mersey who later worked as a cruiser for Minas Basin, rounded out the lot.

50 This is confirmed by Wilfrid Creighton, the Provincial Forester at the time. Interview with Wilfrid Creighton, 1994.

51 This is a point made by Kenny, 'A New Dependency.'

52 Some of the local pulp and paper mills were in a poor state. The Sheet Harbour pulp and paper mill was such a case. Operations were seasonal. When the water was low, the mill closed, and the workers turned to the land and sea for their livelihoods. This may have been an attraction to the company, which employed docile and stable long-term workers. In spite of this, however, the plant had deteriorated and was technologically obsolete. The plant sold groundwood pulp, which faced increasing competition in the international market. When a specialist came up to check the situation, he informed headquarters that 'they are making something that looks like hay up here.'

53 The province's broader tourism development agenda is treated in Ian McKay, *The Quest of the Folk: Antimodernism and Cultural Selection in Twentieth Century Nova Scotia* (Montreal: McGill-Queen's University Press, 1994).

54 MG 1, vol. 2490, no. 6, PANS.

55 This was the original size of lease. In 1940 the Nova Scotia government expropriated 178,648 acres for the establishment of the Cape Breton Highlands National Park. For more details, see Sandberg, 'Forest Policy in Nova Scotia.'

56 When the Oxford people first approached the firm, they were said to have requested the loudest person in the firm. Charlie Burchill, a senior Liberal lawyer, fit the bill. It was said of him that he did not talk but bark.

57 Mersey had in fact tried to buy the Big Lease on two occasions in the 1930s, but Oxford refused to sell. This does not mean that it was unwilling to sell – the price may simply not have been right. Mersey may well have construed this as a refusal to sell on Oxford's part. See Sandberg, 'Forest Policy in Nova Scotia.'

58 Before they left, they asked Chisholm what was the one most important factor behind his success. He responded that it was the power deal that he had received from the state of Maine: two-mill power per KW hour for so many years. 'Sitting on 2 mill power, why shouldn't I make money?' was Chisholm's reply. Mersey's power deal in the late 1920s was also generous, at five-mill power per KW hour. This agreement, in Bigelow's opinion, was 'a disservice to the people of Nova Scotia.'

59 More details on Clauson's actions are provided in Dietrich Soyez, 'Stora Lured Abroad? A Nova Scotian Case Study in Industrial Decision-Making and Persistence,' *Operational Geographer* 16 (1988): 11-4.

60 This stumpage rate was higher than in Alberta, but the conditions in Alberta were much less favourable than those in Nova Scotia. The wood was poorer, the transportation network was less extensive, the wood was more distant from the mill, fire protection was poorer, and winter conditions were more severe.

61 These fears were later realized. See Clancy, 'The Politics of Pulpwood Marketing in Nova Scotia.'

62 John Bigelow to J.A. Gillis, 28 July 1959, RG 30-3/14/883, SFXUA.

63 John Bigelow, 'Better Utilization of Nova Scotia's Forests,' 5, RG 30-3/14/201, SFXUA. See also Bigelow, 'The Utilization and Marketing Problem.'

64 Rev. J.A. Gillis to Hon. E.D. Haliburton, 2 May 1960, RG 30-3/14/950, SFXUA.

65 Peter Clancy, 'The Politics of Pulpwood Marketing in Nova Scotia.'

66 John Bigelow to J.A. Gillis, 28 July 1959, RG 30-3/14/883, SFXUA.

67 John Bigelow, 'Continuous Income through Forest Management,' 8-10, RG 30-3/14/190-200, SFXUA.

68 Some of the foresters in the DLF, however, had an intricate knowledge of the forests of eastern Nova Scotia and constructive ideas on the utilization of its forests, based on detailed forest inventories, forest-harvesting schedules, spruce budworm monitoring, and considerable rapport with Stora's first Swedish foresters. See Chapter 4 on Lloyd Hawboldt.

69 *Halifax Mail Star*, 14 December 1959.

70 Roy E. George, *The Life and Times of Industrial Estates Limited* (Halifax: Institute of Public Affairs, Dalhousie University, 1974).

71 Anthony Lamport, *Common Ground: 25 Years of Voluntary Planning in Nova Scotia* (Halifax: Government Services, 1988). See also Peter Clancy, 'Concerted Action on the Periphery? Voluntary Economic Planning in "The New Nova Scotia,"' *Acadiensis* 26, 2 (1997): 3-30.

72 See also Geoffrey Stevens, *Stanfield* (Toronto: McClelland and Stewart, 1973), 147.

73 John Bigelow, 'Better Utilization of Nova Scotia's Forest,' 2-3, RG 30-3/14/202-3, SFXUA.

74 This trend was reinforced by the elitist and technical biases of the science and profession of forestry at the time. See Chapter 2 on Otto Schierbeck.

75 John Bigelow, 'Report to Mr. F.W. Walsh on the Meeting with the Quebec Forestry Commission re Pulpwood Prices and Cooperative Marketing,' 1, RG 30-3/14/1032, SFXUA.

76 Kenny, 'A New Dependency.'

77 John Bigelow, 'Better Utilization of Nova Scotia's Forest,' RG 30-3/14/201, SFXUA.

78 There are parallels with the efforts of the first Chief Forester of Nova Scotia, whose advice with respect to the terms of Mersey's access to Crown wood was also ignored. See Chapter 2 on Otto Schierbeck.

79 Based on two documents by John R. Bigelow, 'The Utilization and Marketing Problem of Farm Woodlot Timber,' 27 February 1939, RG 20, vol. 870, no. 12, PANS, and 'A Program of Co-operative Marketing of Woodlot Products for Nova Scotia,' *Forestry Chronicle* 15, 2 (1939): 119-21. Bigelow was at the time Forest Products Representative, Marketing Division, Department of Agriculture, Nova Scotia.

Chapter 4: Lloyd Hawboldt

1 This account is based on interviews conducted during the summers of 1992-6. On 10 April 1997, Hawboldt died at the age of eighty. For obituaries, see David Dwyer, 'Lloyd Hawboldt,' *Forestry Chronicle* 73, 5 (1997): 610-1; L. Anders Sandberg and Peter Clancy, 'Lloyd Hawboldt,' *Bulletin – Entomological Society of Canada* 29, 3 (1997): 89-91; and Gary Saunders, '"The Best Boss I Ever Had": Remembering Lloyd Stanley Hawboldt,' *Atlantic Forestry Review* 1 (1997): 2-3.

2 Michael Howlett, 'The 1987 National Forest Sector Strategy and the Search for a Federal Role in Canadian Forest Policy,' *Canadian Public Administration* 32 (1989): 545-63; Michael Howlett and Jeremy Rayner, 'Do Ideas Matter? Policy Network Configurations and Resistance to Policy Change in the Canadian Forest Sector,' *Canadian Public Administration* 38 (1995): 382-410; Alan Miller, 'The Role of Citizen-Scientist in Nature Resource Decision-Making: Lessons from the Spruce Budworm Problem in Canada,' *Environmentalist* 13, 1 (1993): 47-59.

3 The literature questioning this commitment is voluminous.

4 Aldo Leopold, *A Sand County Almanac* (New York: Ballantyne Books, 1970 [1949]), 258.

5 Ibid., 258-9.

6 For the various concepts of ecology that opposed the development of type A forestry, see Curt Meine, 'The Oldest Task in Human History,' in Richard L. Knight and Sarah F. Bates, eds., *A New Century for Natural Resource Management* (Washington, DC: Island Press, 1995), 7-35. On the increasing popularity of research on pesticides in entomology after the Second World War (at the expense of biological controls), see John H. Perkins, *Insects, Experts, and the Insecticide Crisis: The Quest for New Pest Management Strategies* (New York: Plenum Press, 1982).

7 This problem of marrying economic, ecological, and equity concerns is still largely unresolved in resource management.

8 Lloyd Hawboldt, 'Bessa Selecta (Meigen) as a Parasite of Gilpinia Hercyniae (Hartig),' MSc diss., McGill University, 1946. Hawboldt did extremely well in the program, and he was the college's first choice for a $1,500 annual scholarship to continue graduate work at Cornell University. Being in debt, and newly married, he declined the scholarship.

9 Vernon R. Vickery, 'William H. Brittain,' *Bulletin – Entomological Society of Canada* 29, 2 (1997): 73-6; and Robin Stewart, 'E. Melville Duporte: A MacDonald Great,' *MacDonald Journal* August 1983: 12-3.

10 Among other distinctions, Pickett was accorded a prominent place in Rachel Carson's *Silent Spring* (Cambridge, MA: Riverside Press, 1962). On this point, see also Kermode Parr, '"If You Poison Us, Do We Not Die?"' *Atlantic Advocate* November 1962: 21-4. On the fruit orchard spray program that gained Pickett international fame, see Norman Creighton, 'Less Spray, More Dollars,' *Atlantic Advocate* August 1963: 70-2. Perhaps not coincidentally, Pickett had three or four MacDonald College graduates working in his laboratory.

11 Wilfrid Creighton, *Forestkeeping: A History of the Department of Lands and Forests in Nova Scotia, 1926-1969* (Halifax: DLF, 1988), 77.

12 During the war, plywood was in high demand for the construction of Mosquito bombers. Cut into bolts, yellow birch was also put on a lathe and peeled into sheets of veneer.

13 The employment of a forest pathologist was breaking new ground in Canadian forest biology. Although the Canadian Forest Service had a few forest pathologists employed, it focused its research in forest biology on tree insects. This had serious limitations.

14 Kostjukovits's influence on Hawboldt will be covered in more detail in a later section.

15 Hawboldt's publications on the birch dieback included 'Dieback of Birch,' Nova Scotia Department of Lands and Forests, *Annual Report 1944*; 'Dieback of Birch,' Nova Scotia Department of Lands and Forests, *Annual Report 1945*; 'Aspects of Yellow Birch Dieback in Nova Scotia,' *Journal of Forestry* 45, 6 (1947): 414-22; with Arthur J. Skolko, 'Investigation of Yellow Birch Dieback in Nova Scotia in 1947,' *Journal of Forestry* 46, 9 (1948): 659-71; 'Yellow Birch Investigations,' Nova Scotia Department of Lands and Forests, *Annual Report 1948:* 63; *'Dieback' of Yellow Birch,* Nova Scotia Department of Lands and Forests, Bulletin 4, 1951; and *Climate and Birch 'Dieback,'* Nova Scotia Department of Lands and Forests, Bulletin 6, 1952.

16 Lloyd S. Hawboldt, 'Forestry in Nova Scotia,' *Canadian Geographical Magazine* 51 (1955): 80.

17 Otherwise, its profile tended to ebb and flow. Apart from the secretary-manager, all help was voluntary. Consequently, that person's vigour determined what took place in the association's name. Hawboldt remembers one particularly active officer, the president of the Dominion Atlantic Railway, who used to host annual dinners at the Lord Nelson Hotel in Halifax.

18 These were many of the same barren wildlands that B.E. Fernow, in his reconnaissance of 1910, had labelled as 'natural' and inevitable. Bernhard Fernow, *The Forest Resources of Nova Scotia* (Ottawa: Commission of Conservation, 1913).

19 Some of the reasons for this are covered in Chapter 2 dealing with Otto Schierbeck. For one such protest, see *Chronicle Herald* 31 August 1954.

20 For Nova Scotia, see L. Anders Sandberg and Peter Clancy, 'Property Rights, Small Woodlot Owners, and Forest Management in Nova Scotia,' *Journal of Canadian Studies* 31, 1 (1996): 25-47. For the northeastern United States, see Richard Judd, *Common Lands, Common People: The Origins of Conservation in Northern New England* (Cambridge, MA: Harvard University Press, 1997).

21 Lloyd S. Hawboldt, *Blueberries: Cash from Idle Acres* (Halifax: DLF, 1954).

22 Lloyd S. Hawboldt, *Christmas Trees Are a Crop*, Bulletin 12 (Halifax: DLF, 1953).

23 Lloyd S. Hawboldt and Gordon R. Maybee, *The Nova Scotia Christmas Tree Trade*, Bulletin 14 (Halifax: DLF, 1955).

24 Biological controls were tried successfully on occasion. Once on a Christmas tree plantation infested by the tussock caterpillar, Hawboldt took diseased insects, ground them up, sprayed them on the whole stand, and got rid of the infestation.

25 The same condition prevailed in the lumber industry, in which poor-quality lumber had already driven the price structure to the lowest common denominator. See Chapter 3 on John Bigelow.

26 In fact, when the Nova Scotia government transferred forest marketing to the DLF in 1944, it faded into obscurity almost immediately. For more on this, see Chapter 3 on John Bigelow.

27 Lloyd S. Hawboldt and Richard M. Bulmer, *The Forest Resources of Nova Scotia* (Halifax: DLF, 1958).

28 This is described in Lloyd S. Hawboldt and Simon N. Kostjukovits, *Forest Regulation in Nova Scotia, Part 1: Site Quality Normal Yield Tables for Softwoods,* Bulletin 20 (Halifax: DLF, 1961).

29 Still, as late as 1963, when Hawboldt sent a copy of a paper outlining his forest mensuration technique for Nova Scotia for comment to several Canadian foresters, he received no replies! The paper in question was Lloyd S. Hawboldt, Simon N. Kostjukovits, and Richard M. Bulmer, 'Mensuration for Management,' presented to the Forest Mensuration Committee, Annual Meeting of the Canadian Institute of Forestry, Halifax, 1963.
30 Hawboldt and Bulmer, *The Forest Resources of Nova Scotia*, 64.
31 Lloyd S. Hawboldt, 'The Spruce Budworm,' mimeo, 1955, available in the legislative library in Halifax.
32 See also Hector A. Richmond, *Forever Green: The Story of One of Canada's Foremost Foresters* (Lantzville, BC: Oolichan, 1983), 185.
33 See Chapter 6 on David Dwyer.
34 In retrospect, Hawboldt likened the Hotel Nova Scotia process to the preparation of the background studies on the effluent situation at Scott's sulphate pulp mill at Boat Harbour. An engineer at work on the scheme claimed that the raw wastes would be organically broken down and turned back to pure water. It has since developed into one of the most polluted sites in the province. In each case, the chosen outcome was instrumental in the decision to invest. Each outcome was the easiest and the best for the company involved. Such revisions of forest inventories were common at the time, and they even received sanction from forestry science. Gordon Baskerville's comments on New Brunswick are instructive. There several reviews of allowable harvest levels were made in the 1960s, 'but it always seemed that with the appropriate adjustment in the growing stock for cull, proper merchantability rules, and a good rotation age, the level of harvest currently desired turned out to be just about right.' Gordon Baskerville, 'The Forestry Problem: Adaptive Lurches or Renewal,' in Lance H. Gunderson, Crawford S. Holling, and Stephen S. Light, eds., *Barriers and Bridges to the Renewal of Ecosystems and Institutions* (New York: Columbia University Press, 1995), 54. Instructive comparisons can be made with the controversy over the Mersey cruise of Crown lands between the company and Otto Schierbeck. See Chapter 2 on Schierbeck.
35 Peter Clancy and L. Anders Sandberg, 'Crisis or Opportunity: The Political Economy of the Stora Mill Closure,' *Business Strategy and the Environment* 4 (1995): 208-19.
36 Hans Akesson, 'The Nova Scotia Pulp Limited Story: Forest Management for the Revitalization of the Eastern Nova Scotia Forests, Nova Scotia Pulp Limited, and the Nova Scotia Economy,' paper presented to the Standing Forest Management Committee, Canadian Institute of Forestry, 63rd Annual Meeting, Victoria, BC, 1971.
37 L.S. Hawboldt, 'Extension and Forest Biology Division,' 21 November 1963, RG 30-3/14/37, SFXUA.
38 Ibid.
39 This social support was an integral part in Swedish woodlot owners coming 'to accept' mandatory forest legislation. This point is made in L. Anders Sandberg, 'Swedish Forestry Legislation in Nova Scotia: The Rise and Fall of the Forest Improvement Act, 1965-1986,' in Douglas Day, ed., *Geographical Perspectives on the Maritime Provinces* (Halifax: St. Mary's University, 1988), 184-96.
40 See Chapter 5 on Donald Eldridge and Chapter 7 on Richard Lord.
41 For more details, see Chapter 8 on Mary Guptill.
42 The 'budworm battles' are chronicled in Elizabeth May, *Budworm Battles: The Fight to Stop the Aerial Insecticide Spraying of the Forests of Eastern Canada* (Halifax: Four East Publications, 1982). Hawboldt is given his due in this account (47, 58, 87, 89), though the broader historical context is missing.
43 He wrote another document to explain the rationale. See 'Toward "Budworm-Proofing" the Forests of Nova Scotia,' written in 1976 and reprinted in this book.
44 Lloyd S. Hawboldt and Gary L. Saunders, *A Guide to Forest Practices* (Halifax: DLF, 1966).
45 Nova Scotia, the Provincial Forest Practices Improvement Board, *The Trees around Us* (Halifax: Queen's Printer, 1980). On the history of the Forest Improvement Act, see Glyn Bissix and L. Anders Sandberg, 'The Political Economy of Nova Scotia's Forest Improvement Act, 1962-1986,' in L. Anders Sandberg, ed., *Trouble in the Woods: Forest Policy and Social Conflict in Nova Scotia and New Brunswick* (Fredericton: Acadiensis Press, 1992), 168-97.

46 See Sandberg, 'Swedish Forestry Legislation,' and Bissix and Sandberg, 'The Political Economy.'
47 Allison D. Pickett, 'The Philosophy of Orchard Insect Control,' *Report of the Entomological Society of Ontario, 1949,* 40.
48 It is a credit to Hawboldt that his ideas were by no means inflexible. In retirement, he seemed to be more sensitive to the positions of other cultures and non-human species in the forest. His comment on wildlife management is instructive: 'Wildlife management, wildlife conservation or whatever, means different things to different persons. The truly aesthetic value of wildlife to the artist, the writer, the photographer and the outing family, is not comparable to the "conservation" of the wildlife biologist based on carrying capacities, kill statistics and hunter/success ratios. There is a philosophical gap between the "value" of the cock ruffed grouse drumming on a fallen tree and the limp "prize" dangling from the hunter's fingers. The purpose of wildlife management? The aboriginal people will provide one answer – quite a different one is provided by the sportsman and yet another by the nature lover.' Quoted in Wilfrid Creighton, 'An Open Letter to Dr. Donald Dodds,' unpublished, 9-10.
49 Mersey's woodland manager, by contrast, praised Hawboldt for *The Forest Resources of Nova Scotia*, writing in a personal note to Hawboldt, 'I find it interesting that the one true forester in Nova Scotia is an entomologist.'
50 Barney W. Flieger, 'Forest Protection from the Spruce Budworm in New Brunswick,' *Report of the Entomological Society of Ontario, 1953,* 16. Flieger later served as the dean of the Faculty of Forestry at the University of New Brunswick.
51 And, while his notion of 'budworm-proofing' the forest is typically rejected by industrial foresters, there is a broad range of studies that supports this notion. Moreover, the 'silvicultural hypothesis' has never been put to a real test in North America. For an excellent review of its history and debate, see Alan Miller and Paul Rusnock, 'The Rise and Fall of the Silvicultural Hypothesis in Spruce Budworm ... Management in Eastern Canada,' *Forest Ecology Management* 61 (1993): 171-89.
52 This is not unique to Hawboldt's ecological approach. It is also often part of the present discourse on forest sustainability and ecosystem management. For a recent statement, see the 'Special Issue on Sustainability' of the *Journal of Canadian Studies* 31, 1 (1996).
53 This document is dated 10 June 1976. It appeared as an internal document to the Nova Scotia Department of Lands and Forests. A copy is available in the legislative library in Halifax.
54 J.D. Tothill, 'Notes on the Outbreaks of Spruce Budworm, Forest Tent Caterpillar, and Larch Sawfly in New Brunswick,' *Acadian Entomological Society Proceedings* (1922): 177.
55 Lloyd S. Hawboldt, 'The Spruce Budworm,' mimeo, 1955. A copy is available in the legislative library in Halifax.

Chapter 5: Donald Eldridge

1 This account is based on two interviews and our own primary research in the newspaper and archival records. We are indebted to Donald Eldridge for giving generously of his time. We are also thankful for his pointing us to an earlier transcript of an interview with Glyn Bissix and to Glyn for so graciously and readily providing it for us. Bissix's transcript proved to be invaluable, since Eldridge's illness and passing on 7 June 1995 prevented further interviews. We would also like to thank former Minister of Natural Resources Don Downe, who provided copies of Eldridge's unpublished reports as Commissioner of Forest Enhancement. We have included excerpts from these reports in 'From the *Annual Reports*, 1988-90, Commissioner of Forest Enhancement' (hereafter Exhibit 4). An obituary of Eldridge, detailing his various positions and commitments on various committees, is provided in the *Chronicle Herald* of 9 June 1995.
2 See, for example, Ernest R. Forbes and Delphin A. Muise, eds., *The Atlantic Provinces in Confederation* (Toronto: University of Toronto Press, 1993), Chapters 11-13.
3 This stems from the 1937 provincial election, referred to as the 'Woodpecker Election,' when it was revealed that Anderson was cutting wood illegally off land in Guysborough County. The opposition Conservatives did much to exploit this incident but still lost the election to the Liberals. For more information on Anderson, see Chapter 3 on John

Bigelow. See also Murray Beck, *The Politics of Nova Scotia, Vol. 2: 1896-1988* (Tantallon, NS: Four East Publications, 1988).

4 The assessors were clearly not trained at the time. Many also consciously set higher assessment rates for larger than for smaller properties. By 1994 the rate was twenty-five cents per acre, which had stabilized land prices. Eldridge noted, however, that even then Municipal Affairs used the tax regime to try to induce people into forest management. There was a 'forest resource' rate and an 'other' rate. The former was twenty-five cents, and the latter was higher for not being under forest management. The 'resource' rate applied to only about 10 percent of Nova Scotia forest lands.

5 For more on Clarence Mason, see Wilfrid Creighton, 'An Open Letter to Dr. Donald Dodds,' unpublished.

6 A quitclaim deed is often issued where land ownership patterns are uncertain and confused, a situation for which Nova Scotia qualified. The quitclaim deed passes any title that a granter may have to a piece of land to a seller without professing that the title is valid.

7 Wilfrid Creighton, *Forestkeeping* (Halifax: DLF, 1988), 93.

8 Transcript of 'Don Eldridge Interview with Glyn Bissix,' (hereafter Interview with Don Eldridge), 21 September 1987, 17.

9 For another interesting anecdote, see Walter Webber's account in Canadian Institute of Forestry, Nova Scotia Section, *Report of the 36th Annual Meeting, 1989*.

10 Eldridge lamented the loss of the Small Tree Act and the absence of any forest legislation: 'Right now, we are without any laws altogether, a terrible thing. There are only the Forest Guidelines for Crown land and for those on management plans. Today, if you draw a line from New Glasgow to Halifax, there are very few small sawmills east of the line.' Interview with Don Eldridge.

11 Harry Thurston, 'Prest's Last Stand: Keeping Kafka and the Bureaucrats out of the Acadian Forest,' *Harrowsmith* August-September 1983: 22-31.

12 This may have had a longer history as well. In 1950 the company began a shift in emphasis to sales of building material, hardware, plumbing, electrical, and municipal supplies. Today, the Eddy Group Limited is one of the largest wholesale distribution companies in the Maritimes. See 'Special Issue,' *Atlantic Construction Journal* September 1993: E1-12.

13 Eldridge, in his capacity as Commissioner of Forest Enhancement, strongly advised the government of Nova Scotia to buy these lands. See Exhibit 4.

14 Bob Douglas pioneered the sale of Nova Scotia forest lands for recreational purposes. For a closer examination of this phenomenon, see Kell Antoft, 'Symptom or Solution: The Nova Scotia Land Holdings Disclosure Act of 1969,' in L. Anders Sandberg, ed., *Trouble in the Woods: Forest Policy and Social Conflict in Nova Scotia and New Brunswick* (Fredericton: Acadiensis Press, 1992), 198-211; Land Research Group, 'Whither Our Land? Who Owns Nova Scotia? And What Are They Doing with It?' *New Maritimes* 8, 6 (1990): 14-25.

15 Donald Eldridge, 'Executive Director's Report,' Nova Scotia Forest Products Association, *43rd Annual Meeting*, 1977, 45.

16 In unveiling the access road program, DLF minister Haliburton observed, 'Because of the cooperative attitude of your membership, and the agreement to work together, for the common good of the industry, you have been able to work out a proposal that the Government can accept.' Press Release, 21 April 1967.

17 A detailed set of provincial regulations accompanied the program. See 'Access Road Construction Assistance Regulations,' Nova Scotia Forest Products Association (Truro, NS, 1967).

18 Clarence Porter, 'Special Committee Report,' *38th Annual Meeting*, Nova Scotia Forest Products Association, 1971, 21-4.

19 For background, see Peter Clancy, 'The Politics of Pulpwood Marketing in Nova Scotia, 1960-1986,' in Sandberg, ed., *Trouble in the Woods*, 142-67.

20 Nova Scotia Forest Products Association, 43rd Annual Meeting, *Report* (Truro, NS, 1977), 44.

21 Ibid., 45-6.

22 See *Chronicle Herald* 27 May and 21 July 1992.

23 'Forestry Group Pondering New Headquarters,' *Chronicle Herald* 20 February 1989.

24 Douglas MacDonald, *The Politics of Pollution in Canada* (Toronto: McClelland and Stewart, 1991).

25 For an interpretation of the shaping of Nova Scotia environmentalism, see Rod Bantjes, 'Hegemony and the Power of Constitution: Labour and Environmental Coalition Building in Maine and Nova Scotia,' *Studies in Political Economy* 54 (1997).
26 Elizabeth May, *Budworm Battles: The Fight to Stop the Aerial Insecticide Spraying of the Forests of Eastern Canada* (Halifax: Four East Publications, 1982).
27 Julia McMahon, 'The New Forest in Nova Scotia,' and Aaron Schneider, 'Underdeveloping Nova Scotia: Forests and the Role of Corporate Counter-Intelligence,' in Gary Burrill and Ian McKay, eds., *People, Resources, and Power* (Fredericton: Acadiensis Press, 1987).
28 Nova Scotia, Royal Commission on Forestry, *Report* (Halifax: Queen's Printer, 1984).
29 Bill Devall, ed., *Clearcut: The Tragedy of Industrial Forestry* (San Francisco: Sierra Club/Earth Island Press, 1993).
30 Peter Clancy and L. Anders Sandberg, 'Formulating Standards for Sustainable Forest Management in Canada,' *Business Strategy and the Environment* 6, 4 (1997); and Peter Clancy, 'The Politics of Stewardship: Certification for Sustainable Forest Management in Canada,' in L. Anders Sandberg and Sverker Sörlin, eds., *Sustainability – The Challenge* (Montreal: Black Rose Books, 1998), 108-20.
31 John F. Reiger, *American Sportsmen and the Origin of Conservation* (New York: Winchester Press, 1975).
32 Interview with Don Eldridge, 21.
33 Ibid., 3.
34 Nova Scotia Forest Products Association, 46th Annual Meeting, *Report* (Truro, NS, 1980), 84.
35 Interview with Don Eldridge, 21.
36 Nova Scotia Forest Products Association, 44th Annual Meeting, *Report* (Truro, NS, 1980), 75-6.
37 Interview with Don Eldridge, 22.
38 Ibid.
39 For a more detailed account of the context and history of the Forest Improvement Act, see L. Anders Sandberg, 'Swedish Forestry Legislation in Nova Scotia: The Rise and Fall of the Forest Improvement Act, 1965-1986,' in Douglas Day, ed., *Geographical Perspectives on the Maritime Provinces* (Halifax: Saint Mary's University, 1988), 184-96.
40 Interview with Don Eldridge, 24.
41 Ibid., 21. See also Nova Scotia Department of Lands and Forests, 'Submission to the Royal Commission on Forestry' (Halifax, DLF, 1983).
42 Extensive excerpts from these hitherto unpublished reports are presented in Exhibit 4.
43 Interview with Don Eldridge, 3.
44 Eldridge reinforces this point in the document "Excerpts from the Annual Reports, 1988-90, Commissioner of Forest Enhancement" reprinted in this book.
45 Interview with Don Eldridge, 4.
46 A history and a different interpretation of the Group Venture program can be found in Chapter 6 on David Dwyer.
47 Interview with Don Eldridge, 18.
48 Ibid., 1.
49 Ibid., 5.
50 Ibid., 8.
51 Ibid., 4-5.
52 Ibid., 8.
53 Ibid., 5.
54 Ibid., 8.
55 Ibid., 15.
56 Ibid., 7.
57 Ibid., 14.
58 Ibid., 16-7.
59 Commissioner of Forest Enhancement, *Interim Report*, 1990, 8. For a contrasting view, provided by one of the 'opportunists,' see Peter Clancy, 'On the Axe's Edge: The Graham Langley Saga – Pulp and Punishment in the Nova Scotia Woods,' *New Maritimes* 11, 5 (1993): 6-16.
60 Commissioner of Forest Enhancement, *Interim Report*, 1990, 9.

61 In several newspaper interviews late in his term, Eldridge drew public attention to the need for the office to continue. See 'Retiring Commissioner Looks to Forestry's Future,' *Chronicle Herald* 13 June 1990, and 'Forestry Watchdog Faces Uncertain Fate,' *Chronicle Herald* 23 March 1991.
62 Nova Scotia Commissioner of Forest Enhancement, *Second Annual Report*, 1989, 9.
63 *Interim Report*, 1990, 13.
64 *Interim Report*, 1990, 17.
65 *First Annual Report*, 1987, 14.
66 *Interim Report*, 1990, 13.
67 *Third Annual Report*, 1990, 12.
68 *Second Annual Report*, 1989, 15-16.
69 *Third Annual Report*, 1990, 15.
70 *Interim Report*, 1990, 15.
71 *Third Annual Report*, 1990, 17.
72 Ibid., 7.

Chapter 6: David Dwyer

1 Jack Westoby, *The Purpose of Forests: Follies of Development* (Oxford: Basil Blackwell, 1987), 269.
2 This is described in Lloyd S. Hawboldt and Simon N. Kostjukovits, *Forest Regulation in Nova Scotia, Part 1: Site Quality Normal Yield Tables for Softwoods*, Bulletin 20 (Halifax: DLF, 1961).
3 G. David Dwyer, *Twenty Years of Forestry on the Antrim Woodlot, 1951-1971* (Halifax: DLF, 1974).
4 G. David Dwyer, 'Ground Stereopair Photographs,' *Forestry Chronicle* 32 (1956): 309-12.
5 Lloyd S. Hawboldt and Richard M. Bulmer, *The Forest Resources of Nova Scotia* (Halifax: DLF, 1958), 68.
6 G. David Dwyer, 'A Study of Blowdown in Nova Scotia,' unpublished paper, Faculty of Forestry, University of New Brunswick, 1958.
7 G. David Dwyer, 'Woodlands Shaped by Past Hurricanes,' *Forest Times* November 1979.
8 Provincial Parks Act, c.7, *Statutes of Nova Scotia* (Halifax: Queen's Printer, 1959).
9 Beaches Protection Act, c.2, *Statutes of Nova Scotia* (Halifax: Queen's Printer, 1960).
10 Nova Scotia, Royal Commission on the Prices of Pulpwood and Other Forest Products, *Report* (Halifax, 1964).
11 Scott Maritimes Pulp Ltd. Act, c.15, *Statutes of Nova Scotia* (Halifax: Queen's Printer, 1965).
12 One perspective from the point of view of a leading sawmiller is found in Harry Thurston, 'Prest's Last Stand: Keeping Kafka and the Bureaucrats out of the Acadian Forest,' *Harrowsmith* August-September 1983: 22-31.
13 Haliburton's political memoirs can be found in *My Years With Stanfield* (Hantsport, NS: Lancelot Press, 1972).
14 Provincial Forest Practices Improvement Board, *The Trees around Us* (Halifax: Queen's Printer, 1980).
15 R. Edward Bailey and G.E. Mailman, 'Land Capability for Forestry in Nova Scotia' (Truro, NS: Department of Lands and Forests, 1972).
16 'Revised Normal Yield Tables for Nova Scotia Softwoods,' Forest Research Report 22 (Truro, NS: Department of Lands and Forests, 1990).
17 Lloyd S. Hawboldt, *Christmas Trees Are a Crop* (Halifax: DLF, 1953).
18 G. David Dwyer, *Twenty Years of Forestry*.
19 W.E. Hiley, *A Forestry Venture* (London: Faber and Faber, 1964).
20 See Exhibit 5, 'On the Antrim Woodlot.'
21 There was still a large block of unproductive Crown land in the Kemptville Barrens, in the back of Yarmouth, Digby, and Shelburne Counties.
22 G. David Dwyer, 'Stanley Management Unit: Forest Production from the Barrens,' *Lands and Forests Review* 2, 1 (1975).
23 L. Anders Sandberg, 'Swedish Forestry Legislation in Nova Scotia: The Rise and Fall of the Forest Improvement Act, 1965-1986,' in Douglas Day, ed., *Geographical Perspectives on the Maritime Provinces* (Halifax: Saint Mary's University, 1988).

24 Canadian Institute of Forestry, Nova Scotia Section, 'A Forest Policy for Nova Scotia' (Halifax: CIF, 1971).
25 'Guidelines to Sections 9, 10, 11 and 12 of the Forest Improvement Act,' printed in *The Trees around Us*.
26 The board published a pamphlet to mark this event. See Provincial Forest Practices Improvement Board, *Our Forests: A New Look at Nova Scotia's Forest Improvement Act* (Halifax: DLF, 1976). It was written by Dwyer but appeared at a time 'when names disappeared from publications.'
27 The federal Agricultural and Rural Development Administration (ARDA) originated in 1961 to enhance marginal farm and rural land economies across the country. The third agreement in Nova Scotia (ARDA 3, 1971-5) supported pilot projects in forest improvement.
28 Dwyer to Eldridge, 17 December 1984; Eldridge to Dwyer, 18 December 1984; Dwyer to Eldridge, 25 January 1985; Dwyer to Johnson, 1 November 1985.
29 Dwyer to Streatch, 4 February 1986.
30 For details on the CESO program, and a portrait of Dwyer's work, see Jim Lotz, *Sharing a Lifetime of Experience: The CESO Story* (Lawrencetown, NS: Pottersfield Press, 1997).
31 G. David Dwyer, 'Report of Assistance Rendered at Emafini Forest Estate,' Project 10005 (Swaziland, December 1989).
32 G. David Dwyer, 'Report of Assistance Rendered at International Timber Corporation, Indonesia,' Project 11565 (June 1991).

Chapter 7: Richard Lord

1 Richard A. Lord, *Land Use and Rural Settlement Patterns in Northeastern New Brunswick* (Fredericton: Faculty of Forestry, University of New Brunswick, 1964), 60-1.
2 Ibid., 62.
3 Richard A. Lord, *A Study of Private Forest Landownership in Soulanges and Huntingdon Counties, Quebec* (Montreal: MacDonald College, Department of Woodlot Management, 1965).
4 Nova Scotia Pulp Limited Agreement Act, c.9, *Statutes of Nova Scotia* (Halifax: Queen's Printer, 1958).
5 Wendell Coldwell, 'Nova Scotia Woodlot Owners' Association,' December 1960, RG 30-3/3/14/558-64, SFXUA.
6 Rev. J.A. Gillis, 'The Nova Scotia Woodlot Owners' Association: Some Facts about Its Background and Program,' n.d., RG 30-3/14/555-57, SFXUA.
7 Wendell Coldwell to NSWOA members, 15 February 1962, RG 30-3/14/76-8, SFXUA.
8 Nova Scotia, Royal Commission on the Prices of Pulpwood and Other Forest Products, *Report* (Halifax: Queen's Printer, 1964).
9 Department of Extension, St. Francis Xavier University, 'Research on the Organization and Education of Woodlot Owners' (Antigonish, NS, 1966).
10 For details, see Alexander A. MacDonald, 'General Report on Woodlot Owner Organization in Nova Scotia,' Nova Scotia Federation of Agriculture, mimeo, 1967.
11 Press release, Nova Scotia Woodlot Owners Association, 13 May 1969.
12 Richard A. Lord, 'Report of the Contracting Agency to the Nova Scotia Woodlot Owners Association,' 31 January 1970, 2.
13 Lord to Burgess, 28 April 1969, RG 30-3/14, SFXUA.
14 Burgess to Lord, 23 May 1969, ibid.
15 Richard A. Lord, 'Report of the Contracting Agency.'
16 John Smith, *Proceedings*, Nova Scotia Woodlot Owners and Operators Association, 8th Annual Meeting, 1977.
17 Canadian Council of Forest Ministers, *Canada's Forest Heritage* (Ottawa: Forestry Canada, n.d.), 17.
18 'Wood Products Marketing,' *Forestry Chronicle* 26, 1 (1950): 3.
19 'Special Woodlot Issue,' *Forestry Chronicle* 27, 3 (1951).
20 For an example, see H.J. Malsberger, 'Farm Woodland Management in the South,' *Forestry Chronicle* 26, 4 (1950).
21 Details are available in Nova Scotia, Royal Commission on the Price of Pulpwood and Other Forest Products, *Report* (Halifax: Queen's Printer, 1964); and Canada, Restrictive

Trades Practices Commission, *Report Concerning the Purchase of Pulpwood in Certain Districts of Eastern Canada* (Ottawa, 1958).

22 The Woodlands Department of NSPL provided each potential supplier with a loose-leaf manual of rules and practices. See *Forestry Guide*, April 1961.

23 Lloyd S. Hawboldt and Simon N. Kostjukovits, 'Forest Regulation for Nova Scotia, Part 3: A Procedure for Forest Management Planning,' Bulletin 22 (Halifax: DLF, 1961).

24 Lloyd S. Hawboldt, Simon N. Kostjukovits, and Richard M. Bulmer, 'Mensuration for Management,' paper presented to the Annual Meeting of the Canadian Institute of Forestry, Halifax, 1963.

25 Lloyd S. Hawboldt, 'Supply – Can We Keep Up?' *Lands and Forests Newsletter* 1, 3 (1973): 1.

26 R.M. Nacker et al., *Small Private Woodlands in Nova Scotia* (Ottawa: Forest Economics Research Institute, 1972); and G. David Dwyer, *Twenty Years of Forestry on the Antrim Woodlot, 1951-1970* (Halifax: DLF, 1974).

27 Robert H. Burgess, 'Forestry Incentive Programs for Increased Fibre Production in Nova Scotia,' Department of Lands and Forests, talk to the Canadian Pulp and Paper Association, mimeo, 1970, 4.

28 Richard Lord, *Proceedings*, Nova Scotia Woodlot Owners and Operators Association, Annual Meeting, 1977, 8. See also his keen analysis in the document 'Forest Management Situation on Private Holdings,' PANS.

29 Meeting of 9 November 1970; Topshee to Comeau, 11 February 1971, RG 30-3/14, SFXUA.

30 The group included Bowater Mersey (pulp and newsprint – Liverpool, NS); Scott Maritimes (sulphate pulp – Pictou, NS); Nova Scotia Forest Industries (market pulp and newsprint – Port Hawkesbury, NS); and Canexel (hardboard – Chester, NS).

31 C.H. Sproule, 'Opening Remarks, Meeting on Marketing Boards,' Nova Scotia Forest Products Association, 26 May 1970, 1-2.

32 Richard Lord to County Woodlot Owner Associations, August 1970, SFXUA, 1667. See also 1635, 1638.

33 VP is a consultative body linking provincial business (and, to a far lesser degree, labour) groups in an advisory capacity to the provincial government. Established by the Stanfield government in 1963, its original 'planning' mandate was abandoned by 1970, and the organization was attempting to demonstrate to a sceptical Liberal government its capacity as a consensus-seeking forum. For more detail, see Peter Clancy, 'Concerted Action on the Periphery? Voluntary Economic Planning in "the New Nova Scotia",' *Acadiensis* 26, 2 (1997): 3-30.

34 NSWOA, 'Submission to the Forestry Sector Committee of Nova Scotia Voluntary Economic Planning,' Halifax, May 1970, SFXUA, 1626.

35 'Actually two votes were held: for the first vote, the Nova Scotia Woodlot Owners Association plan was approved by a majority of one vote. It was then argued that government members, of which there were two, should not be given a vote (they had voted during the first ballot). This position was upheld and a subsequent vote turned down the [NSWOA] resolution by one vote.' Michael Kirby, 'The Nova Scotia Pulpwood Marketing Board Case,' Halifax, Dalhousie University, unpublished case study, 1979, 9.

36 Further details of this crucial series of events can be found in Peter Clancy, 'The Politics of Pulpwood Marketing in Nova Scotia, 1960-1985,' in L. Anders Sandberg, ed., *Trouble in the Woods: Forest Policy and Social Conflict in Nova Scotia and New Brunswick* (Fredericton: Acadiensis Press, 1992), 168-97.

37 *Proceedings*, 3rd Annual Meeting, NSWOA, January 1972.

38 Clancy, 'Pulpwood Marketing in Nova Scotia,' 55.

39 See, for example, Rick Lord to Hon. Benoit Comeau, 7 April 1972, RG 30-3/14/p.1792, SFXUA.

40 Richard Lord, *Proceedings*, Nova Scotia Woodlot Owners Association, Annual Meeting, February 1974, 4.

41 C.H. Sproule, 'Opening Remarks, Meeting on Marketing Boards,' 26 May 1970.

42 Robert G. Murray, Report to the Nova Scotia Forest Products Association, Annual Meeting, February 1972, 5.

43 Robert H. Burgess to Hon. George A. Snow, 18 March 1970.

44 NSWOA, 'Submission to Hon. G.I. Smith, Premier, and Hon. G.A. Snow, Minister of Lands and Forests,' 11 March 1970, 10 pp., DLF files, vol. 880, no. 7.
45 Robert H. Burgess to Hon. George A. Snow, 18 March 1970, DLF files, vol. 880, no. 7.
46 NSWOA, 'Proposal for the Recruitment, Training, and Employment by the NSWOA of Crews of Silviculture and Harvesting Technicians for Forestry Operations on Privately Owned Woodlots in Nova Scotia,' submitted to the Joint DREE-Nova Scotia Task Force, 16 May 1974, 8.
47 Department of Lands and Forests, 'A Report on the West Pictou Pilot Project,' 1977, DLF files, vol. 898, no. 11.
48 'NSWOOA Honors Richard Lord,' *Forest Times* 8, 1 (1986): 1.
49 Richard A. Lord, 'An Analysis of the Private Forestry Sector,' submission to the Royal Commission on Forestry, 19 April 1983.
50 Lloyd S. Hawboldt, *Christmas Trees Are a Crop*, Bulletin 12 (Halifax: DLF, 1953), 24.
51 Lloyd S. Hawboldt and Gordon R. Maybee, *The Nova Scotia Christmas Tree Trade*, Bulletin 14 (Halifax: DLF, 1955).
52 Judged in terms of content, the DLF extension bulletin of 1953 compares well with the more recent publications. It is in the absence (and later presence) of organizational infrastructure that the Christmas tree improvement efforts of the 1950s contrast with those of the 1980s.
53 Duncan K. MacLellan, *A Study of Selected New Brunswick, Nova Scotia, Regional, and National Forestry Sector Interest Groups* (Halifax: Dalhousie School of Public Administration, 1986).
54 Matthew Wright, *Forest Times* 14, 6 (1982).
55 Matthew Wright, 'A Growers Notebook,' *Forest Times* 15, 4 (1993): 12.
56 The time of writing marks thirty years from the approval of the ARDA contract under which the organizing campaign was financed.
57 Presentation to the Annual Meeting, Nova Scotia Section, Canadian Institute of Forestry, Halifax, 22 September 1977, MG 1, vol. 2862, no. 10, PANS.

Chapter 8: Mary Guptill

1 In this sense, the turn-of-the-century pioneers formed the first generation, and the interwars graduates formed the second. The veterans' bulge of the late 1940s announced the third generation, and the graduates of the age of the environment constitute the fourth. If the 1972 Stockholm Conference marks the beginning of modern environmentalism, the 1976 budworm spray controversies signalled its arrival in Maritime forestry.
2 Franklin O. Carroll, John Freemuth, and Les Aim, 'Women Forest Rangers,' *Journal of Forestry* 94, 1 (1996): 38-41.
3 'Canada-Nova Scotia Subsidiary Agreement – Forestry,' 28 June 1977, 23.
4 Ibid., 24.
5 Mary Guptill, 'Nova Scotia's Group Ventures: Process and Progress,' Seminar on Extension Activities for Owners of Small Woodlots, 28 September 1987, 2.
6 Coalition of Nova Scotia Forest Interests, *A New Forest Strategy for Nova Scotia* (Halifax: DLF, 1996).
7 'Canada-Nova Scotia Subsidiary Agreement – Forestry,' Ottawa, 28 June 1997, 3.
8 'Canada-Nova Scotia Forest Resource Development Agreement,' Halifax, 31 August 1982.
9 'Canada-Nova Scotia Cooperation Agreement for Forestry Development 1989-91,' Halifax, 15 December 1989; 'Canada-Nova Scotia Cooperation Agreement for Forestry Development, 1991-95,' Halifax, 10 January 1992.
10 Earlier chapters have detailed the considerable difference in perspective and support for the Group Venture program by influential figures within the Nova Scotia DLF. See the chapters on Donald Eldridge and David Dwyer in particular.
11 Mary E. Guptill, 'Presentation to the Royal Commission on Forestry,' transcript, 20 January 1983, 137.
12 See the document 'Typical Woodlot Management Plan,' developed by La Forêt Acadienne Ltée, in this book.
13 La Forêt Acadienne Ltée, 'Presentation to the Royal Commission on Forestry,' 20 January 1983, 3.

14 Ibid., 2-3.
15 Ibid., 3.
16 See Map 6, 'Sample Woodlot and Shareholder Property Pattern,' in the document 'Typical Woodlot Management Plan.'
17 Much of that pasture land should never have been cleared. But the early bias of settlement policy was against the forests. This is reflected, as Guptill points out, in the term 'spruce up,' pulling out the young spruce on pasture lands in order to 'improve' the land. The term lives on today.
18 Guptill, 'Presentation to the Royal Commission,' 120.
19 An example of how Guptill and her colleagues developed woodlot management plans is provided in 'Typical Woodlot Management Plan' in this book.
20 Guptill, 'Nova Scotia's Group Ventures,' 3-4.
21 For a contrast, see the chapters on Bigelow and Lord, who took on this task squarely but unsuccessfully.
22 In a monopsony situation, the buyer of an industrial input, such as pulpwood, possesses the economic power to dictate the price to the seller.
23 Guptill, 'Nova Scotia's Group Ventures,' 7.

Chapter 9: Conclusion

1 Bruno Latour, 'Visualization and Cognition: Thinking with Eyes and Hands,' in Henrika Kuklick and Elizabeth Long, eds., *Knowledge and Society: Studies in the Sociology of Culture Past and Present*, vol. 6 (Greenwich, CT: JAI Press, 1986), 1-40.
2 For a general example of this view of the professions, see Ivan Illich et al., *Disabling Professions* (London: Marion Boyers, 1977); for the case of foresters in particular, see Henry E. Lowood, 'The Calculating Forester: Quantification, Cameral Science, and the Emergence of Scientific Forestry Management in Germany,' in Tore Frängsmyr, J.L. Heilbron, and Robin E. Rider, eds., *The Quantifying Spirit in the Eighteenth Century* (Berkeley: University of California Press, 1991), 315-42. For a trenchant critique, see Nancy Langston's powerful account of the failure of modern forestry's 'immutable mobile' to manage the old-growth forests of the western United States. Nancy Langston, *Forest Dreams, Forest Nightmares: The Paradox of Old Growth in the Inland West* (Seattle: University of Washington Press, 1995). For another critique in the context of India, see K. Sivaramakrishnan, 'The Politics of Fire and Forest Regulation in Colonial Bengal,' *Environment and History* 2 (1996): 145-94. Rick Jonasse, 'The Forester's Eye: Technology, Techniques, and Perceptions in Early American Forestry,' *Alternatives* 21, 3 (1995): 32-7.
3 Billie DeWalt, 'Using Indigenous Knowledge to Improve Agriculture and Natural Resource Management,' *Human Organization* 53, 2 (1994): 128.
4 A.E. Balloch, 'Science Enters the Forest,' 23 April 1966, RG 30-3/14/461, 459, SFXUA.
5 Forestry made a much earlier appearance in New Brunswick, Ontario, and Quebec, and even in British Columbia. For details on New Brunswick, see R. Peter Gillis and Thomas R. Roach, *Lost Initiatives: Canada's Forest Industries, Forest Policy, and Forest Conservation* (New York: Greenwood, 1986), Chapter 4. For Ontario, see Richard S. Lambert and A. Paul Pross, *Renewing Nature's Wealth* (Toronto: Government of Ontario, 1967).
6 John Bigelow, 'Better Utilization of Nova Scotia's Forests,' RG 30-3/14/203, 3, SFXUA.
7 Gillis and Roach, *Lost Initiatives*, Chapter 1.
8 Bigelow, 'Better Utilization,' 5.
9 The same applied to New Brunswick in the 1940s. Gordon L. Baskerville, 'The Forestry Problem: Adaptive Lurches of Renewal,' in Lance H. Gunderson, Crawford S. Holling, and Stephen S. Light, eds., *Barriers and Bridges to the Renewal of Ecosystems and Institutions* (New York: Columbia University Press, 1995), 37, 45-7.
10 Baskerville attributes changing inventories to the fact that 'computers were not commonly available to perform the cumbersome calculations' necessary to make more precise calculations. Given that inventories in the past have frequently been contested, determined politically, and then justified scientifically, one wonders whether the refined computer models of the present might not be subject to the same pressures. Baskerville, 'The Forestry Problem.'

11 The prevailing orthodoxy is leaning toward the biocides, mainly because chemical insecticides are not accepted by the public.
12 Allan Miller and Paul Rusnock, 'The Rise and Fall of the Silvicultural Hypothesis in Spruce Budworm ... Management in Eastern Canada,' *Forest Ecology Management* 61 (1993): 171-89.
13 Although Elizabeth May's *Budworm Battles* is instructive in providing a record of the environmentalists' fight to stop aerial insecticide spraying of the budworm in the late 1970s, there is little more than a recognition of the long-standing opposition of the Department of Lands and Forests on this issue. Although Hawboldt is mentioned in the text (and generously, perhaps strategically, but mistakenly given the title Dr.), and several quotations are drawn from Rachel Carson's *Silent Spring*, there is no reference to the different ecological paradigms that competed with the chemical spray option from the 1940s onward. Elizabeth May, *Budworm Battles: The Fight to Stop Aerial Insecticide Spraying of the Forests of Eastern Canada* (Halifax: Four East Publications, 1982).
14 For one recent rationalization in today's climate of environmental attacks, see Hamish Kimmins, *Balancing Act: Environmental Issues in Forestry* (Vancouver: UBC Press, 1992).
15 For an excellent example, see Baskerville, 'The Forestry Problem.'
16 The groups were dominated by sports hunters and anglers. The Mi'kmaq community and other subsistence users were routinely excluded.
17 Bigelow, 'Better Utilization,' 5.
18 R.H. Burgess, 'Forest Industry Incentive Programs for Increased Fibre Production in Nova Scotia,' Canadian Pulp and Paper Association Annual Meeting, 1970, transcript, 10.
19 Peter Clancy, 'Crossroads in the Forest: Change is in the Air for Nova Scotia Woodlot Owners,' *New Maritimes* 9, 5 (1991): 5-8, and 'On the Axe's Edge: The Graham Langley Saga – Pulp and Punishment in the Nova Scotia Woods,' *New Maritimes* 11, 5 (1993): 6-16.
20 This inevitability trap has been raised against our previous writings, which one reviewer characterized as 'thoroughly pessimistic' for overestimating the scale of entrenched power structures and thereby the obstacles to reform.
21 Indeed, James C. Scott, in a recent account, uses the history of scientific forestry as a metaphor for the ways in which state bureaucracies and large commercial firms have more generally manipulated forms of knowledge into simplified and legible immutable mobiles. James C. Scott, *Seeing like a State: How Certain Schemes to Improve the Human Condition Have Failed* (New Haven: Yale University Press, 1999).
22 Joanna Beyers, 'The Forest Unbundled: Canada's National Forest Strategy and Model Forest Program, 1992-1997,' PhD diss., Faculty of Environmental Studies, York University, 1998.
23 James D. Proctor, 'Whose Nature? The Contested Moral Terrain of Ancient Forests,' in William Cronon, ed., *Uncommon Ground: Rethinking the Human Place in Nature* (New York: W.W. Norton, 1996), 269-97.
24 Similarly, there are difficulties defining what constitutes an ancient or old-growth forest area. In the colonial context, environmental historians have by now shown convincingly that the pre-European settlement forests were subject to large-scale change from environmental processes as well as actions by First Nations peoples. See, for example, William Cronon, *Changes in the Land: Indians, Colonists, and the Ecology of New England* (New York: Hill and Wang, 1982); Stephen Pyne, *Burning Bush: A Fire History of Australia* (North Sydney: Allen and Unwin, 1991), 15-152.
25 For a brilliant book addressing this question, see William Cronon, ed., *Uncommon Ground: Rethinking the Human Place in Nature* (New York: W.W. Norton, 1996); for Canadian and Scandinavian perspectives, see L. Anders Sandberg and Sverker Sörlin, eds., *Sustainability – the Challenge: People, Power, and the Environment* (Montreal: Black Rose Books, 1998), and Ari Aukusti Lehtinen, 'Northern Natures: A Study of the Forest Question Emerging within the Timber Line Conflict in Finland,' *Fennia* 169, 1 (1991): 57-169.

Bibliography

Akesson, Hans. 'The Nova Scotia Pulp Limited Story: Forest Management for the Revitalization of the Eastern Nova Scotia Forests, Nova Scotia Pulp Limited, and the Nova Scotia Economy.' Paper presented to the Standing Forest Management Committee, Canadian Institute of Forestry, 63rd Annual Meeting, Victoria, BC, 1971.

Allison, Graham. *The Essence of Decision*. Boston: Little Brown, 1971.

Antoft, Kell. 'Symptom or Solution: The Nova Scotia Land Holdings Disclosure Act of 1969.' In *Trouble in the Woods*, ed. L. Anders Sandberg, 198-211. Fredericton: Acadiensis Press, 1992.

Aplet, Gregory H., et al., eds. *Defining Sustainable Forestry*. Washington, DC: Island Press, 1993.

Atlantic Construction Journal. 'Special Issue' (September 1993): E1-12.

Bailey, R. Edward, and G.E. Mailman. 'Land Capability for Forestry in Nova Scotia.' Truro, NS: Department of Lands and Forests, 1972.

Balloch, A.E. 'Science Enters the Forest.' 23 April 1966. RG 30-3/14/461, 459, SFXUA.

Bantjes, Rod. 'Hegemony and the Power of Constitution: Labour and Environmental Coalition-Building in Maine and Nova Scotia.' *Studies in Political Economy* 54 (1997).

Baskerville, Gordon L. 'The Forestry Problem: Adaptive Lurches of Renewal.' In *Barriers and Bridges to the Renewal of Ecosystems and Institutions,* ed. Lance H. Gunderson, Crawford S. Holling, and Stephen S. Light, 37-102. New York: Columbia University Press, 1995.

Beck, J. Murray. *The Politics of Nova Scotia, Vol. 2, 1896-1988*. Tantallon, NS: Four East Publications, 1988.

Beyers, Joanna. 'The Forest Unbundled: Canada's National Forest Strategy and Model Forest Program, 1992-1997.' PhD diss. York University, 1998.

Bigelow, John R. 'Better Utilization of Nova Scotia's Forests.' RG 30-3/14/203, 3, SFXUA.

–. 'Continuous Income through Forest Management.' Mimeo. SFXUA.

–. 'Cooperative Marketing of Pulpwood.' RG 20, vol. 870, no. 12, PANS.

–. 'Report to Mr. F.W. Walsh on the Meeting with the Quebec Forestry Commission re Pulpwood Prices and Cooperative Marketing.' RG 30-3/14/1032, SFXUA.

–. 'The Utilization and Marketing Problem of Farm Woodlot Products in Nova Scotia.' 27 February 1939, RG 20, vol. 870, no. 12, PANS.

'Bigelow, John Robert, Sr.' Obituary. *Chronicle Herald* 18 December 1997.

Bissix, Glyn, and L. Anders Sandberg. 'The Political Economy of Nova Scotia's Forest Improvement Act, 1962-1986.' In *Trouble in the Woods: Forest Policy and Social Conflict in Nova Scotia and New Brunswick*, ed. L. Anders Sandberg, 168-97. Fredericton: Acadiensis Press, 1992.

Brock, Peter, and G. David Dwyer. 'On the Antrim Woodlot.' Transcript, CBC Television, 'Land and Sea.' October, 1971.

Bulmer, Richard M., and Lloyd S. Hawboldt. *The Forest Resources of Nova Scotia*. Halifax: DLF, 1958.

Burgess, Robert H. 'Forestry Incentive Programs for Increased Fibre Production in Nova Scotia.' Presentation to the Canadian Pulp and Paper Association, 1970. DLF, transcript.

Burrill, Gary, and Ian McKay, eds. *People, Resources, and Power: Critical Perspectives on Underdevelopment and Primary Industries in the Atlantic Region*. Fredericton: Acadiensis Press, 1987.

'A Business Meeting in Hell.' *Forest Crusader* July 1934: 8–9, 18, 21.

Canada. Atlantic Canada Opportunities Agency. *Canada-Nova Scotia Cooperation Agreement for Forestry Development, 1989-91*. Halifax, 1989.

–. Atlantic Canada Opportunities Agency. *Canada-Nova Scotia Cooperation Agreement for Forestry Development, 1991-95*. Halifax, 1992.

–. Department of Natural Resources. *State of the Forest Reports*. Ottawa: Department of Natural Resources, 1991.

–. Department of Regional Economic Expansion. *Nova Scotia Subsidiary Agreement – Forestry*. Ottawa, 1977.

–. Regional Economic Expansion. *Canada-Nova Scotia Forest Resource Development Agreement*. Halifax, 1982.

–. Restrictive Trades Practices Commission. *Report Concerning the Purchase of Pulpwood in Certain Districts of Eastern Canada*. Ottawa, 1958.

Canada Lumberman 1 October 1929: 40.

Canadian Council of Forest Ministers. *Canada's Forest Heritage*. Ottawa: Forestry Canada, n.d.

–. *Compendium of Canadian Forestry Statistics, 1994*. Ottawa: CCFM, 1995.

Canadian Council of Forest Ministers. *Sustainable Forests, a Canadian Commitment*. Hull: CCFM, 1992.

Canadian Institute of Forestry, Nova Scotia Section. *A Forest Policy for Nova Scotia*. Halifax: CIF, 1971.

Canadian Pulp and Paper Association. Woodlands Section *Proceedings, 1927-1938*.

Carroll, Franklin O., John Freemuth, and Les Aim. 'Women Forest Rangers.' *Journal of Forestry* 94, 1 (1996): 38-41.

Carson, Rachel. *Silent Spring*. Cambridge, MA: Riverside Press, 1962.

CIF-NSS. 'The Way We Were.' *Forestry Chronicle* 71, 6 (1995): 792.

Clancy, Peter. 'Concerted Action on the Periphery? Voluntary Economic Planning in "the New Nova Scotia."' *Acadiensis* 26 (1997): 3-30.

–. 'Crossroads in the Forest: Change Is in the Air for Nova Scotia Woodlot Owners.' *New Maritimes* 9 (1991): 5-8.

–. 'On the Axe's Edge: The Graham Langley Saga – Pulp and Punishment in the Nova Scotia Woods.' *New Maritimes* 11, 5 (1993): 6-16.

–. 'The Politics of Pulpwood Marketing in Nova Scotia, 1960-1985.' In *Trouble in the Woods: Forest Policy and Social Conflict in Nova Scotia and New Brunswick*, ed. L. Anders Sandberg, 142-67. Fredericton: Acadiensis Press, 1992.

–. 'The Politics of Stewardship: Certification for Sustainable Forest Management in Canada.' In *Sustainability – the Challenge: People, Power, and the Environment*, ed. L. Anders Sandberg and Sverker Sörlin, 108-20. Montreal: Black Rose Books, 1998.

Clancy, Peter, and L. Anders Sandberg. 'Crisis or Opportunity: The Political Economy of the Stora Mill Closure.' *Business Strategy and the Environment* 4 (1995): 208-19.

–. 'Formulating Standards for Sustainable Forest Management in Canada.' *Business Strategy and the Environment* 6 (1997): 206-17.

Coalition of Nova Scotia Forest Interests. *A New Forest Strategy for Nova Scotia*. Halifax: DLF, 1996.

Coldwell, Wendell. 'Nova Scotia Woodlot Owners Association.' December 1960. RG 30-3/3/14/558-64, SFXUA.

–. 'Speech to Nova Scotia Woodlot Owners Association Members.' 15 February 1962. RG 30-3/14/76-8, SFXUA.

Colpitts, Nancy. 'Sawmills to National Park: Alma, New Brunswick, 1921-1947.' In *Trouble in the Woods: Forest Policy and Social Conflict in Nova Scotia and New Brunswick*, ed. L. Anders Sandberg, 90-109. Fredericton: Acadiensis Press, 1992.

Compendium of Canadian Forestry Statistics, 1994. Ottawa: Canadian Council of Forest Ministers, 1995.

Creighton, Wilfrid. *Forestkeeping: A History of the Department of Lands and Forests in Nova Scotia, 1926-1969.* Halifax: DLF, 1988.

–. 'Less Spray, More Dollars.' *Atlantic Advocate* August 1963: 70-2.

–. 'An Open Letter to Dr. Donald Dodds.' Unpublished.

–. Personal interview with the authors. 4 August 1994.

Cronon, William, ed. *Uncommon Ground: Rethinking the Human Place in Nature.* New York: W.W. Norton, 1996.

–, ed. *Changes in the Land: Indians, Colonists, and the Ecology of New England.* New York: Hill and Wang, 1982.

DeBonis, Jeff. 'Natural Resource Agencies: Questioning the Paradigm.' In *A New Century of Natural Resources Management*, ed. R. Knight and S. Bates, 159-70. Washington, DC: Island Press, 1995.

Department of Extension, St. Francis Xavier University. 'Research on the Organization and Education of Woodlot Owners.' Antigonish, NS, 1966.

Devall, Bill, ed. *Clearcut: The Tragedy of Industrial Forestry.* San Francisco: Sierra Club/Earth Island Press, 1993.

DeWalt, Billie. 'Using Indigenous Knowledge to Improve Agriculture and Natural Resource Management.' *Human Organization* 53, 2 (1994): 128.

Dodds, Donald. *Challenge and Response: A History of Wildlife and Wildlife Management in Nova Scotia.* Halifax: Department of Natural Resources, 1993.

Drengson, Alan, and Duncan Taylor, eds. *Ecoforestry: The Art of Sustainable Forest Use.* Gabriola Island, BC: New Society Publishers, 1997.

Dunk, Thomas. '"Is It Only Forest Fires that Are Natural?" Boundaries of Nature and Culture in White Working Class Culture.' In *Sustainability – the Challenge: People, Power, and the Environment,* ed. L. Anders Sandberg and Sverker Sörlin, 157-66. Montreal: Black Rose Books, 1998.

–. 'Talking about Trees: Environment and Society in Forest Workers' Culture.' *Canadian Review of Sociology and Anthropology* 31 (1994): 14-34.

Dunlap, Thomas R. *DDT: Science, Citizens, and Public Policy.* Princeton: Princeton University Press, 1981.

Dupuis, Melanie. 'In the Name of Nature: Ecology, Marginality, and Rural Land Use Planning during the New Deal.' In *Creating the Countryside: The Politics of Rural and Environmental Discourse*, ed. Melanie Dupuis and Peter Vandergeest. Philadelphia: Temple University Press, 1996.

Dwyer, G. David. 'Ground Stereopair Photographs.' *Forestry Chronicle* 32 (1956): 309-12.

–. 'Lloyd Hawboldt.' *Forestry Chronicle* 73, 5 (1997): 610-1.

–. 'Report of Assistance Rendered at Emafini Forest Estate.' Project 10005, Swaziland, December 1989.

–. 'Report of Assistance Rendered at International Timber Corporation, Indonesia.' Project 11565, Indonesia, June 1991.

–. 'Stanley Management Unit: Forest Production from the Barrens.' *Lands and Forests Review* 2, 1 (1975): 1, 4.

–. 'A Study of Blowdown in Nova Scotia.' Unpublished paper. Faculty of Forestry, University of New Brunswick, October 1958.

–. *Twenty Years of Forestry on the Antrim Woodlot, 1951-1970.* Halifax: DLF, 1974.

–. 'Woodlands Shaped by Past Hurricanes.' *Forest Times* November 1979.

Eldridge, Donald. 'Commissioner of Forest Enhancement.' *Annual Report.* Halifax, 1987-1990.

–. 'Executive Director's Report.' Nova Scotia Forest Products Association, 43rd Annual Meeting (1977), 45.

–. Interview with Glyn Bissix. Transcript. 21 September 1987.

–. 'Obituary.' *Chronicle Herald* 9 June 1995.

Ellefson, Paul E. 'Politics and Policymaking: A Teaching Challenge in Forestry.' *Journal of Forestry* 91, 3 (1993): 24-7.

Evenden, Matthew. 'The Laborers of Nature: Economic Ornithology and the Role of Birds as Agents of Biological Pest Control in North American Agriculture, ca. 1880-1930.' *Forest and Conservation History* 39 (1995): 172-83.
Fellows, Edward S. 'Popular, Professional Views Often Clash.' *Daily Gleaner* 6 February 1993.
Fensom, Ken G. *Expanding Forestry Horizons: A History of the Canadian Institute of Forestry – Institut Forestier du Canada, 1908-1969*. Montreal: CIF, 1972.
Fernow, Bernhard. *The Forest Conditions of Nova Scotia*. Ottawa: Commission of Conservation, 1912.
Finegold, Kenneth, and Theda Skocpol. *State and Party in America's New Deal*. Madison: University of Wisconsin Press, 1995.
Flieger, Barney W. 'Forest Protection from the Spruce Budworm in New Brunswick.' *Report of the Entomological Society of Ontario, 1953*.
Forbes, Ernest R., and Delphin A. Muise. *The Atlantic Provinces in Confederation*. Toronto: University of Toronto Press, 1993.
'Forestry Group Pondering New Headquarters.' *Chronicle Herald* 20 February 1989: A16.
'Forestry Watchdog Faces Uncertain Fate.' *Chronicle Herald* 23 March 1991: B1.
Franklin, J.F. 'Toward a New Forestry.' *American Forests* 95 (1989): 37-44.
Frome, Michael. *The Forest Service*. New York: Praeger Publishers, 1971.
Galbraith, John Kenneth. *The New Industrial State*. New York: Signet, 1978.
George, Donald F. 'The View from the Finish Line.' Unpublished paper. Antigonish, NS, n.d.
George, Roy E. *The Life and Times of Industrial Estates Limited*. Halifax: Institute of Public Affairs, Dalhousie University, 1974.
Gillis, Peter. 'Limits of Federal-Provincial Cooperation, 1920-1936: Ernest Herbert Finlayson and Canadian Forestry.' In *Canada's Timber Supply*, ed. David G. Brand. Chalk River, ON: Forestry Canada, 1991.
Gillis, R. Peter, and Thomas R. Roach. *Lost Initiatives: Canada's Forest Industries, Forest Policy, and Forest Conservation*. New York: Greenwood Press, 1986.
Gillis, Rev. J.A. 'The Nova Scotia Woodlot Owners' Association: Some Facts about Its Background and Program.' N.d. RG 30-3/14/555-57, SFXUA.
Gonzalez, Ellice. *Changing Roles for Micmac Men and Women: An Ethnocultural Analysis*. National Museum of Man Mercury Series. Canadian Ethnology Service Paper 72. Ottawa: National Museums of Canada, 1981.
Grumbine, Edward. 'Policy in the Woods.' In *Clearcut: The Tragedy of Industrial Forestry*, ed. Bill Devall, 253-62. San Francisco: Sierra Club Books, 1993.
Guha, Ramachandra. 'The Malign Encounter: The Chipko Movement and Competing Visions of Nature.' In *Who Will Save the Forests? Knowledge, Power, and Environmental Destruction*, ed. Tariq Banuri and Frédérique Apffel-Marglin. London: Zed, 1993.
Guptill, Mary E. 'Nova Scotia's Group Ventures: Process and Progress.' Seminar on Extension Activities for Owners of Small Woodlands, 28 September 1987.
–. 'Presentation to the Royal Commission on Forestry,' 20 January 1983. Transcript.
–. 'Typical Woodlot Management Plan.' La Forêt Acadienne Ltée, n.d.
Haliburton, Edward Douglas. *My Years with Stanfield*. Windsor: Lancelot Press, 1972.
'Hardwood Stands of Nova Scotia: Recent Controversy Regarding Methods of Manufacturing Sheds Light on Hardwood Resources of Maritime Province.' *Canada Lumberman* 15 April 1929: 34.
Hawboldt, Lloyd S. 'Aspects of Yellow Birch Dieback in Nova Scotia.' *Journal of Forestry* 45, 6 (1947): 414-22.
–. 'Bessa Selecta (Meigen) as a Parasite of Gilpinia Hercyniae (Hartig).' MSc diss. McGill University, 1946.
–. *Blueberries: Cash from Idle Acres*. Halifax: DLF, 1954.
–. *Christmas Trees Are a Crop*. Nova Scotia Department of Lands and Forests Bulletin 12. Halifax: DLF, 1953.
–. *Climate and Birch 'Dieback.'* Nova Scotia Department of lands and Forests Bulletin 6. Halifax: DLF, 1952.

–. 'Dieback of Birch.' Nova Scotia Department of Lands and Forests, *Annual Report 1944* and *Annual Report 1945*. Halifax: Government Services, 1944, 1945.
–. *'Dieback' of Yellow Birch*. Nova Scotia Department of Lands and Forests Bulletin 4. Halifax: DLF, 1951.
–. 'Forestry in Nova Scotia.' *Canadian Geographical Magazine* 51 (1955): 80.
–. 'The Spruce Budworm.' Mimeo. 1955. Halifax, legislative library.
–. 'Supply – Can We Keep Up?' *Lands and Forests Newsletter* 1, 3 (1973).
–. 'Yellow Birch Investigations.' Nova Scotia Department of Lands and Forests, *Annual Report 1948*. Halifax: Government Services, 1948.
Hawboldt, Lloyd S., and Arthur J. Skolko. 'Investigation of Yellow Birch Dieback in Nova Scotia in 1947.' *Journal of Forestry* 46, 9 (1948): 659-71.
Hawboldt, Lloyd S., and Gary L. Saunders. *A Guide to Forest Practices*. Halifax: DLF, 1966.
Hawboldt, Lloyd S., and Gordon R. Maybee. *The Nova Scotia Christmas Tree Trade*. Nova Scotia Department of Lands and Forests Bulletin 14. Halifax: DLF, 1955.
Hawboldt, Lloyd S., and Richard M. Bulmer. *The Forest Resources of Nova Scotia*. Halifax: DLF, 1958.
Hawboldt, Lloyd S., and Simon N. Kostjukovits. 'Forest Regulation in Nova Scotia, Part 1: Site Quality Normal Yield Tables for Softwoods.' Nova Scotia Department of Lands and Forests Bulletin 20. Halifax: DLF, 1961.
–. 'Forest Regulation in Nova Scotia, Part 3: A Procedure for Forest Management Planning.' Nova Scotia Department of Lands and Forests Bulletin 22. Halifax: DLF, 1961.
Hawboldt, Lloyd S., Simon N. Kostjukovits, and Richard M. Bulmer. 'Mensuration for Management.' Paper presented to the Forest Mensuration Committee, Annual Meeting of the Canadian Institute of Forestry, Halifax, 1963.
Hiley, W.E. *A Forestry Venture*. London: Faber and Faber, 1964.
Hiller, James. 'The Origin of the Pulp and Paper Industry in Newfoundland.' *Acadiensis* 11 (1982): 42-68.
–. 'The Politics of Newsprint: The Newfoundland Pulp and Paper Industry, 1915-1939.' *Acadiensis* 19 (1990): 3-39.
Hirt, Paul. *A Conspiracy of Optimism: Management of the National Forests since World War Two*. Lincoln: University of Nebraska Press, 1994.
Hornborg, Alf. 'Environmentalism, Ethnicity, and Sacred Places: Reflections on Modernity, Discourse, and Power.' *Canadian Review of Sociology and Anthropology* 31 (1994): 245-67.
–. 'Mi'kmaq Environmentalism: Local Initiatives and Global Projections.' In *Sustainability – the Challenge: People, Power, and the Environment,* ed. L. Anders Sandberg and Sverker Sörlin, 202-11. Montreal: Black Rose Books, 1998.
How, Douglas. *A Very Private Person: The Story of Izaak Walton Killam and His Wife Dorothy*. N.p.: n.p., 1976.
Howlett, Michael. 'The 1987 National Forest Sector Strategy and the Search for a Federal Role in Canadian Forest Policy.' *Canadian Public Administration* 32 (1989): 545-63.
Howlett, Michael, and Jeremy Rayner. 'Do Ideas Matter? Policy Network Configurations and Resistance to Policy Change in the Canadian Forest Sector.' *Canadian Public Administration* 38 (1995): 382-410.
Illich, Ivan, et al. *Disabling Professions*. London: Marion Boyers, 1977.
Innis, Harold. 'Complementary Report.' In *Report of the Royal Commission Provincial Economic Inquiry* (Halifax: King's Printer, 1934), 131-230.
'Is Reforestation Bunk?' *Forest Crusader* April 1934: 18.
Janzen, Russ. 'Hegemony and Genealogy: Managerialist Discourse in the Forests of British Columbia.' In *Sustainability – the Challenge: People, Power, and the Environment,* ed. L. Anders Sandberg and Sverker Sörlin, 149-56. Montreal: Black Rose Books, 1998.
Johnson, Ralph S. 'Forest Legislation.' Panel Discussion for Nova Scotia Section of Canadian Institute of Forestry. New Glasgow, NS, 7 September 1964.
–. *Forests of Nova Scotia*. Halifax: Four East Publications, 1986.
Johnstone, Kenneth. *Timber and Trauma: 75 Years with the Federal Forestry Service*. Ottawa: Supply and Services, 1991.

Jonasse, Rick. 'The Forester's Eye: Technology, Techniques, and Perceptions in Early American Forestry.' *Alternatives* 21, 3 (1995): 32-7.
Judd, Richard. *Common Lands, Common People: The Origins of Conservation in Northern New England*. Cambridge, MA: Harvard University Press, 1997.
Juhlin-Dannfelt, M. 'Forestry Conditions in Canada and Sweden from a Swedish Forester's View-Point.' *Illustrated Canadian Forest and Outdoors* 20 (1924): 531-2.
Kaufman, Herbert. *The Forest Ranger: A Study in Administrative Behaviour*. Baltimore: Johns Hopkins University Press, 1960.
Kennedy, John de Navarre. *History of the Department of Munitions and Supply*. Ottawa: King's Printer, 1950.
Kenny, James. 'A New Dependency: State, Local Capital, and the Development of New Brunswick's Base Metal Industry, 1960-1970.' *Canadian Historical Review* 78 (1997): 1-39.
Kettela, Edward. 'Byron Wentworth Flieger.' *Bulletin – Entomological Society of Canada* 29, 2 (1997): 78-9.
Kimmins, Hamish. *Balancing Act: Environmental Issues in Forestry*. Vancouver: UBC Press, 1992.
Kirby, Michael. 'The Nova Scotia Pulpwood Marketing Board Case.' Unpublished case study. Halifax: Dalhousie University, 1979.
Klyza, Christopher McGrory. *Who Controls the Public Lands? Mining, Forestry, and Grazing Policies, 1879-1990*. Chapel Hill: University of North Carolina Press, 1996.
Knight, Richard, and Sarah Bates. *A New Century for Natural Resources Management*. Washington, DC: Island Press, 1995.
La Forêt Acadienne Ltée. 'Presentation to the Royal Commission on Forestry.' Unpublished. January 1983.
Laidlaw, A.F. *The Campus and the Community: The Global Impact of the Antigonish Movement*. Montreal: Harvest House, 1961.
Lambert, Richard S., and A. Paul Pross. *Renewing Nature's Wealth*. Toronto: Government of Ontario, 1967.
Lamport, Anthony. *Common Ground: 25 Years of Voluntary Planning in Nova Scotia*. Halifax: Government Services, 1988.
Land Research Group. 'Whither Our Land? Who Owns Nova Scotia? And What Are They Doing with It?' *New Maritimes* 8, 6 (1990): 14-25.
Langston, Nancy. *Forest Dreams, Forest Nightmares: The Paradox of Old Growth in the Inland West*. Seattle: University of Washington Press, 1995.
Larson, Magali Sarfatti. *The Rise of Professionalism: A Sociological Analysis*. Berkeley: University of California Press, 1977.
Latour, Bruno. 'Visualization and Cognition: Thinking with Eyes and Hands.' In *Knowledge and Society: Studies in the Sociology of Culture Past and Present*, vol. 6, ed. Henrika Kuklick and Elizabeth Long, 1-40. Greenwich, CT: JAI Press, 1986.
Leavitt, Robert M. *Maliseet Micmac: First Nations of the Maritimes*. Fredericton: New Ireland Press, 1995.
Lehtinen, Ari Aukusti. 'Northern Natures: A Study of the Forest Question Emerging within the Timber Line Conflict in Finland.' *Fennia* 169, 1 (1991): 57-169.
Leopold, Aldo. *A Sand County Almanac*. 1949. New York: Ballantyne Books, 1970.
Long, H.D. *Report on Nova Scotia Plots*. Ed. George A. Mulloy. 1943. RG 39, vol. 148, file 47361, NAC.
Lord, Richard A. 'An Analysis of the Private Forestry Sector.' Submission to the Nova Scotia Royal Commission on Forestry, 19 April 1983.
–. 'Forest Management in Private Holdings.' Submission to the Nova Scotia Royal Commission on Forestry, Halifax, 1983.
–. 'Forest Management Situation on Private Holdings.' Paper presented to the Nova Scotia Section, CIF, 1977. MG 1, vol. 2862, no. 10, PANS.
–. *Land Use and Rural Settlement Patterns in Notheastern New Brunswick*. Fredericton: Faculty of Forestry, University of New Brunswick, 1964.
–. *Proceedings*. Nova Scotia Woodlot Owners Association, Annual Meeting, February 1974.
–. *Proceedings*. Nova Scotia Woodlot Owners Association, 3rd Annual Meeting, January 1972.

–. *Proceedings*. Nova Scotia Woodlot Owners and Operators Association, Annual Meeting, 1977.

–. 'Report of the Contracting Agency to the Nova Scotia Woodlot Owners Association.' Department of Extension, St. Francis Xavier University, Antigonish, 1970.

–. *A Study of Private Forest Landownership in Soulanges and Huntingdon Counties, Quebec*. Thesis, MacDonald College, 1965.

Lotz, Jim. *Sharing a Lifetime of Experience: The CESO Story*. Lawrencetown, NS: Pottersfield Press, 1997.

Loucks, O.L. *A Forest Classification for the Maritime Provinces*. Fredericton: Canada, Department of Forestry, 1962.

Lowood, Henry E. 'The Calculating Forester: Quantification, Cameral Science, and the Emergence of Scientific Forestry Management in Germany.' In *The Quantifying Spirit in the Eighteenth Century*, ed. Tore Frängsmyr, J.L. Heilbron, and Robin E. Rider, 315-42. Berkeley: University of California Press, 1991.

MacDonald, Alexander A. 'Exploitation of the Private Owners of Forest Resources in Nova Scotia.' Unpublished paper, 1975.

–. 'General Report on Woodlot Owner Organization in Nova Scotia.' Nova Scotia Federation of Agriculture, 1967.

–. 'Policy Formulation Process: Nova Scotia Dairy Marketing, 1933-1978.' Unpublished report, 1980.

MacDonald, Douglas. *The Politics of Pollution*. Toronto: McClelland and Stewart, 1991.

McKay, Ian. *The Quest of the Folk: Antimodernism and Cultural Selection in Twentieth Century Nova Scotia*. Montreal: McGill-Queen's University Press, 1994.

MacLellan, C. Roger. 'Dr. Allison Deforest Pickett.' *Bulletin –Entomological Society of Canada* 29, 2 (1997): 76-8.

MacLellan, Duncan K. *A Study of Selected New Brunswick, Nova Scotia, Regional, and National Forestry Sector Interest Groups*. Halifax: Dalhousie School of Public Administration, 1986.

Mahood, Ian, and Ken Drushka. *Three Men and a Forester*. Vancouver: Harbour Publishing, 1990.

Malsberger, H.J. 'Farm Woodland Management in the South.' *Forestry Chronicle* 26, 4 (1950): 302-7.

Maser, Chris. *The Redesigned Forest*. Toronto: Stoddard, 1990.

May, Elizabeth. *At the Cutting Edge: The Crisis in Canada's Forests*. Toronto: Key Porter, 1998.

–. *Budworm Battles: The Fight to Stop the Aerial Insecticide Spraying of the Forests of Eastern Canada*. Halifax: Four East Publications, 1982.

Meine, Curt. 'The Oldest Task in Human History.' In *A New Century for Natural Resources Management*, ed. Richard Knight and Sarah Bates. Washington, DC: Island Press, 1995.

Miller, Alan. 'The Role of Citizen-Scientist in Nature Resource Decision-Making: Lessons from the Spruce Budworm Problem in Canada.' *Environmentalist* 13, 1 (1993): 47-59.

Miller, Alan, and Paul Rusnock. 'The Ironical Role of Science in Policy-Making: The Case of the Spruce Budworm.' *International Journal of Environmental Studies* 43 (1993): 239-51.

–. 'The Rise and Fall of the Silvicultural Hypothesis in Spruce Budworm (Choristoneura fumiferana) Management in Eastern Canada.' *Forest Ecology Management* 61 (1993): 171-89.

Moore, Patrick. *Pacific Spirit*. Vancouver: Terra Bella, 1995.

Mullen, Eric, and Millie Evans. *In the Mersey Woods*. Liverpool, NS: Bowater Mersey, 1989.

Mulloy, George A. 'Nova Scotia's Great Opportunity: A Province that Grows Prolific Crops of Forests and Is Rapidly Re-Establishing Lost Areas.' *Illustrated Canadian Forest and Outdoors* 25 (1929): 93-4.

Murray, Robert G. 'Report to the Nova Scotia Forest Products Association.' Annual Meeting, February 1972.

Nacker, R.M., et al. *Small Private Woodlands in Nova Scotia*. Ottawa: Forest Economics Research Institute, 1972.

Neary, Peter. 'The Bradley Report on Logging Operations in Newfoundland, 1934.' *Labour/Le Travail* 16 (1985): 193-232.

Nelles, H. Vivian. *The Politics of Development: Forests, Mines, and Hydro-Electric Power in Ontario, 1849-1941*. Toronto: Macmillan, 1974.
Nova Scotia. Commissioner of Forest Enhancement. *Annual Reports*, 1987-90.
–. Department of Lands and Forests. Annual Reports, 1925-6, 1930-1, and 1931-2. Halifax: Queen's Printer.
–. –. *Forests and Forestry in Nova Scotia*. Halifax: Queen's Printer, 1930.
–. –. *The Imperial Economic Conference Report on Labour Conditions in the Woods, Forest Inventory, Duration of Supply, Forest Industries, and the English Market*. Halifax: Queen's Printer, 1932.
–. –. 'A Report on the West Pictou Pilot Project.' 1977. DLF files, vol. 898, no. 11.
–. –. 'Revised Normal Yield Tables for Nova Scotia Softwoods.' Forest Research Report 22. Truro, NS, March 1990.
–. –. *Submission to the Royal Commission on Forestry*. Halifax: DLF, 1983.
–. Provincial Forest Practices Improvement Board. *Our Forests: A New Look at Nova Scotia's Forest Improvement Act*. Halifax: DLF, 1976.
–. –. *The Trees around Us*. Halifax: Queen's Printer, 1980.
–. *Public Accounts, 1933-34*. Halifax: Queen's Printer, 1934.
–. Royal Commission on Forestry. *Report*. Halifax: Queen's Printer, 1984.
–. Royal Commission on the Prices of Pulpwood and Other Forest Products. *Report*. Halifax: Queen's Printer, 1964.
–. Royal Commission Provincial Economic Inquiry. *Report*. Halifax: King's Printer, 1934.
–. *Statutes of Nova Scotia*. 'Beaches Protection Act, c.2.' Halifax: Queen's Printer, 1960.
–. *Statutes of Nova Scotia*. 'Nova Scotia Pulp Limited Agreement Act, c.9.' Halifax: Queen's Printer, 1958.
–. *Statutes of Nova Scotia*. 'Provincial Parks Act, c.7.' Halifax: Queen's Printer, 1959.
–. *Statutes of Nova Scotia*. 'Scott Maritimes Pulp Ltd. Act, c.15.' Halifax: Queen's Printer, 1965.
Nova Scotia Forest Products Association. 'Access Road Construction Assistance Regulations.' Truro, NS, 1967.
Nova Scotia Members of the Canadian Society of Forest Engineers. *Forestry, Economy, and Post-War Reconstruction in Nova Scotia*. Annual Report. Halifax: DLF, 1944.
Nova Scotia Museum. *The Natural History of Nova Scotia, Vol. 1: Topics and Habitats*, and *Vol. 2: Theme Regions*. Halifax: Nimbus, 1996.
'Nova Scotia Will Send Lumber Envoy.' *Canada Lumberman* 15 March 1933: 24.
Nova Scotia Woodlot Owners Association. Press release. 13 May 1969.
–. 'Proposal for the Recruitment, Training, and Employment by the Nova Scotia Woodlot Owners Association of Crews of Silviculture and Harvesting Technicians for Forestry Operations on Privately Owned Woodlots in Nova Scotia.' Paper submitted to the Joint DREE-Nova Scotia Task Force, 16 May 1974.
–. 'Submission to Hon. G.I. Smith, Premier, and Hon. G.A. Snow, Minister of Lands and Forests.' 11 March 1970. DLF files, vol. 880, no. 7.
–. 'Submission to the Forestry Sector Committee of Nova Scotia Voluntary Economic Planning.' Halifax, May 1970. SFXUA 1626.
'Nova Scotia Woodlot Owners Association Honors Richard Lord.' *Forest Times* 8, 1 (1986): 1.
Öckerman, Anders. 'Culture versus Nature in the History of Swedish Forestry.' In *Sustainability – the Challenge: People, Power, and the Environment*, ed. L. Anders Sandberg and Sverker Sörlin, 72-9. Montreal: Black Rose Books, 1998.
Ontario Forest Policy Panel, Diversity. *Forests, People, Communities: A Comprehensive Forest Policy for Ontario*. Toronto: Ministry of Natural Resources, 1993.
Parenteau, William, and L. Anders Sandberg, 'Conservation and the Gospel of Economic Nationalism: The Canadian Pulpwood Question in Nova Scotia and New Brunswick, 1918-1925.' *Environmental History Review* 19, 2 (1995): 57-84.
Parker, Mike. *Woodchips and Beans: Life in the Early Lumberwoods of Nova Scotia*. Halifax: Nimbus, 1992.
Parr, Kermode. '"If You Poison Us, Do We Not Die?"' *Atlantic Advocate* November 1962: 21-4.

Patterson, Stephen. 'Indian-White Relations in Nova Scotia, 1749-61.' *Acadiensis* 28 (1993): 23-59.

Perkins, John H. *Insects, Experts, and the Insecticide Crisis: The Quest for New Pest Management Strategies*. New York: Plenum Press, 1982.

Pickett, Allison D. 'The Philosophy of Orchard Insect Control.' *Report of the Entomological Society of Ontario,* 1949.

Pickett, S.T.A., and Richard S. Ostfeld. 'The Shifting Paradigm in Ecology.' In *A New Century for Natural Resources Management*, ed. Richard Knight and Sarah Bates, 261-78. Washington, DC: Island Press, 1995.

Pinkett, Harold T. *Gifford Pinchot, Private and Public Forester.* Urbana: University of Illinois Press, 1970.

'Politics and Conservation: Is Our Political System Faulty?' *Forest Crusader* April 1934: 26.

Porter, Clarence. 'Special Committee Report.' Nova Scotia Forest Products Association 38th Annual Meeting, 1971.

Proctor, James D. 'Whose Nature? The Contested Moral Terrain of Ancient Forests.' In *Uncommon Ground: Rethinking the Human Place in Nature*, ed. William Cronon. New York: W.W. Norton, 1996.

Pross, A. Paul. 'The Development of Professions in the Public Service: The Foresters of Ontario.' *Canadian Public Administration* 10 (1969): 376-404.

Pyne, Stephen. *Burning Bush: A Fire History of Australia.* North Sydney: Allen and Unwin, 1991.

–. *Fire in America: A Cultural History of Wildland and Rural Fire*. Princeton: Princeton University Press, 1982.

–. 'Flame and Fortune.' *New Republic* 8 August 1994: 19-20.

Raddall, Thomas. *The Dreamers*. Porter's Lake, NS: Pottersfield Press, 1986.

Radforth, Ian. *Bush Workers and Bosses: Logging in Northern Ontario, 1900-1980.* Toronto: University of Toronto Press, 1987.

Rajala, Richard. *Clearcutting the Pacific Rain Forest: Production, Science, and Regulation*. Vancouver: UBC Press, 1998.

Raphael, Ray. *Tree Talk: The People and Politics of Timber.* Covelo, CA: Island Press, 1981.

–. *More Tree Talk: The People, Politics, and Economics of Timber.* Washington, DC: Island Press, 1994.

Reiger, John F. *American Sportsmen and the Origin of Conservation*. New York: Winchester Press, 1975.

Renfrew, Stewart. 'The Commission of Conservation.' *Douglas Library Notes* 19, 3-4 (1971): 17-26.

'Retiring Commissioner Looks to Forestry's Future.' *Chronicle Herald* 13 June 1990.

Richmond, Hector A. *Forever Green: The Story of One of Canada's Foremost Foresters*. Lantzville, BC: Oolichan, 1983.

Roach, Thomas, and Richard Judd. 'A Man for All Seasons: Frank John Dixie Barnjum, Conservationist, Pulpwood Embargoist, and Speculator!' *Acadiensis* 20 (1991): 129-44.

Rodgers III, Andrew D. *Bernhard Eduard Fernow: A Story of North American Forestry.* Princeton: Princeton University Press, 1951.

Sacouman, James. 'Underdevelopment and the Structural Origins of Antigonish Movement Co-operatives in Eastern Nova Scotia.' In *Underdevelopment and Social Movements in Atlantic Canada*, ed. Robert Brym and James Sacouman, 109-26. Toronto: New Hogtown Press, 1979.

Sandberg, L. Anders. 'The Forest Landscape in Maritime Canadian and Swedish Literature.' In *A Few Acres of Snow,* ed. Paul Simpson-Housley and Glen Norcliffe, 109-21. Toronto: Dundurn, 1992.

–. 'Swedish Forestry Legislation in Nova Scotia: The Rise and Fall of the Forest Improvement Act, 1965-1986.' In *Geographical Perspectives on the Maritime Provinces*, ed. Douglas Day, 184-96. Halifax: Saint Mary's University, 1988.

–, ed. *Trouble in the Woods: Forest Policy and Social Conflict in Nova Scotia and New Brunswick*. Fredericton: Acadiensis Press, 1992.

Sandberg, L. Anders, and Peter Clancy. 'Forestry in a Staples Economy: The Checkered

Career of Otto Scheirbeck, Chief Forester, Nova Scotia, Canada, 1925-1933.' *Environmental History* 2, 1 (1997): 74-95.
–. 'Lloyd Hawboldt.' *Bulletin – Entomological Society of Canada* 29, 3 (1997): 89-91.
–. 'Property Rights, Small Woodlot Owners, and Forest Management in Nova Scotia.' *Journal of Canadian Studies* 31 (1996): 25-47.
Sandberg, L. Anders, and Sverker Sörlin, eds. *Sustainability – the Challenge: People, Power, and the Environment.* Montreal: Black Rose Books, 1998.
Sandberg, L. Anders, and William Parenteau. 'From Weapons to Symbols of Privilege: Political Cartoons and the Rise and Fall of the Pulpwood Embargo Debate in Nova Scotia, 1923-1933.' *Acadiensis* 26, 2 (1997): 31-58.
Schierbeck, John. Personal interview with the authors. Ashcroft, BC, 4 November 1993.
Schierbeck, Otto. 'Fra Amerika.' *Tidskrift for Skovvaesen* 18 (1906): 63-7.
–. 'Nova Scotia Has Fine Young Tree to Replace Forest Primeval of Evangeline's Time.' *Canada Lumberman* 15 September 1928: 43.
–. 'Nova Scotia Is Looking Forward to Greater Overseas Trade in Timber.' *Canada Lumberman* 15 July 1932: 20.
–. 'Selective Thinning.' *Forestry Chronicle* 12 (1936): 368, 372-3.
–. 'The Sins of the Fathers Shall Be Visited upon the Children unto the Third and Fourth Generation.' RG 20, vol. 751, no. 3, PANS, 2.
–. 'Working Plan and Experimental Forest.' RG 20, vol. 746, PANS.
Schneider, Aaron. 'Underdeveloping Nova Scotia: Forests and the Role of Corporate Counter-Intelligences.' In *People, Resources, and Power*, ed. Gary Burrill and Ian MacKay. Fredericton: Acadiensis Press, 1987.
Schrecker, Ted. 'Of Invisible Beasts and the Public Interest: Environmental Cases and the Judicial System.' In *Canadian Environmental Policy: Ecosystems, Politics, and Process*, ed. Robert Boardman, 83-105. Toronto: Oxford University Press, 1992.
Scott, James C. *Seeing like a State: How Certain Schemes to Improve the Human Condition Have Failed.* New Haven: Yale University Press, 1999.
Sisam, John W.B. *Forestry Education at Toronto.* Toronto: University of Toronto Press, 1961.
–. 'Historical Highlights, Canadian Institute of Forestry.' *Forestry Chronicle* April 1983: 55.
Sivaramakrishnan, K. 'The Politics of Fire and Forest Regulation in Colonial Bengal.' *Environment and History* 2 (1996): 145-94.
Smith, C. Ray, and David R. Witty. 'Conservation, Resources, and Environment: An Exposition and Critical Evaluation of the Commission of Conservation, Canada.' *Plan 11* (1970, 1972).
Smith, John. *The Acquisition of Forest Reserves.* Halifax: DLF, 1975.
–. *Proceedings.* Nova Scotia Woodlot Owners and Operators Association, 8th Annual Meeting, 1977.
'Some Handicaps that Must Be Overcome in Canadian Lumber Export.' *Canada Lumberman* 1 October 1933: 45.
Soyez, Dietrich. 'Stora Lured Abroad? A Nova Scotian Case Study in Industrial Decision-Making and Persistence.' *Operational Geographer* 16 (1988): 11-4.
'Special Issue on Sustainability.' *Journal of Canadian Studies* 31 (1996).
'Special Woodlot Issue.' *Forestry Chronicle* 27, 3 (1951).
Spicer, Stanley T. *Maritimers Ashore and Afloat.* Vol 2. Hantsport, NS: Lancelot, 1994.
Sproule, C.H. 'Opening Remarks.' Nova Scotia Forest Products Association Meeting on Marketing Boards, 26 May 1970.
Steen, Harold. *The U.S. Forest Service: A History.* Seattle: University of Washington Press, 1976.
Stevens, Fred. 'The Section's First Sixty Years.' *Pulp and Paper Canada* 78 (1977): 22.
Stevens, Geoffrey. *Stanfield.* Toronto: McClelland and Stewart, 1973.
Stewart, Robin. 'E. Melville Duporte: A MacDonald Great.' *MacDonald Journal* August 1983: 12-3.
Swift, Jamie. *Cut and Run: The Assault on Canada's Forests.* Toronto: Between the Lines, 1983.
–. 'Cut and Run: The MacAlpine Case.' *Harrowsmith* October-November 1983: 38-45, 107, 112.

Thurston, Harry. 'Prest's Last Stand: Keeping Kafka and the Bureaucrats out of the Acadian Forest.' *Harrowsmith* August-September 1983: 22-31.

Thyssen, Aksel. *Danske Forstkandidater*. København: Danske Forstkandidaters Forening, 1956.

Tothill, J.D. 'Notes on the Outbreaks of Spruce Budworm, Forest Tent Caterpillar, and Larch Sawfly in New Brunswick.' *Proceedings*. Acadian Entomological Society, 1922.

University of New Brunswick Forestry Association. *The Fiftieth Anniversary of the Faculty of Forestry at the University of New Brunswick, 1908-1958*. Fredericton: UNB Forestry Association, 1958.

Van den Bosch, R. *The Pesticide Conspiracy*. Garden City, NY: Doubleday, 1978.

Vickery, Vernon. 'William H. Brittain.' *Bulletin – Entomological Society of Canada* 29, 2 (1997): 73-6.

Walsh, F. Waldo. *We Fought for the Little Man*. Moncton, NB: Co-op Atlantic, 1978.

Webber, Walter. *Report of the 36th Annual Meeting, 1989*. Canadian Institute of Forestry, Nova Scotia Section.

Westoby, Jack. *The Purpose of Forests: Follies of Development*. Oxford: Basil Blackwell, 1987.

White, Bruce A. *Elbert Hubbard's 'The Philistine: A Periodical of Protest' (1895-1915)*. Lanham, MD: University Press of America, 1989.

Wicken, William. 'Heard It from Our Grandfathers: Mi'kmaq Treaty Traditions and the Syliboy Case of 1928.' *UNB Law Journal* 44 (1995): 145-61.

–. 'Mi'kmaq and Waustukwiuk Treaties.' *UNB Law Journal* 43 (1994): 241-53.

–. '26 August 1726: A Case Study in Mi'kmaq-New England Relations in the Early 18th Century.' *Acadiensis* 28 (1993): 5-22.

'Will Specialize in Forestry Activities.' *Canada Lumberman* July 1933: 19.

'Wood Products Marketing.' *Forestry Chronicle* 26, 1 (1950): 3.

Wright, Matthew. *Forest Times* 14, 6 (1982).

–. 'A Grower's Notebook.' *Forest Times* 15, 4 (1993): 12.

Index

Set in Stone Serif by Brenda and Neil West, BN Typographics West

Printed and bound in Canada by Friesens

Copy editor: Dallas Harrison

Indexer: Thiam Jin Tan